T0314246

Offshore Mechanics

Offshore Mechanics

Structural and Fluid Dynamics for Recent Applications

Madjid Karimirad
Queen's University Belfast
UK

Constantine Michailides
Cyprus University of Technology
Cyprus

Ali Nematbakhsh
University of California, Riverside
USA

This edition first published 2018
© 2018 John Wiley & Sons Ltd

The right of Madjid Karimirad, Constantine Michailides and Ali Nematbakhsh to be identified as the authors of this work has been asserted in accordance with law.

Registered Office(s)
John Wiley & Sons, Inc., 111 River Street, Hoboken, NJ 07030, USA
John Wiley & Sons Ltd, The Atrium, Southern Gate, Chichester, West Sussex, PO19 8SQ, UK

Editorial Office
The Atrium, Southern Gate, Chichester, West Sussex, PO19 8SQ, UK

For details of our global editorial offices, customer services, and more information about Wiley products visit us at www.wiley.com.

Wiley also publishes its books in a variety of electronic formats and by print-on-demand. Some content that appears in standard print versions of this book may not be available in other formats.

Library of Congress Cataloging-in-Publication Data

Names: Karimirad, Madjid, author. | Michailides, Constantine, 1978– author. |
    Nematbakhsh, Ali, 1984– author.
Title: Offshore mechanics : structural and fluid dynamics for recent applications / by Madjid Karimirad,
    Queen's University Belfast, UK, Constantine Michailides, Cyprus University of Technology,
    Cyprus, Ali Nematbakhsh, University of California, Riverside USA.
Description: First edition. | Hoboken NJ, USA : Wiley, 2018. | Includes bibliographical references
    and index. |
Identifiers: LCCN 2017043079 (print) | LCCN 2017044707 (ebook) | ISBN 9781119216643 (pdf) |
    ISBN 9781119216636 (epub) | ISBN 9781119216629 (cloth)
Subjects: LCSH: Offshore structures–Hydrodynamics.
Classification: LCC TC1665 (ebook) | LCC TC1665 .K38 2018 (print) | DDC 627/.98–dc23
LC record available at https://lccn.loc.gov/2017043079

Cover design by Wiley
Cover image: © BenGrasser/Gettyimages

Set in 10/12pt Warnock by SPi Global, Pondicherry, India

Printed and bound by CPI Group (UK) Ltd, Croydon, CR0 4YY

10 9 8 7 6 5 4 3 2 1

Contents

About the Authors

Dr Madjid Karimirad is Senior Lecturer (Associate Professor) in Marine and Coastal Engineering in the School of Natural and Built Environment, Queen's University Belfast (QUB), UK, since March 2017. Prior to joining QUB, he was Scientist at MARINTEK (Norwegian Marine Technology Research Institute) and SINTEF Ocean, Trondheim, Norway. Dr Karimirad has been researching in the field of marine structures and offshore technology with more than 10 years of work and research experience. He has a strong background in academia and industry, including postdoctoral and PhD research, working expertise, and competencies in the offshore oil and gas business. Dr Karimirad got his PhD in 2011 in Marine Structures from the Norwegian University of Science and Technology (NTNU). He has been employed by CeSOS (Centre for Ships and Ocean Structure), a Centre of Excellence (CoE) in Norway. Dr Karimirad served as postdoctoral academic staff at CeSOS, and his postdoctoral was part of the NOWITECH (Norwegian Research Centre for Offshore Wind Technology) program. Dr Karimirad obtained his MSc (2007) in Mechanical Engineering/Ships Structures and his BSc (2005) in Mechanical/Marine Engineering from the Sharif University of Technology, Tehran. In addition, he has worked in industry as Senior Engineer at the Aker-Kvaerner EPCI Company. Offshore renewable energy (ORE) structures and oil/gas technologies are among his interests, and he has carried out different projects focusing on these issues. Dr Karimirad has served as editor and reviewer for several international journals and conferences; also, he has been appointed as topic and session organizer of the International Conference on Ocean, Offshore & Arctic Engineering (OMAE). His knowledge covers several aspects of offshore mechanics, hydrodynamics, and structural engineering. The work and research results have been published in several technical reports, theses, book chapters, books, scientific journal articles, and conference proceedings.

Dr Constantine Michailides is a Lecturer in Offshore Structures in the Department of Civil Engineering and Geomatics, Cyprus University of Technology, Cyprus, since July 2017. He holds a BSc, MSc, and PhD in Civil Engineering from Aristotle University of Thessaloniki (AUTh), Greece. He did a PhD in marine and offshore engineering in hydroelasticity of floating structures and wave energy production; for his PhD research, he awarded by AUTh and International Society of Offshore and Polar Engineering (ISOPE). In May 2013, he joined Centre for Ships and Ocean Structures (CeSOS) and Centre for Autonomous Marine Operations and Systems (AMOS) in the Department of Marine Technology at the Norwegian University of Science and Technology as

Postdoctoral Researcher until January 2016, when he was appointed as Senior Lecturer in the Department of Maritime and Mechanical Engineering, Liverpool John Moores University, UK. Dr Michailides has performed research on numerical analysis, experimental testing, and structural-field monitoring of offshore and coastal structures and systems. His research focuses on global and local numerical analyses of offshore structures and systems (oil and gas, and renewable energy), fluid–structure interaction, hydrodynamics and hydroelasticity, ocean energy devices, offshore wind technology, offshore combined energy systems, physical model testing, structural health monitoring, optimization, genetic algorithms, and floating bridges and tunnels. Recent interests include computational fluid dynamics (CFD) and reliability analysis for offshore engineering applications. Dr Michailides is a member of research and professional bodies in marine and offshore engineering technology like ISOPE, International Ship and Offshore Structures Congress (ISSC), Technical Chamber of Greece, and Engineering and Physical Sciences Research Council (EPSRC) Peer Review Associate College. He authored more than 40 research papers in peer-reviewed journals and conference proceedings in the field of marine and offshore engineering.

Dr Ali Nematbakhsh is an expert in the field of computational fluid dynamics (CFD) and numerical modeling, with 9 years of experience in these fields. He is currently Project Scientist at University of California, Riverside. He received his PhD in 2013 from Worcester Polytechnic Institute in Massachusetts in the field of Mechanical Engineering. During his PhD, he wrote original code to model nonlinear hydrodynamic loads on floating wind turbines. Right after his PhD, he was invited to be a Postdoctoral Researcher at Center of Ships and Ocean Structures (CeSOS) at Norwegian University of Science and Technology, where he further developed and used his numerical model to simulate different offshore structures, including floating wind turbines, wave energy converters, barges, and monopiles. Dr Nematbakhsh also has extensive knowledge in high-performance computing using parallel CPUs and GPUs, which enable him to simulate complex and nonlinear offshore engineering problems. Dr Nematbakhsh is a member of the American Society of Mechanical Engineers. His research has been published in more than 15 papers in international conference proceedings and scientific journals.

Preface

Offshore industry has seen rapid development in recent years. New marine structures have emerged in different fields such as offshore oil and gas, marine renewable energy, sea transportation, offshore logistics and sea food production. As a result, new concepts and innovative offshore structures and systems have been proposed for use in the oceans. An obvious need exists for a book which provides the capabilities and limitations of theories and numerical analysis methods for performing dynamic analysis for the case of recent applications in offshore mechanics. This book covers the needs for the required analysis and design of offshore structures and systems. Particular emphasis has been given to recent applications in offshore engineering. This includes ship-shaped offshore structures, fixed-bottom and floating platforms, ocean energy structures and systems (wind turbines, wave energy converters and tidal turbines) and multipurpose offshore structures and systems. Theoretical principles are introduced, and simplified mathematical models are presented. Practical design aspects for various offshore structures are presented with handy design guides and examples. Each example is followed with an analytical or a numerical solution. Additionally, special attention has been paid to present the subject of computational fluid dynamics (CFD) and finite element methods (FEM) that are used for the high-fidelity numerical analysis of recent applications in offshore mechanics. The book provides insight into the philosophy and power of numerical simulations and an understanding of the mathematical nature of the physical problem of the fluid–structure interaction, with focus on offshore applications.

The book helps students, researchers and engineers with a mid-level engineering background to obtain insight on theories and numerical analysis methods for the structural and fluid dynamics of recent applications in offshore mechanics. The main key feature of the book is using "new" applications for describing the theoretical concepts in offshore mechanics. Furthermore, the present book not only covers traditional methodologies and concepts in the field of offshore mechanics, but also includes new approaches such as novel CFD and FEM techniques. Nowadays, due to the rapid increase of computational resources, offshore industry is using various advanced CFD and FEM tools to design offshore structures. Therefore, qualified graduated students and engineers need to be familiar with both traditional methodologies and new methods applied in offshore mechanics proper for recent applications. The book helps engineers and researchers in the field of offshore mechanics to become familiar with recently applied trends and methodologies.

This book covers the fundamental knowledge of offshore mechanics by teaching the reader how to use numerical methods for design of different concepts in offshore engineering. Recent methodologies for hydrodynamic and structural analysis of offshore structures are introduced and explained. The authors believe that a graduate student or an engineer in offshore industry should be well familiar with these concepts. The book is intended for graduate students, researchers, faculty members and engineers in the fields of offshore engineering, offshore renewable energy (wind energy, wave energy and tidal energy), marine structures, ocean and coastal engineering, fluid dynamics and mechanical engineering. The readers of the book must have basic offshore engineering knowledge and interest related to the analysis and design of recent applications in offshore mechanics. The presented theories and applications are developed in a self-contained manner, with emphasis on fundamentals, concise derivations and simple examples. Some of the main key words covered in this book are as follows:

> Offshore mechanics; structural dynamics; fluid–structure interaction; hydrodynamics; fluid dynamics; ocean energy devices; offshore wind; multipurpose floating platforms; offshore structures; wave energy converters; floating wind turbines; combined wave and wind energy; finite element method (FEM); computational fluid dynamics (CFD)

The book consists of the following chapters:

1) Preliminaries
2) Offshore Structures
3) Offshore Environmental Conditions
4) Hydrodynamic and Aerodynamic Analyses of Offshore Structures
5) Fundamentals of Structural Analysis
6) Numerical Methods in Offshore Structural Mechanics
7) Numerical Methods in Offshore Fluid Mechanics
8) Mooring and Foundation Analysis

Acknowledgements

Dr Karimirad would like to thank his family (in particular, Soosan and Dorsa) for their support which helped him to finish this important task.

Dr Michailides would like to thank his wife, Christina, and his son, Evangelos, for inspiring him; and also his mentor, Professor Demos Angelides, for teaching him about marine structures.

Dr Nematbakhsh would like to thank his wife, Nafiseh, for her support during writing this book.

Dr Madjid Karimirad, Queen's University Belfast (QUB), Belfast, UK

Dr Constantine Michailides, Cyprus University of Technology (CUT), Limassol, Cyprus

Dr Ali Nematbakhsh, University of California, Riverside (UCR), USA

October 2017

1

Preliminaries

Compared to inland structures, offshore structures have the added difficulty of being placed in the ocean environment. Hence, offshore structures are subjected to complicated loads and load effects. Important factors affect the design, functionality, structural integrity and performance of offshore structures, including but not limited to: fluid–structure interaction, intense dynamic effects, nonlinear loadings, extreme and harsh weather conditions and impact pressure loads. Offshore industry has seen rapid development in recent years. This includes the emergence of new marine structures in different areas such as offshore petroleum, marine renewable energy, sea transportation, offshore logistics and seafood production. As a result, new concepts and innovative offshore structures and systems have been proposed for use in the oceans.

An obvious need exists for a book providing the limitations and capabilities of theories and numerical analysis methods for structural and fluid dynamic analysis of recent applications in offshore mechanics. This book attempts to provide a comprehensive treatment of recent applications in offshore mechanics for researchers and engineers. The book covers important aspects of offshore structure and system analysis and design. Its contents cover the fundamental background material for offshore structure and system applications. Particular emphasis has been paid to the presentation of recent applications from the required theory and their applicability. The book covers recent applications in a broad area. This includes ship-shaped offshore structures, recent fixed-bottom and floating oil and gas platforms, ocean energy structures and systems (wind turbines, wave energy converters, tidal turbines and hybrid platforms), multipurpose offshore structures and systems, submerged tunnels and floating bridges for transportation purposes and aquacultures (fish farms).

Many of the applications of the theoretical principles are introduced, and several exercises as well as different simplified mathematical models are presented for recent applications in offshore engineering. In this book, practical design aspects of the aforementioned offshore structures are presented with handy design guides and examples, simple description of the various components for their robust numerical analysis and their functions. Additionally, special attention has been paid to present the subjects of computational fluid dynamics (CFD) and finite element methods (FEM) along with the high-fidelity numerical analysis of recent applications in offshore mechanics.

The book makes available an insight into the philosophy and power of numerical simulations and an understanding of the mathematical nature of the fluid and structural dynamics, with focus on offshore mechanics applications. The current book helps

Offshore Mechanics: Structural and Fluid Dynamics for Recent Applications, First Edition.
Madjid Karimirad, Constantine Michailides and Ali Nematbakhsh.
© 2018 John Wiley & Sons Ltd. Published 2018 by John Wiley & Sons Ltd.

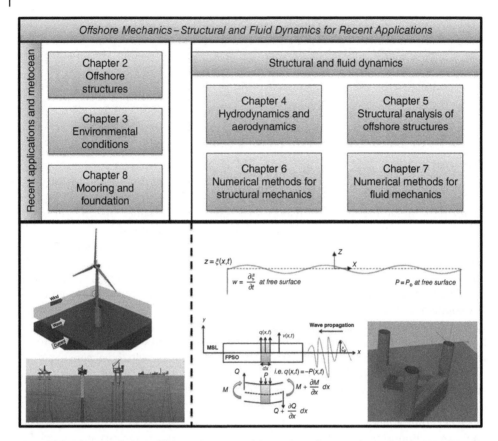

Figure 1.1 Schematic layout, different chapters and their roles in forming the present book, *Offshore Mechanics: Structural and Fluid Dynamics for Recent Applications.*

students, researchers and engineers with mid-engineering background gain good insights on theories and numerical analysis methods for structural and fluid dynamics for the cases of recent applications in offshore mechanics. Figure 1.1 presents the schematic layout of the book and shows different chapters as well as their roles in shaping this book.

The key features of the book are using "new" applications for describing the theoretical concepts in offshore mechanics, and covering *both* traditional and recent methodologies used in offshore structure modelling. Most of the books currently available in the field of offshore mechanics are based on using traditional oil, gas and ship industry examples to explain the fundamentals of offshore mechanics. Therefore, the reader becomes familiar with the basic concepts very well, but his or her viewpoint will remain limited to the traditional applications. This book tries to address this limitation by covering some recent applications, such as: offshore wind farms, ocean energy devices, aquaculture, floating bridges and submerged tunnels.

Furthermore, the current book not only covers traditional methodologies and concepts in the field of offshore mechanics, but also includes new approaches such as CFD and FEM techniques. The material in this book will help graduate students get needed knowledge in offshore industry for recent applications. Currently, due to the rapid increase in speed of computational resources, offshore industry is using various advanced CFD and FEM tools such as ANSYS and ABAQUS to analyse offshore structures. Therefore, qualified graduated students and engineers need to be familiar with both traditional methodologies and new methods applied in offshore mechanics proper for recent applications.

Structural fluid mechanics of offshore structures, the theories applied to recent applications and proper case studies to explain analytical and numerical methods make the core of this book. The hydrodynamic, stochastic dynamics and structural analyses are the book's focus. What makes this book distinct from similar available books is that it covers recent applications in offshore industry by providing suitable examples. Simplified examples help students, researchers and engineers to understand the subjects and know how to use proper methods.

This book will help engineers and researchers in the field of offshore mechanics to become familiar with new trends and methodologies that have been applied recently. Different new offshore concepts such as offshore energy harvesters, floating bridges, submerged tunnels, multipurpose platforms, hybrid floaters as well as fish farms are going to play important roles in the future of offshore industry. Furthermore, new numerical techniques such as advanced CFD and FEM methods are currently used in industry.

We believe that the new offshore concepts that are now the focus of academic investigations gradually will be adopted by industry and probably result in greater popularity of this book. This book helps readers to learn the basic concepts of offshore mechanics not only by traditional standard applications, but also by applying these concepts for new structures in offshore engineering. In addition, it introduces the fundamentals of new numerical techniques that are emerging in offshore industry.

The book covers the fundamentals of offshore mechanics by teaching the reader how to use these concepts for traditional and (more specifically) current demands in offshore industry. The examples, given throughout the book, are for offshore structures that have been recently designed or are currently under development. For example, different offshore wind farms have been installed in Europe in recent years, and several projects are ongoing for harvesting energy from waves. We believe that a graduate student or an engineer in offshore industry should be well familiar with these concepts.

The methodologies for hydrodynamic and structural analyses of offshore structures are introduced and explained in this book. By learning the basics of the new methodologies, the reader has enough background to further expand his or her knowledge based on the needs in a specific industry. Throughout the chapters, special attention is given to familiarize the reader with numerical methods. These numerical methods cover both structural and hydrodynamic analysing of offshore structures.

This book is intended for graduate students, researchers, faculty members and engineers in the fields of: offshore structural engineering, offshore renewable energy (wind energy, wave energy and tidal energy), marine structures, ocean and coastal engineering, fluid dynamics and mechanical engineering. Its reading level can be considered as introductory or advanced. However, readers must have basic offshore engineering knowledge and interest related to the analysis and design of recent applications in offshore mechanics. The presented theories and applications are developed in a self-contained manner, with an emphasis on fundamentals, concise derivations and simple examples.

The book has eight chapters. The first chapter introduces the book and explains its scope and objectives. The second chapter covers offshore structures, explaining different concepts such as ship-shaped, oil and gas platforms (bottom-fixed and floating), ocean energy devices (e.g. wind turbines, wave energy, ocean tidal turbines and hybrid platforms), multipurpose floaters, submerged tunnels, floating bridges and aquaculture and fish farms. The third chapter covers metocean and environmental conditions; in particular, wind, wave and current conditions, joint distribution of wave and wind, oceanography, bathymetry, seabed characteristics, extreme environmental conditions and environmental impacts of offshore structures. The fourth chapter explains the wave, wind and current kinematics as well as aerodynamic and hydrodynamic loads. This covers coupled hydrodynamic and aerodynamic analysis for offshore structures. Chapter 5 covers structural analysis and fundamental structural mechanics. This includes beam theories, stress–strain relation as well as buckling, bending, plate and plane theories and similar basic theories useful for studying the structural integrity of offshore and marine structures. In Chapter 6, the stress analysis, dynamics analysis, multibody formulation, time-domain and frequency-domain simulations, finite element methods, nonlinear analysis, extreme response calculation as well as testing and validation of offshore structures are discussed. The seventh chapter is dedicated to computational methods for fluid mechanics covering potential theories (i.e. a panel method covering radiation and diffraction as well as excitation forces). Computation fluid dynamics (CFD) is the core of this chapter, and different practical theories are included in this chapter. The eighth chapter covers mooring and foundation as well as theories related to soil mechanics and soil–foundation interaction.

The objective of the present book is to help the readers on different levels – namely, knowledge, comprehension, application, analysis, synthesis and evaluation – whenever they are dealing with physical problems that exist in offshore mechanics, especially with recent applications. As a result, the readers of the book will be able to: (a) exhibit learned material by recalling facts, terms and basic concepts; (b) demonstrate understanding of facts and basic concepts; (c) solve problems by applying acquired knowledge, facts, techniques and rules in a different way; (d) examine and split any possible information into parts by identifying motives or causes, making inferences and finding evidence to support solution methods; (e) compile information in a different way by combining elements in a new pattern; and (f) present and defend opinions by making judgments about relevant information based on a set of criteria.

In Chapter 2, we will review and present important information for different types of offshore structures, and we will try to identify an outline of their numerical analysis needs and the methods that have been used up to now. The types that will be presented are ship-shaped offshore structures, oil and gas offshore platforms, offshore wind turbines, wave energy converters, tidal energy converters, multipurpose offshore structures and systems, submerged floating tunnels and aquaculture and fish farms. For all the types of recent applications of offshore structures, categorization and basic design aspects are presented.

In Chapter 3, we will present important information about the generation and the process of propagation of different environmental conditions that may affect the structural integrity of recent applications of offshore structures. Different environmental processes like the wave, wind, current, scour and erosion are described appropriately in connection with possible effects that they have on all the different types of offshore structures that are examined. Moreover, the effect of joint analysis on wind and wave is presented. Finally, insight is presented about the estimation of extreme environmental conditions that have straightforward relation with the survivability of offshore structures.

In Chapter 4, we deal with the three dominant excitation loading conditions that influence the lifetime of offshore structures: the wave, tidal and wind loadings. Wave kinematic theories that exist for addressing regular and irregular waves are presented. Moreover, methods for estimating the wave loads induced by inviscid flows in members of offshore structures are presented, too. In addition, tide and current kinematic methods are presented with emphasis on methods for estimating the current loads on offshore structures. Wind kinematic methods that have application for the design of offshore structures are presented along with numerical methods for estimating the wind loadings. Finally, fundamental topics of the required aerodynamic analysis for the design of offshore wind turbines are presented. Emphasis is on presenting numerical methods for estimating the aforementioned environmental loadings and on how these loads are used compared to different numerical tools.

In Chapter 5, some of the important principles of statics and dynamics and how these are used to determine the resultant internal loadings in an offshore structure are initially presented. Furthermore, the concepts of normal and shear stress are introduced along with the strains induced by the deformation of the body. Moreover, important information about the appropriate development of structural elements of offshore structures is presented. Beams and plates, and methods for developing numerical models with the use of these types of structural elements for the structural analysis of offshore structures, are presented.

In Chapter 6, numerical methods that are used in offshore engineering for the structural response dynamic analysis of different types of offshore structures are presented. Dynamic loadings dominate the response of offshore structures. Numerical methods for the development of numerical models and tools for the dynamic analysis of offshore structures in both frequency and time domain are presented. Also, special cases where a multibody approach is needed or nonlinear phenomena exist, and numerical methods for handling these special cases, are presented. Methods for estimating or predicting numerically the extreme response values of different components of offshore structures (e.g. mooring lines, pontoons of a semisubmersible platform and tower of a wind turbine) are presented. Finally, the fundamental required process for the development of a physical model test of an offshore structure is presented.

In Chapter 7, the different possible numerical methods that exist in offshore fluid mechanics are presented. Initially, the bases of potential flow theory models are presented and explained. Afterwards, a comprehensive presentation of CFD-based models in offshore engineering is presented. Details about the discretization of the Navier–Stokes equation on rectangular structures' grids, with details about the advection, viscous and pressure terms and mass conservation equation, are presented. Possible numerical methods for solving the Navier–Stokes equations, incorporating the Poisson equation, the effects of free surface and the volume of fluid method, are presented. Moreover, the discretization of the Navier–Stokes equation in a mapped coordinate system (which can be used for different types of moving offshore structures) is presented. Finally, methods for discretization of level set function and of reinitialization of the equation of motion are presented in connection with use for the numerical analysis of offshore structures.

Chapter 8 presents the effects of different possible foundation systems that are used in offshore engineering. Initially, different mooring line systems are described, with emphasis on catenary and taut mooring systems; the appropriate numerical modelling of these mooring line systems is presented and explained. Afterwards, fundamental theories for the numerical analysis of soil in offshore areas are presented, with focus on possible soil–structure interaction effects that should be taken into account. Finally, the chapter presents design aspects for the case of foundations that are used in offshore engineering, like piles, caissons, direct foundations and anchors.

2

Offshore Structures

2.1 Ship-shaped Offshore Structures

Oil and gas demand will not decrease in the near future unless substantial changes and developments happen in renewable and sustainable energy technologies. Oil prices may increase and change world economics again. This leads oil companies and countries to be keen to explore new offshore fields in deeper water, in harsher conditions and in areas with longer distance to shore.

Offshore oil and gas have seen a movement from deep water (500 to 1500 m) to ultra-deep water (more than 1500 m) (Lopez-Cortijo *et al.*, 2003) in the past decade, and several oil and gas subsea facilities have been installed in offshore sites with water deeper than 2000 m (TOTAL, 2015). This is mainly due to energy supply development limits and policies, as well as the world economy and increased energy consumption. To develop these offshore fields, ship-shaped structures are widely used, particularly in sites with no (or limited) pipeline infrastructure. However, the application of ship-shaped offshore structures is not limited to deep water, and these structures may be considered in near-shore oil and gas terminals (Paik and Thayamballi, 2007).

In moderate-depth and relatively shallow offshore sites (less than 500 m in depth), bottom-fixed platforms (e.g. jackets) have been widely used for oil and gas development. However, they are not feasible in deep water (500+ m depth). Floating-type offshore structures (e.g. semisubmersibles) are considered for deep-water and ultra-deep-water areas. In Section 2.2, fixed-type and floating-type offshore oil and gas platforms are discussed. The current section is dedicated to ship-shaped offshore structures.

Several alternatives exist for production, storage and offloading. The produced oil and gas should be transported to shore for further processing and use. All types of floating platforms (e.g. spar, semisubmersible and tension leg platform) need systems and infrastructures, such as pipelines and other associated facilities, to store and transport the products. The processed oil may be stored in platforms and transported via pipelines to shore.

Before the oil and gas can be transferred to shore or stored for offloading, the product should be processed. Normally, the rig flow is separated into water, gas and oil. The gas may be compressed and stored, or flared (burned to atmosphere). The stored gas can be used for reinjection in later phases of oil production. The water is drained and normally

Offshore Mechanics: Structural and Fluid Dynamics for Recent Applications, First Edition.
Madjid Karimirad, Constantine Michailides and Ali Nematbakhsh.
© 2018 John Wiley & Sons Ltd. Published 2018 by John Wiley & Sons Ltd.

reinjected to enhance oil production in nearby wells. The oil is further processed to obtain necessary crude oil characteristics.

An oil tanker converted to an FPSO (floating, storage, processing and offloading) or FSO (without processing) vessel, or a vessel built for this application, is a very attractive alternative for field development compared to ordinary floating platforms (Devold, 2013). If the structure is equipped with drilling, the unit is called an FDPSO (floating, drilling and production, storage and offloading) vessel.

By using shuttle tankers, the produced and processed oil and gas can be transported to shore from the ship-shaped unit. Thus, the ship-shaped structure is an active unit combining several functions, which reduces infrastructure needed for transporting products to shore. Development of oil and gas fields in deep water and ultra-deep water with limited access to pipelines has been extensively enhanced using economical ship-shaped units.

Numerous types of research about design, engineering, construction, installation and operation covering structural integrity, cost and reliability of these structures have been carried out; nevertheless, the first ship-shaped structures were in place 40 years ago. The first FPSO was built in 1977, and currently more than 270 FPSOs are installed worldwide.

There are similarities and differences between ship-shaped offshore structures and trading tankers. The number of conversions has grown in recent years, and several oil tankers have been converted to FPSOs during that time (Biasotto *et al.*, 2005). The structural geometry of ship-shaped units and oil tankers is similar. Ship-shaped structures (similar to other types of oil and gas platforms) are designed for a specific offshore site with specific environmental conditions for which the design is tailored (ABS, 2014). Oil tankers like other merchant ships can avoid harsh weather or change their heading angle. However, ship-shaped structures are located in a defined location and are subjected to environmental conditions at the site (Hwang *et al.*, 2012).

The other important parameter is that the trading ships are regularly checked, surveyed and repaired (i.e. by dry-docking to maintain the ship in proper condition). But the structural integrity of ship-shaped units is considered to cope with their long-term safety demands. Risk assessment and management plans for field development are initiated very early in the concept selection phase, and the project's feasibility may be questioned regarding converting an oil tanker or designing a new-built unit (Mierendorff, 2011).

Another difference between an oil tanker and FPSO is that the FPSO has a production and process unit. In converted FPSOs, the production/process plant is constructed in modular form and added onto the deck of the oil tanker. Regarding the storage, the newly built ships and converted units are the same, and the processed oil is stored in tanks of the units. During offloading, the oil is pumped to shuttle tankers using a flexible floating discharge hose (Karimirad and Mazaheri, 2007). Figure 2.1 shows a schematic layout of a turret-moored FPSO; the shuttle tanker is moored to the FPSO for offloading.

Although loading and unloading of trading ships are normally performed in protected ports at still-water conditions, the loading and offloading (ballast, transient and fully loaded conditions) of ship-shaped offshore structures are subjected to major loading (Terpstra *et al.*, 2001). Ship-shaped offshore units are installed permanently at a specified location and hence similar to other offshore structures, designed to withstand 100-year environmental conditions. For converting oil tankers, the structural integrity of the hull should be checked with respect to offshore industry standards (compared to

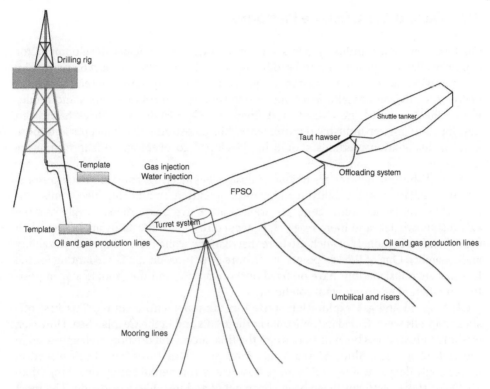

Figure 2.1 Schematic layout of an FPSO offshore oil and gas field; the ship-shaped offshore structure (FPSO) is moored by a turret-mooring system. The shuttle tanker, drilling rig, umbilical and risers are shown as well.

shipbuilding industry standards) to confirm durability and reliability (Ayala-Uraga, 2009), in particular the interfaces between the hull and topside.

Although the shipbuilding industry's standards are more cost-effective, it is not easy to apply these to ship-shaped offshore structures due to differences between the shipbuilding and offshore industries in requirements, background, tradition and culture of staffs. Moreover, there are many interface matters complicating the design, for example topside–hull interactions. These issues should be considered during the design of new-built units or conversion of oil tankers to FPSOs (Paik and Thayamballi, 2007).

Ship-shaped offshore structures are used for storage and processing of natural gas as well. Floating liquefied natural gas (FLNG) production is the only option for some fields, although the cost associated with production, liquefaction, storage and transferring using FLNG can be greater compared to land-based LNG units (Anonymous, 2005). However, this technology helps natural gas to be produced, liquefied, stored and transferred at sea (Shell, 2015), which is required in marginal gas fields and offshore-associated gas resources. There are unique characteristics for LNG-FPSO design, such as:

1) Restricted space
2) Platform motion
3) LNG sloshing in the inner storage tank and offloading system (Gu and Ju, 2008).

2.2   Oil and Gas Offshore Platforms

Offshore petroleum industry has seen continuous technological development for exploring, drilling, processing and producing oil and gas during the past 65 years. The key motivations for such nonstop progress are reducing cost, increasing safety, decreasing environmental impact, increasing remote-control operations and reducing accidents. Hence, new fixed- and floating-type offshore structures have been developed to answer industry requirements. More cost-effective concepts and more efficient installation methods should be developed to overcome offshore industry challenges.

Many offshore structures are unique in design, engineering, construction, transportation, installation, accessibility, maintenance, operation, monitoring, decommissioning and so on. Hence, some concepts are less attractive than others, considering the existing knowledge, and more research is needed to reduce the associated costs (both capital and operational) of such offshore installations while increasing their durability and reliability. One of the key points in offshore structures design is accounting for the fact that such installations have no fixed onshore access, and they should stay in position in different environmental conditions.

Offshore oil and gas exploration started in the nineteenth century. The first offshore oil wells were drilled in California in the 1890s and in the Caspian Sea. However, offshore industry was born in 1947 when the first successful offshore well appeared in the Gulf of Mexico. Since 65 years ago, innovative structures have been placed in increasingly deeper waters and in more hostile environmental conditions. More than 10,000 offshore platforms have been constructed and installed worldwide. The most important deep-water and ultra-deep-water offshore petroleum fields are located in the Gulf of Mexico, West Africa and Brazil, known as the Golden Triangle (Chakrabarti, 2005).

Offshore shallow water reserves have been depleted, and offshore petroleum industry has moved to explore deep water and ultra-deep water. Offshore exploration and production of oil and gas in deep water created new challenges for offshore technology. Several offshore structures have already been installed in deep water. Furthermore, new oil and gas fields have been discovered in ultra-deep water, and offshore petroleum industry has moved to ultra-deep water in the past decade. However, several of these fields are small, and their development requires novel concepts and innovative structures to present competitive and cost-effective solutions.

In general, offshore structures are divided into two main types: fixed and floating. Fixed-type offshore structures are fixed to the seabed using their foundation, while floating-type offshore structures may be (a) moored to the seabed, (b) dynamically positioned by thrusters or (c) allowed to drift freely. However, there is another innovative group of structures that are partially fixed to the seabed, and their stability is ensured using guyed lines and floatation devices. Among those structures are articulated columns, guyed towers and compliant platforms (Johnson, 1980).

Fixed-type offshore structures – for example, jackets, gravity-based structures and jack-ups – have been widely used for oil and gas production in moderate and shallow waters. Figure 2.2 shows a schematic layout of bottom-fixed offshore structures. As water depth increases, fixed-type structures become more expensive, and the installation of bottom-fixed structures in deep water is very challenging.

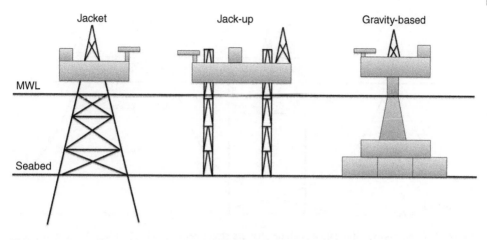

Figure 2.2 Artistic layout of the fixed-type offshore structure.

The guyed tower is an innovative and cheaper alternative to fixed-type structures (Finn, 1976). Guyed towers can deflect easier under wave and wind loads, compared to fixed-type structures. The structure is supported by piles extending from the seafloor and mooring lines attached to the platform helping the structure resist harsh conditions. The Lena (Exxon's Mississippi Canyon 280-A) platform represents the first commercial application of the guyed tower concept for offshore drilling and production platforms. The Lena platform was installed in 305 m water depth (Power *et al.*, 1984). The other famous compliant platforms installed in the Gulf of Mexico are Amerada Hess' Baldpate in 502 m and ChevronTexaco's Petronius in 535 m. Petronius, the world's deepest bottom-fixed oil platform and one of the world's tallest structures, was completed in 2000 (Texaco Press, 2000).

Moving further in deep-water and ultra-deep-water areas using fixed-type offshore structures is not feasible. Moreover, alternative solutions such as compliant towers are becoming expensive; hence, the only practical option is floating structures. Floating offshore structures have been widely considered for developing deep-water oil and gas fields. In Section 2.1, ship-shaped offshore units (i.e. FPSOs) were discussed. In addition to ship-shaped structure, the main floating-type structures are spar, tension leg platform (TLP) and semisubmersible; see Figure 2.3. The world deepest floating oil platform is Perdido, which is a spar platform in the Gulf of Mexico in a water depth of 2438 m. Perdido started production in 2010 (Shell, 2010).

Jacket: Jacket structures, also called templates or lattices, are three-dimensional space frame structures consisting of tubular members (legs and braces) that are welled. Jackets are the most common offshore structures used for offshore petroleum drilling and production. Jackets normally have four to eight legs, which are not normally vertical to increase the stability under environmental loads and corresponding overturning moments. Piles penetrating soil fix the structure to seabed; the piles are also tubular members and are driven (hammered) through the jacket legs into the sea bottom. The pile design is highly affected by soil conditions and seabed characteristics. The jacket structure provides an enclosure for well conductors. The platform

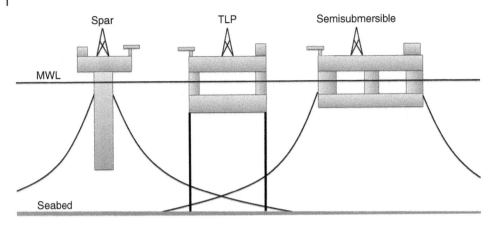

Figure 2.3 Artistic layout of the floating-type offshore structure.

topside (superstructure) consists of 2–3 decks with drilling units and production facilities. Jackets are widely applied in shallow and moderate water depth around the world. However, water depth is not the only decision-making parameter, and in areas with more moderate environmental conditions, it may be possible to use a jacket structure in deeper water depth (i.e. environmental conditions in the Gulf of Mexico are moderate compared to the North Sea).

Jack-up: Jack-up (self-elevating unit) platforms are normally three-legged structures supporting a deck for drilling. The deck is buoyant (i.e. a barge). Jack-ups are designed for exploration drilling and hence move from one site to another. The legs are truss structures and support the platform by standing at the seabed. The legs penetrate the seabed, fit with enlarged sections or footings, or attach to a bottom mat. Typically, jack-ups are not self-propelled and are towed by tugs using the buoyancy of the hull (i.e. the barge). The barge (the buoyant part) enables transportation of all machinery needed for drilling and operations. However, jack-ups may be transported on top of transport barges if needed. After arriving at the offshore oil and gas field, the legs are set on the seabed and the deck is jacked up out of water. During drilling, the deck (i.e. the barge) is stationary and dry. In offshore wind industry, jack-ups are used as service platforms, for example as installation vessels.

Gravity-based: Gravity-based structures (GBSs) are placed on the sea bottom and are kept in place by gravity (using their weight). No additional piles or anchors are needed to hold GBSs in place. GBSs are appropriate for production and storage of petroleum. GBS construction is normally performed in sheltered water close to shore, for example in fjords, as sufficient depth is required. Although there are examples of steel GBS platforms, as GBS platforms require large volume and high weight, they are normally constructed of steel-reinforced concrete, with tanks and cells for ballasting (controlling the buoyancy and weight) of finished installed GBSs. They are transported to the offshore site in upright position and ballasted in the final place to be submerged and sit on the seabed. It may be possible to carry the superstructure (topside) with the GBS (support structure). The soil should resist the huge structure weight. Hence, the seabed properties must be studied to confirm the possibility of an onsite GBS installation. GBSs are prone to scour and sinkage. Scour study and sediment transport analysis should be performed. GBS platforms have been used for offshore wind farms.

Spar: Spar platforms are deep-draft floating cylinders that are designed for drilling, production and storage. The buoyancy of the platform supports the topside and facilities above water. They are stabilized by their ballast and weight. The centre of mass is located much below the centre of buoyancy, which restores and stabilizes the platform. Multiple catenary, slack or taut mooring lines are used to anchor the floating structure to the seabed. The mooring system can be made of wire rope, polyester rope and chain. The main types of spar are classic, truss and cell. The first designs, classic spars, are long cylinders with hard tanks near the top for providing buoyancy. The lower part is called the soft tank, which is used for ballasting. In truss spars, to reduce the weight, cost and hydrodynamic loads, the middle section is replaced by truss structures. To minimize the heave motion, horizontal plates are added. The next generation, the cell spar, consists of multiple stiffened tubes connected by horizontal and vertical plates. Cell spars can be cheaper than the other two designs. Spars cannot be transported upright due to their length. A spar is towed to the offshore site on its side, and afterwards it is ballasted to be upright. Then, the mooring lines are attached to pre-deployed anchors, and the spar is moored to the seabed.

Tension leg platform: A tension leg platform (TLP) is moored by tendons (tension legs or tethers) to the seabed. The difference between weight and excess buoyancy force results in a huge tension force in the tendons which stabilizes the TLPs. TLPs are compliant platforms, and they are restrained in heave, pitch and roll. However, TLPs are compliant in surge and sway. TLPs' natural frequencies in heave, pitch and roll motions are kept above the wave frequencies, while most of the floating platforms are designed to have natural frequencies below wave frequency. In other words, TLPs have high stiffness in heave, roll and pitch due to tethers. The axial stiffness of the tendons controls the heave natural frequency, while pitch and roll also depend on the distance between tendons. Increasing the pipe wall thickness of tethers increases the heave stiffness, and wide spacing between tethers will increase the pitch and roll stiffness (however, this will affect the cost). Installation of TLPs is a challenging operation; the platform is towed to the site and then ballasted so that the tethers are attached. Then, it will be de-ballasted to provide excess buoyancy and provide enough tension in tendons.

Semisubmersible: Semisubmersibles are floating structures made of pontoons and columns. Columns support the deck and provide the area moment of inertia to stabilize the system. The restoring moments are increased by increasing the distance between the columns. The columns are interconnected at the bottom to pontoons (horizontal buoyant structures) which are submerged. Early semisubmersibles were towed by tugs (or self-propelled) to new offshore sites for exploration and drilling. However, new semisubmersibles may be transported by heavy transport ships. Semisubmersibles may be braced by cross elements to resist the prying and racking of wave loads (Chakrabarti, 2005). However, diagonal bracing can be removed to simplify the construction. Braceless semisubmersibles have been widely constructed and installed around the world.

2.3 Offshore Wind Turbines

Offshore renewable energy systems are expected to significantly contribute in the coming years to reach the energy targets worldwide; the leading technology in offshore renewable energy sector that can be considered mature is that for offshore wind turbines (OWTs) (Magagna, 2015). Especially, the bottom-fixed offshore wind turbines

(BFOWTs) have successfully passed into the commercialization and industrialization phases, and there is an increasing trend for using this very promising renewable energy technology in the coming years (EWEA, 2016).

Wind power has been used since ancient years by humans for different types of useful applications. The first onshore wind turbine used for the production of electricity was constructed in 1887 in Glasgow, UK (Price, 2005); in 2012, 510 TWh of electricity were generated by wind turbines globally.

The wind energy E_w is the kinetic energy of air in motion through a surface with area A during time t, and is estimated with Equation 2.1:

$$E_w = \frac{1}{2}mu^2 = \frac{1}{2}Aut\rho_a u^2 = \frac{1}{2}At\rho_a u^3 \qquad\qquad \text{Equation 2.1}$$

where ρ_a is the density of air, and u is the wind speed. The wind power P_w is the wind energy per unit time:

$$P_w = \frac{E_w}{t} = \frac{1}{2}A\rho_a u^3 \qquad\qquad \text{Equation 2.2}$$

Alternatively, compared to onshore wind turbines (that developed majorly since 1900), in the last two decades, the OWTs have been developed significantly. The main advantage for use of OWTs is that, compared to onshore sites, the wind speed in offshore sites are usually higher and as a result more wind resources are available. Moreover, in offshore sites, there are much fewer restrictions regarding land space limitations, and the wind power is more stable. This results in significant increase of the daily operational hours of a wind turbine. Additionally, for most of the large cities that are located near a coastline, OWTs are suitable for large-scale deployment near them, avoiding the need for expensive energy transmission systems. The types of wind turbines that have been deployed in OWT technology are the horizontal axis with two or three blades and the vertical-axis wind turbines. In OWTs, the most commonly used type is the horizontal axis with three blades.

The first offshore wind farm was installed in Denmark in 1991. Since June 2015, there are (fully grid connected in the European Union [EU] countries) 3072 offshore wind turbines in 82 wind farms across 11 countries, with a capacity of 10,393.6 MW (demonstration projects are included) (EWEA, 2015). The majority of the already installed OWTs are placed in the North Sea ($\approx$60%) at an average distance to shoreline equal to 32.9 km and with an average water depth equal to 22.4 m. Around 90% of the already installed OWTs are supported by the monopile fixed-bottom substructure foundation type. Usually, the OWTs are deployed in a farm configuration; the average size of connected wind farms is 368 MW. Nowadays, the offshore installed capacity is growing at a rate of 40% per year. In the UK, 46 GW of offshore projects is registered, and around 10 GW has progressed to consenting, construction or operation; while in the USA, 3.8 GW of OWTs is under development. European countries are the leading players in the OWT market. The two largest OWT farms in the world are the London Array farm (LondonArray, 2016) with capacity 630 MW and the Greater Gabbard farm with capacity 504 MW; both OWT farms are located off the UK coast.

The levelized cost of energy (LCOE) of OWTs in mid-2015 was in the range of 11 ~ 18 €/kWh, while the LCOE of onshore wind turbines is in the range of 6 ~ 11 €/kWh

(Magagna, 2015). In an OWT farm in shallow waters, the wind turbine (tower, nacelle, hub and blades) corresponds to 33% of the LCOE; the support structure equals 24%; the operational and maintenance issues, 23%; the grid connection, 15%; the decommissioning, 3%; and the management, 2% (EWEA, 2016). OWT-related technology is relatively new. It is expected that the costs will reduce in the near future and OWT technology will advance, helping OWTs to be more cost-efficient and competitive. Several steps towards decreasing the LCOE must be performed by researchers jointly with industry. As a result of efforts to decrease the LCOE to an efficient level compared to other renewable energy technologies, the rated power of the wind turbines is continuously increasing (Figure 2.4). Consequently, the demand for different support structures that can withstand the environmental loadings without losing their structural integrity is unavoidable. Apart from decreasing the LCOE of OWTs, environmental concerns should be considered; for example, the effects of deployment of large OWT farms on the behaviour of marine mammals, fishes and seabirds, as well as the effects on the risk of heating the seabirds by the huge wind turbine blades, should be carefully investigated.

Based on the foundation of the support substructure of the wind turbine, OWTs can be divided into the following major categories: (a) fixed-bottom offshore wind turbines (FBOWTs) and (b) floating offshore wind turbines (FOWTs).

Depending on the type of foundation substructure, the FBOWTs can be further divided into: (a) monopile, (b) tripod, (c) jacket and (d) gravity base. In EU countries, the monopile foundation type of FBOWT dominates (91% usage) (EWEA, 2016). In Figure 2.5, the Alpha Ventus monopile OWT supplied by Adwen in the North Sea is presented. Monopiles can be used in either shallow or intermediate waters. For shallow waters, the use of monopiles has several advantages that result in reducing the LCOE of OWTs and making their usage more attractive to companies and investors. The main advantages of using monopiles are the simplicity of their design, the minimum engineering design requirements, the minimum operational requirements for the transition from onshore construction site to offshore operation site and the small installation cost. Usually, monopiles have a cylindrical shape and are thick-walled, with an outer diameter that typically varies between 4 and 6 m. Reflecting the recent trend of increase in the

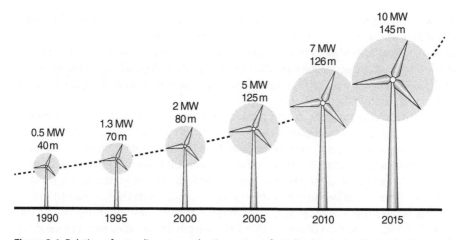

Figure 2.4 Relation of rotor diameter and rating power of wind turbine rotor during the time evolution.

Figure 2.5 Alpha Ventus monopile offshore wind turbine supplied by Adwen in the North Sea. *Source:* Courtesy of SteKrueBe, 2009; this file is licensed under the Creative Commons Attribution-Share Alike 3.0 Unported license.

rating power, the diameter of the monopile FBOWTs needs to be increased. Therefore, the robustness of the applied numerical methods and tools should be validated, and new numerical methods should be developed (Breton, 2009). Typically, approximately 50% of the length of the monopiles is inserted into the soil, depending on the environmental characteristics (e.g. soil properties, wind, wave and current) of the site where the monopile will be installed.

For intermediate water depths, tripod foundation substructures can be used. Often, a tripod consists of a central steel column which is connected to an upper joint with three legs that are tubular pipes and connect the pile sleeve to the upper joint of the column. Moreover, the central column is connected to a lower joint (close to the seabed) with three braces that are tubular pipes and connect the pile sleeve to the lower joint of the central column. Usually the centre-to-centre distance from the column to one of the foundation piles is in the order of $15 \sim 25$ m for wind turbines with rated power less than 5 MW. Typically, approximately 80% of the length of the piles that are used for the tripod foundation are inserted into the soil, depending on the environmental characteristics (e.g. soil properties, wind and wave). The piles are both laterally and axially loaded.

Jackets are also another option that can be used as foundation substructures of FBOWTs. Jacket substructures consist of vertical or inclined legs, horizontal braces and diagonal braces; all of the used elements are usually tubular members. Four legs are used with an inclined vertical direction, and the horizontal braces form a rectangle at the upper level (beginning of the tower of the wind turbine) that becomes larger and larger as we move to the seabed. For the connection of the horizontal and diagonal braces, specific connection types are utilized (e.g. KT type) that have been used extensively in offshore oil and gas industry. Many different combinations of braces and legs can be used in FBOWTs (Byrne, 2003). The foundation piles of the jackets are both laterally and axially loaded.

Finally, gravity base foundations that use their weight to resist the environmental loads can be used as FBOWT substructure. With the use of gravity base foundations, the problems that may exist with the flexibility (e.g. elasticity) of monopiles can be overcome, but on the other hand the cost will increase dramatically since a large amount of concrete is used. Moreover, gravity base foundation substructures require special operations at the seabed, sophisticated site-specific soil analysis and soil–structure interaction analysis. On the other hand, piles are not required for gravity base foundations of OWTs, and this can be considered as the most significant advantage that they have.

Although the cost of onshore wind energy technology has been reduced to a fairly competitive level when compared with traditional power such as thermal power and hydropower, the cost of OWTs is one of the main handicaps for harvesting offshore wind energy. OWTs are more expensive than onshore wind turbines due to the high cost of the (a) supporting structures, (b) sea transportation, (c) installation, (d) maintenance, (e) operation and (f) offshore grids and cables. To reduce the cost of generated power, the development of large-scale wind turbines in deep sea is considered as a feasible and efficient potential direction for offshore wind energy. For deep waters (approximately, water depths greater than 70 m), the use of FOWTs is considered as the most appropriate, mainly from a cost–benefit point of view (Myhr, 2014). FOWTs and their related technology are under intense development.

Depending on the floating support platform configurations, FOWTs can be further divided into (Bagbanci, 2012): (a) spar-buoys, (b) semisubmersibles and (c) TLPs. In Figure 2.6, a semisubmersible-type floating offshore wind turbine is presented.

A spar-buoy is a very simple support substructure of FOWTs that very often consists of a steel cylinder, concrete ballast and mooring line system. Spar-buoys are ballast-stabilized structures that keep the centre of gravity below the centre of buoyancy for suppressing the roll and pitch motions. The small water area of the spar-buoy in combination with the deep draft result in the reduction of the heave hydrodynamic loads (Karimirad and Moan, 2012b). Spar-buoys are relatively simple structures with a minimum number of required welds during their construction. The first deep-water FOWT, which was put in operation in 2009, is the Hywind in the North Sea off the Norwegians coast; Hywind is a spar-buoy FOWT with a 100 m draft that carries on a 2.3 MW wind turbine (Figure 2.7).

Semisubmersible platforms can be characterized as column-stabilized since the columns of the platform provide the required ballast and flotation stability (Karimirad,

Figure 2.6 University of Maine's VolturnUS 1:8 floating wind turbine. *Source:* Courtesy of Jplourde umaine, 2013; this file is licensed under the Creative Commons Attribution-Share Alike 4.0 International license.

Figure 2.7 Hywind floating wind turbine. *Source:* Courtesy of Lars Christopher 2009; this file is licensed under the Creative Commons Attribution-Share Alike 2.0 Generic license.

2016). Alternatively, the columns of the semisubmersible platform can be connected by pontoons without any kind of braces for reducing the fatigue damage of the platform during its lifetime (Karimirad and Michailides, 2015a, 2015b). Semisubmersible FOWTs usually have low draft that results in larger wave-induced motions that may affect the rotor life cycle. Proper design of the mooring lines will lead to insignificant effects of the produced power for the case of semisubmersible floating wind turbines (Michailides, 2015). Semisubmersibles can be constructed and assembled onshore or in a dry dock, which is considered one of the main advantages of this type of floating wind turbine.

Finally, TLPs can be used as supporting structures in FOWTs (Nematbakhsh, 2015). Usually, TLPs consist of one main cylinder, submerged arms and tendons. The submerged arms are used for connecting the main cylinder with the tendons. TLPs are floating structures with no ballast, and the pitch-restoring stiffness is provided by the mooring system (tendons). Due to the smaller wave-induced motions of the TLP and compared to the other possible FOWTs platforms (spar-buoys and semisubmersibles), the fatigue loads in the tower are expected to be much smaller (Bachynski, 2014). Moreover, TLPs can be fully assembled in dry dock, and a low number of welds are required. On the other hand, the tendon tensioning and installation require special attention and great operational efforts.

For both FBOWTs and FOWTs, a steel transition piece is installed at the upper part of the structural member of the foundation substructure where the tower of the wind turbine will be placed. With the transition piece, a rigid connection between the tower of the wind turbine and the support structure of the FBOWT or FOWT is achieved. Usually, the gap between the foundation support structure and the transition piece is filled with grouting concrete; damages have been reported so far for these connections induced by fatigue loads (Dallyn, 2015). Nonlinear local type numerical models have to be developed for the robust numerical analysis of these connections.

A typical horizontal-axis wind turbine consists of the following parts: tower, nacelle, hub and three blades. The hub is attached to the nacelle, and the blades are connected with the hub. The blades with the hub assembly are often called the *rotor*. The blades are made of composite materials and are connected with the hub through bolted connections. The dimensions of the blades depend upon the rated power of the wind turbine and can be considered as flexible elements; their flexibility should be accounted during numerical analysis of the OWTs. The blades, through an appropriate control system, can rotate around their longitudinal axis of rotation to reduce the amount of lift when the wind speed becomes too strong. The hub usually is a ductile cast-iron component that behaves rigidly and has large mass weight (e.g. for a 2 MW wind turbine, the weight of the hub is 8 ~ 10 tons). The hub is covered by a nose cone to protect it from environmental loadings.

The nacelle of the wind turbine is a box-shaped component that sits at the top of the tower and is connected to the hub. The majority of the components of the wind turbine, namely the gearbox, generator and main frame, are placed into the nacelle. The total weight of the nacelle is large (e.g. for a 3 MW wind turbine, the nacelle weighs 70 tons). The large weight of different wind turbine components should be included in the numerical analysis in their real position since intense dynamic effects are expected to exist. Figure 2.8 shows a wind turbine hub being installed.

The tower of the wind turbine is a tubular flexible element. Tower dimensions depend upon the rated power of the wind turbine (e.g. for a 5 MW wind turbine, the height of

Figure 2.8 Wind turbine hub during installation. *Source:* Courtesy of Paul Anderson 2008; this file is licensed under the Creative Commons Attribution-Share Alike 2.0 Generic license.

the tower is approximately 90 m, while the diameter at the base is 6.5 m). Wind turbines for application in offshore areas have rated power of 2 ~ 5 MW; greater wind turbines of up to 10 MW are under development. A 5 MW and a 10 MW reference wind turbine have been proposed for offshore applications and are widely used for benchmark studies from different research groups. For the dynamic analysis of FBOWTs and FOWTs, usually the nacelle and its components are simplified and modelled as mass points with six degrees of freedom. More sophisticated models can be used for the numerical modelling of nacelle components through multibody drivetrain generator models; these models use (as input) shaft loads from different global integrated numerical analysis models and provide (as output) time series of loads for inclusion in the global numerical analysis models of the whole structure.

The loads that should be taken into account for the analysis of FBOWTs and FOWTs are: (a) aerodynamic loads, (b) hydrodynamic loads, (c) wind turbine control loads, (d) earthquake loads (mainly for FBOWTs), (e) ice loads, (f) current loads, (g) tidal loads, (h) soil–structure interaction loads and (i) marine growth loads. Usually, aerodynamic, hydrodynamic and wind turbine controller loads dominate the response of FBOWTs and FOWTs and define their lifetime assessment (Karimirad and Moan, 2012c).

Here, we will examine three different types of aerodynamic loads: (a) steady loads induced by the mean wind speed; (b) periodic loads induced by the wind shear, rotation of the rotor and shadow of tower; and (c) randomly fluctuating aerodynamic loads (e.g. gusts). The steady aerodynamic load (thrust force) T is attributed to the deceleration of the wind speed due to the presence of wind turbine and is calculated with Equation 2.3:

$$T = \frac{1}{2} A \rho_a \left(v_1^2 - v_2^2 \right)$$

Equation 2.3

where v_1 is the wind speed far in front of the rotor, v_2 is the wind speed far behind the rotor and A is the rotor swept area.

Considering a constant thrust force for a wind turbine is an extremely simplified model, since in reality wind speed is not constant over the height of the wind turbine. Therefore, aerodynamic analysis should be carried out to determine the acting wind loads more precisely. Two numerical methods are widely used for the aerodynamic analysis of FBOWTs and FOWTs and can be coupled with the use of finite element method tools, namely the blade element momentum (BEM) and the generalized dynamic wake (GDW) method (Hansen, 1993). Alternatively, other methods that are rarely used are the vortex method, the panel method and the Navier–Stokes solver methods. In general, different methods have been used for different purposes depending on the required accuracy and computational cost of the required analysis.

With regard to the hydrodynamic loads, the Morison equation is often used for the slender members of the support structure. For cases where large-volume structures are involved in FBOWTs and especially in FOWTs (e.g. semisubmersibles and spar-buoys), usually the wave loads are estimated with the use of the linear potential theory and panel models in frequency domain, taking into account the diffraction and radiation effects. The foundation substructures are considered to behave rigidly, and the hydrodynamic coefficients that are calculated with the frequency-domain analysis are used in the subsequent time-domain motion analysis (Karimirad, 2013). The first-order hydrodynamic wave loads dominate the responses of FBOWTs and FOWTs, but also (and depending on the type of support platform) the second-order hydrodynamic wave loads due to the difference- and sum-frequency effects have to be included in the coupled numerical analysis. Second-order difference wave loads can be calculated with the use of the Newman approximation method or the use of the full quadratic transfer function, while sum-frequency loads can be estimated with the use of the full quadratic transfer function (Robertson, 2014). Appropriate methods should be used for the estimation of these types of loads.

Wind turbine control loads need to be addressed for the numerical analysis of FBOWTs or FOWTs. The purpose of the control system in a turbine level is to regulate the start-up, shutdown, fault monitoring and operational condition of a wind turbine. The controller provides input to local dynamic controllers (e.g. generator torque and blade pitch) in order to conform with the modifications to operational or environmental conditions (Jonkman, 2011). For wind speeds between cut-in and rated wind speed, the blade pitch should be kept constant by the controller with the application of some forces, and the generator torque should vary such that the turbine operates as closely as possible to the rated speed. Active control systems should be integrated within the dynamic analysis of FBOWTs and FOWTs.

Different numerical methods can be used for the dynamic analysis of FBOWTs and FOWTs. Mainly, these methods can be categorized into fully coupled (aero-hydro-elastic-servo) methods and uncoupled numerical methods (Bachynski, 2012). A strong interaction between the motions of the different parts of the wind turbine with the loadings exists. Fully coupled integrated numerical methods along with their analysis in the time domain considering the wind turbine controller effects are desired. On the other hand, various types of uncoupled analyses are useful with small computational efforts (compared to fully coupled analysis), especially for investigating the response in local-level analysis (e.g. drivetrain, stress at column pontoon intersections, tendon connection detail to the TLPs and grouting connections).

Experimental investigation of the functionality of either FBOWTs or FOWTs has been conducted and reported so far by various researchers (Müller, 2014). For physical model testing of wind turbines, one particular, highly important uncertainty related to interpretation of the model test results is the scaling effect, since it is not possible to simultaneously scale both the aerodynamic loads according to Reynolds's law and the hydrodynamic loads using Froude's law (Huijs, 2014). Additionally, there are different techniques for the rotor's modelling and for induced thrust force physical modelling. The rotor can be simplified as a disk providing drag force or as a controlled fan providing an active force to mimic the thrust force. For the latter case, the geometrically scaled rotor will result in less corresponding thrust force at model scale as compared to a full-scale rotor (Martin, 2012). Hence, a redesign of the rotor blade is then necessary to achieve the correct thrust curve. Usually, the blade pitch angle of the wind turbine is fixed, but it can be manually adjusted prior to the tests. Recently, active pitch control mechanisms of blades similar to what is expected for the full-scale pitch-regulated wind turbine are used in physical model testing of OWTs.

2.4 Wave Energy Converters

In the coming years, offshore renewable energy systems are expected to significantly contribute to reaching energy security targets worldwide; significant opportunities and benefits have been identified in the area of ocean wave energy. Ocean waves are an extremely abundant and promising resource of alternative and clean energy; it is estimated (Panicker, 1976; Thorpe, 1999) that the potential from ocean wave power resource worldwide is estimated to be in the magnitude of terawatts (TW), a quantity that is comparable with the present power consumption globally. Compared to other forms of offshore renewable energy, ocean wave energy has significant benefits: higher energy density that enables the energy devices to extract more power from a smaller volume, limited negative environmental impacts, long operational time and more predictable energy. Solar energy has a typical intensity of $0.1 \sim 0.3 \, kW/m^2$ per horizontal surface of the earth, while wind intensity is typically equal to $0.5 \, kW/m^2$ per area perpendicular to wind direction. In ocean surface, the average power flow intensity is typically $2 \sim 3 \, kW/m^2$ per area perpendicular to the wave direction.

In deep water, the ocean average (mean) energy E per unit area of horizontal sea surface (J/m^2) is:

$$E = \rho g H_{mo}^2 / 16 = \rho g \int_0^\infty S(f) df = \rho g \int_{\beta 1}^{\beta 2} \int_0^\infty S(f,\beta) df d\beta \qquad \text{Equation 2.4}$$

where ρ is the density of the water (kg/m^3), g is the acceleration of gravity (m/sec^2), H_{mo} is the significant wave height of the sea state and S(f) is the wave spectrum (m^2/Hz). In cases where the waves are propagating in a combination of different directions, $S(f,\beta)$ is the direction energy spectrum, β is the direction of wave incidence and $\beta_1 < \beta < \beta_2$ are the directions that are contributing to the estimation of the spectrum. In cases where waves from all directions are contributing, then $\beta_1 = -\pi$ and $\beta_2 = \pi$. This energy is equally partitioned between kinetic energy, due to the motion of the water, and potential energy.

In deep waters, the wave energy flux P (kW/m) is equal to:

$$P = \frac{\rho g^2}{64\pi} H_{mo}^2 T_e \approx 0.5 H_{mo}^2 T_e \qquad \text{Equation 2.5}$$

where T_e is the wave energy period (sec). For example, and based on Equation 2.5, for waves that are propagating with $H_{m0} = 2\,m$ and $T_e = 6\,sec$, the corresponding wave energy flux is equal to $P = 12\,kW/m$.

The energy of the waves is transported as the waves are propagating; their energy transport velocity is the well-known group velocity c_g (m/sec):

$$P = Ec_g \qquad\qquad \text{Equation 2.6}$$

The possibility of converting wave energy into usable energy has inspired a lot of researchers and inventors since the end of the nineteenth century. Many different types of wave energy converters (WECs) have been inspired and proposed, more than 1000 patents were registered by 1990 (McCormick, 1981) and since then the number has widely increased (Michailides, 2015). The first patent was registered in 1799 in France by a father and son named Girard (Ross, 1995). Unfortunately, the technology of WECs is not mature enough for large-scale commercial deployment. WECs span a wide range of different proposed concepts so far and do not converge on a particular concept for commercialization and industrialization. Four key words, with nearly equal importance, can be considered as the leading design parameters for WECs: (a) functionality, (b) survivability, (c) structural integrity and (d) maintenance. The research community has to focus on all four parameters equally, using appropriate integrated numerical analysis methods and tools as well as physical model testing.

Around 170 developers are associated with the wave energy sector, with 45% of them in Europe. Already 46 companies have reached their concepts to the level of testing their prototype in open real sea conditions. The possibility of converting the ocean energy to electric power with the use of WECs has been approved for installation of prototypes totalling around 125 MW (76 MW in Europe, 43 MW in Oceania, 4 MW in Asia and 1.5 MW in USA). Meanwhile, a lot of projects (with total capacity around 500 MW) were awarded funding by the EU or by national programs globally. The LCOE of WECs in mid-2015 was in the range of 50~65 €/kWh, which can be considered very high.

A very basic parameter that characterizes the functionality of a WEC is the capture width, which is the width of the wave front that contains the same amount of power as that absorbed by the WEC. The CW can be calculated by dividing the produced power of WEC, P_{WEC}, by the power available in the sea:

$$CW = \frac{P_{WEC}}{P_{iw}} \qquad\qquad \text{Equation 2.7}$$

where the power of incident waves of frequency ω and amplitude A may be written as:

$$P_{iw} = \frac{\rho_w g^2 A^2}{4\omega}\left(1 + \frac{2\kappa d}{\sinh(2\kappa d)}\right)\tanh(\kappa d) \qquad\qquad \text{Equation 2.8}$$

where κ is the wave number, d is the water depth, g is the acceleration of gravity, ρ_w is the water density and A is the wave amplitude.

The very large number of proposed WECs can be categorized depending on the: (a) water depth and location of application, (b) type and size consideration of WECs and (c) working principle in which wave energy can be absorbed.

Based on the location where the WEC is applied, WECs can be categorized as: (a) shoreline WECs, (b) near-shore WECs and (c) deep-water offshore WECs. Shoreline WECs are fixed structures that are placed close to the utility network; the main disadvantage of this type of WECs is that as the waves are approximating the shore, their energy decreases. Near-shore WECs are usually deployed in shallow waters and typically are attached to the seabed; as for the case of shoreline WECs, wave energy phenomena (e.g. shoaling) are observed and result in a decrease of the available wave energy. For a specific type of WECs (e.g. oscillating bodies), the decrease of the produced power due to shoaling phenomena is estimated to be in the order of 35% (Michailides, 2015). To date, 64% of the WECs are designed for offshore sites; most of the installations have been performed at near-shore sites.

Based on the type and size considerations or, alternatively, on the horizontal (longitudinal) dimension and orientation, the WECs can be categorized as: (a) line absorbers and (b) point absorbers. If the WEC has horizontal direction so that its value is comparable to or even larger than typical wavelengths, the WEC is called a line absorber (attenuator or terminator); when the larger horizontal dimension of the WEC is parallel to the wave propagation direction, then the WEC is categorized as an attenuator. However, when the wave propagation is normal compared to the longer horizontal direction of the WEC, the WEC is categorized as a terminator. WECs are categorized as point absorbers when they have horizontal dimensions that are small compared to the wavelength of the incident waves.

Based on the working principle in which energy can be absorbed (Falcão, 2010) WECs can be categorized as: (a) oscillating water columns, (b) oscillating bodies and (c) overtopping structures (Figure 2.9). Oscillating water columns can be either fixed onshore structures or floating structures with partly submerged parts that consist of a chamber with an internal free surface that is forced to oscillate by the incident waves. The oscillation of the internal free surface forces the air above that is inside the chamber to flow through a turbine that usually drives a generator (e.g. turbine).

Oscillating bodies present a major category of WECs that up to now enumerate a very large number of proposed WECs. Usually, this type of WEC is based on the harnessing of the relative motion(s) between (a) two or more oscillating bodies and (b) an oscillating body and the sea bed, and converting this relative motion(s) into power by using a power take-off (PTO) mechanism. This category of WECs can be further categorized as: (a) single-body heaving buoys, (b) two-body systems operating in one degree of freedom, (c) fully submerged heaving systems, (d) pitching devices, (e) bottom-hinged systems and (f) multiple-body systems.

In overtopping structures, usually a large quantity of sea water is captured in a reservoir where the mean water level is higher compared to the surrounding sea mean water level. In such cases, the potential energy that is stored in the reservoir is converted to useful energy through conventional low-head hydraulic turbines. Very comprehensive and useful reviews with regards to the WEC technology can be found in Falcão (2010) and Falnes (2007).

For all the aforementioned categories of WECs, the produced power is generated with the use of a PTO that is placed into the WEC. In general, the PTO is the mechanical interface that converts the motion of the WEC itself, or different parts of the WEC, into useful power. The basic categories of the PTOs that are used in WEC technology are: (a) air turbines, (b) low- and high- head water turbines, (c) high-pressure oil driven hydraulic motors and (d) linear electrical generators.

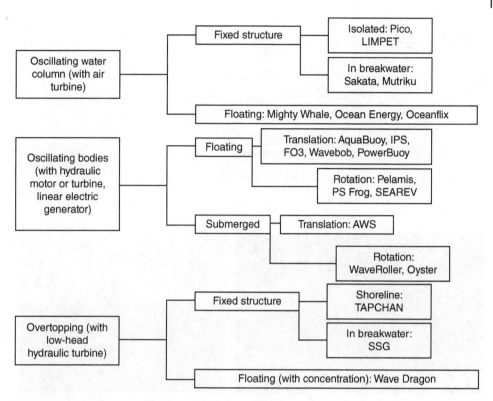

Figure 2.9 Categorization of WECs based on the working principle in which energy can be absorbed. *Source:* Falcão (2010).

Air turbines are used by different type of WECs, but usually they are used by the oscillating water columns. Using air turbines, the flow of air drives the turbine that is directly connected to the generator; the main advantage of using the air is that the slow velocities of incident waves are transformed to high air flow rates very easily. Since the conventional air turbines do not cover efficiently the demands of wave energy production, self-rectifying air turbines have been studied for adoption in WEC technology. Several types of air turbines have been proposed so far; the Wells turbine is the most frequently used in oscillating water columns technology because of its ability to rotate in the same direction irrespective of the direction of the air flow.

Usually, low-head conventional hydraulic turbines are used in the overtopping WEC type, while high-head turbines (e.g. Pelton) can be efficiently used in the oscillating-bodies WEC type. An important advantage of this type of PTO is that the efficiency is approximately 0.9 (the maximum efficiency is equal to 1.0).

High-pressure oil driven hydraulic motors are employed in a large number of WECs. Usually, the motions of the oscillating body due to waves are turned into the flow of a liquid in a hydraulic circuit that includes a hydraulic cylinder, a high-pressure accumulator, a low-pressure accumulator and a hydraulic motor (Figure 2.10). Different challenges exist for this type of PTO and require further research regarding fluid containment, the sealing of fluid, efficiency, maintenance, the end-stop issue and energy storage. Falcão (2008) gives a comprehensive description about the PTO's equipment and the basic quantities of their hydraulic function.

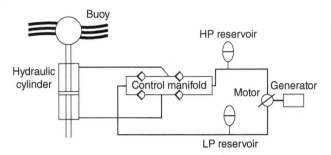

Figure 2.10 A schematic representation of a hydraulic PTO.

Figure 2.11 Prototypes of (a) Pelamis wave energy converter (WEC) on site at the European Marine Energy Test Centre and (b) Wave Dragon WECs. *Source:* (a) Courtesy of P123 (2008); this file is licensed under the Public Domain license. (b) Courtesy of Friis-Madsen (2008); this file is licensed under the Creative Commons Attribution 3.0 Unported license.

Electrical linear generators within WECs can be used as an alternative PTO system. The main components of this type of PTO are the translator that is connected with the oscillating body and on which magnets are mounted, and the stator that contains windings. As the translator moves relative to the stator, an electric current is induced to the stator, and consequently electricity is produced. Rhinefrank (2006) and Mueller (2002) describe the equipment, the basic quantities and the functionality of the electrical generators in the WEC-related technology.

A very comprehensive review with regard to the mechanical PTO equipment for WECs can be found in Salter (2002). In Figure 2.10, a schematic representation of a hydraulic PTO is presented. The motion of the buoy applies a force to the hydraulic cylinder that moves in its axial direction; this motion has, as a result and through a control manifold, a high pressure, a low pressure and a motor, the generation of power.

Very promising WECs from all categories according to the working principle in which energy can be absorbed are briefly described in this section. The Pelamis (Figure 2.11a) (Yemm, 2000) is an attenuator WEC type consisting of slack mooring lines and four cylindrical bodies (modules) that are interconnected with special hinged joints. The PTOs are placed into these joints and can be categorized as high-pressure oil-driven hydraulic generators. Great efforts in numerical and experimental campaigns have been performed for the last 20 years for Pelamis. Pelamis prototype experiments were performed at sea off the Scotland coast in 2004.

Another promising WEC is the Oyster (Whittaker, 2007), which is a bottom-hinged WEC operating in pitching mode as an inverted pendulum. Oyster consists of a flap that has a piercing surface and spans the whole water depth. The wave-induced rotation of the flap is converted into useful energy by means of a hydraulic system that uses sea water and powering an onshore Pelton turbine. Prototype experiments of Oyster were performed at sea, also off the Scotland coast, in 2009.

Moreover, Wave Dragon (Figure 2.11b) (Kofoed, 2006) is an overtopping WEC that consists of two wave reflectors focusing the waves towards the ramp, a doubly curved ramp, a water storage reservoir and a set of low-head turbines. Prototype experiments of Wave Dragon were performed in Nissum Bredning, Denmark, in 2003 and tested for several years.

Mighty Whale (Washio, 2000) is a floating-type oscillating water column; it consists of a floating barge platform that contains three air chambers at the front and buoyancy tanks. Wells air turbines are installed in all the chambers and generate power. Prototype experiments of Mighty Whale have been performed in Gokasho Bay, Japan, since 1998 and tested for several years.

The two critical parameters for the numerical analysis of WECs are the appropriate estimation and inclusion of the wave loads in the equation of motion of the WECs as well as the inclusion of PTO's motion effects into the equation of motion (Michailides, 2014). On one hand, the numerical analysis of WECs is performed to estimate the power absorption of PTOs and consequently the produced power of the WEC. On the other hand, the numerical analysis is used to predict the wave-induced motions of the WEC's different components as well as to estimate the structural response and behaviour of the WEC system and its components for further Ultimate Limit State (ULS) and Fatigue Limit State (FLS) design checks. For the design of WEC mechanical components or hydraulic components (such as the PTO systems), usually a hierarchical analysis method is required. For this method, a global analysis is performed firstly with the use of a simplified model (normally with one degree of freedom) of these components, followed by an analysis with a detailed multibody simulation or finite element model.

Depending on the purpose of the analysis, hydrodynamic loads can be modelled as integrated force/moment, distributed force/moment or distributed pressure, and structural components of WECs. WECs can be modelled as rigid bodies, flexible beams or shell finite elements for structural response analysis. The hydrodynamic wave loads on the WECs can be estimated using the potential flow theory or Morison's equation. When applying the potential flow theory, the hydrodynamic coefficients of added mass and potential damping as well as the first- and second- order wave excitation hydrodynamic loads are obtained in the frequency domain by using (usually) a panel model, and then applied in time-domain simulation tools for dynamic analysis in which viscous effect can also be included as drag forces on the structural components. Fully nonlinear potential flow models and/or CFD models and tools can be used to capture nonlinear phenomena for extreme wave conditions. Afterwards, these nonlinear phenomena must be included appropriately in the solution of the equation of motion in the frequency or time domain. It must be noted that for overtopping WECs, the linear potential theory cannot cover their analysis, and more sophisticated nonlinear numerical methods and tools have to be used even for preliminary studies and estimations. For

WECs that consist of multiple bodies, the hydrodynamic interaction between the different bodies must be estimated appropriately and included in the solution of the equation of motion. WEC oscillating bodies are characterized by flexibility, since they are usually constructed by connecting multiple bodies with connectors. Hydroelasticity is a very important factor for this type of WECs, and appropriate numerical methods have to be developed to calculate the forces that the PTO exerts in rigid as well as in flexible generalized degrees of freedom (Michailides, 2012, 2013). The hydroelastic effects have connection with both the functionality and survivability of WECs and consequently with the lifetime of the WECs.

With regard to the PTO and its inclusion in the equation of motion, usually for preliminary studies and assessments of WECs, the PTO is simplified and considered to behave linearly; and consequently, it consists of a linear damper and a linear spring. The approximation of linear wave loads with linear PTO allows the linear frequency domain analysis to be adopted with low computational time. However, this analysis can be considered as very initial, since the PTO for most of the WEC cases behaves as a nonlinear damper and time-domain analysis is required.

2.5 Tidal Energy Converters

Tidal energy and its related technology can play a major role in the required energy sustainability of the future. In general, tidal energy results from the gravitational and centrifugal forces between earth, moon and sun. Tides are usually generated by the rise and fall of the ocean surface, depending on the gravitational pull of moon and sun on earth, which is larger in magnitude on the nearest side of earth, which is closer to moon. At the same time, centrifugal tidal forces are generated by the rotation of earth and moon around each other, which is larger in magnitude on the side of earth, which is closer to moon (Mazumder, 2005). The aforementioned tidal phenomena with different intensity occur very frequently (e.g. twice every 24 hours, 50 min, 28 sec) and at predefined times. Tidal energy consists of gravitational (vertical) water motion that can be characterized as rise and fall of the water and of horizontal water motion that can be characterized as current (Clark, 2007). The rise and fall of the water will generate potential energy, while the horizontal current will generate kinetic energy. The tidal energy is the sum of the potential and kinetic energy (Lemonis, 2004).

Tides are an extremely abundant resource of alternative and clean energy in the oceans; it is estimated that worldwide the tidal resources are 3 TW, with 1 TW of them being technically usable. The very basic advantage of tidal energy compared to the other forms of renewable energy is that tidal energy is highly predictable, with daily, bi-weekly, biannual and annual cycles of prediction over a longer time span. Moreover, tidal energy is less affected by the weather conditions compared to offshore wind and ocean wave energy technologies and their related technologies. Also, the environmental impact of the tidal energy converters (TECs) that are used for tidal energy extraction is relatively low. Contrary to the high predictability of tidal energy, the production of tidal energy into useful electric power is still not mature enough; this fact poses different challenges and gives uncertainty to this technology (Rourke, 2010). The LCOE of TECs in mid-2015 was in the range of 33 ~ 46 €/kWh (Magagna, 2015), which can be considered very high. More than 100 companies are associated and invested in the tidal energy sector, with more than 50% of them in Europe. Apart from EU countries, activities in tidal energy sector can be found in Australia, eastern Asia, Canada and the USA. TECs that are using ocean currents, and more specifically the kinetic energy of tides, represent the commonest type found in research and development (R&D) in tidal energy sector, reaching 76%. Since 2008, 11,000 MWh of electricity was generated by TECs in the UK. In recent years, a lot of projects in tidal energy sector have been announced in EU countries, with an installed capacity that will reach 1500 MW (Denny, 2009).

TECs can be categorized into two major types with their own related technology: (a) tidal barrage converters that make use of the potential energy part of the tides and (b) tidal current turbines that convert the kinetic energy part of the tidal energy into useful power (Baker, 1991).

Tidal barrage converters have many similarities with the hydroelectric technology, with the difference that for the case of tidal barrages, the water flow can be conducted in two opposite directions and not only in one. Usually, tidal barrages are built in areas with tidal variation that exceeds 5 m water level height, and they utilize the potential energy of this difference in head between ebb tide and flood tide. These special environmental conditions that are independent of the weather conditions can be found in locations where, due to geological and ecological conditions, a large volume of water flow exists (Baker, 1991). Usually, tidal barrages consist of: (a) turbines, (b) sluice gates, (c) embankments and (d) ship locks. The turbines that are used are either unidirectional or bidirectional; they are bulb, straflo, rim or tubular type turbines. Tidal barrages can be further categorized as: (a) single-basin tidal barrages that can use ebb, flood or two-way generation methods to produce electricity; and (b) double-basin tidal barrages for matching the delivery of the electricity to consumer demands.

The largest operating tidal barrage converter is the La Rance facility in Brittany, France (Charlier, 2007). The length of the barrage is 720 m with a surface area of the estuary that is equal to 22 km^2; 24 reversible bulb turbines with rated power of 10 MW each are placed inside the barrage walls and convert the tidal potential energy to electricity. The total generating capacity of the La Rance facility is 240 MW. Other operating tidal barrage converters are in Canada, Russia, China, the UK, the USA, Norway, Argentina, Ireland, Iran, South Korea and Australia. Figure 2.12 presents the tidal barrage in La Rance, France.

The tidal current converters convert the kinetic energy part of the tides to electricity. Most of the large-scale demonstration projects use horizontal-axis tidal current turbines. The technology of tidal current converters has a lot of similarities with the wind turbine technology; the main differences are related to the nature of the physical problem, since in tidal current converters the flow speed is slower compared to the air and the density of the fluid is higher compared to the density of the air depending upon the location (e.g. 832 times higher) (Bryden, 2004).

The tidal current converters can be categorized as (Bryden, 1998): (a) horizontal-axis axial-flow tidal turbines (HATTs), (b) vertical-axis cross-flow tidal turbines (VATTs) and (c) oscillating hydrofoil/reciprocating turbines. In terms of power efficiency, for the same amount of water volume, the HATTs dominate compared to the other types. For the case of HATTs, the turbine blades rotate around an axis that is parallel to the direction of the current; meanwhile, for VATTs, the turbine blades rotate around an axis that is perpendicular to the direction of the current. Usually, HATTs and VATTs can be either shrouded or open and consist of two or three blades, a hub, a gearbox, a generator and a support structure. The

(a)

(b)

 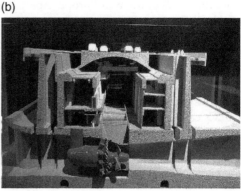

Figure 2.12 (a) La Rance, France, and (b) scale model of the power station. *Source:* (a) Courtesy of P123 2008; this file is licensed under the Creative Commons Attribution-Share Alike 3.0 Unported license. (b) Courtesy of Unknown 2002; this file is licensed under the Creative Commons Attribution-Share Alike 3.0 Unported license.

shrouded turbines are used to accelerate the incoming flow for achieving larger efficiency by the use of a shaped infrastructure. The incoming flow of the tides pushes the blades to rotate around their axis of rotation and also forces the hub to rotate. The hub is connected with a power shaft and transmits the generated torque. The torque through a gearbox is transmitted to a generator for power production that is transmitted on land through undersea cables. All the parts (apart from the blades and hub) are placed in an underwater nacelle that is mounted on the support substructure. Instead of using the rotational motion of the blades, the oscillating hydrofoil/reciprocating tidal turbines use the up-and-down motion of the hydrofoils as the tidal stream flows. The hydrofoils are usually shaped like airplane wings and are connected with a shaft or a piston for power production. Usually, the oscillation of the hydrofoil/reciprocating is controlled with the modification of the hydrofoils' angle of attack with regard to the direction of the incoming tide.

Tidal current converters and related technologies can be considered early mature, since there is a small number of commercial projects. However, tidal current turbines are expected to rapidly develop and industrialize in the coming years; the mean installed capacity of projects under development is 10 MW, and HATTs technology is being used. The projects are placed in Australia, Canada, France, India, Korea and the UK (Rourke, 2010). Moreover, prototype sea tests have been performed for different types of tidal current turbines so far at the European Marine Energy Centre in Scotland, UK, since 2005. The most advanced and developed TEC is the MCT SeaGen with a turbine capacity of 750 kW that was installed in the Bristol Channel in 2003. A second device with 1.2 MW capacity is being prototype tested in Strangford Lough, Ireland (Denny, 2009). The MCT can be deployed in water depths in the range of 20–40 m, with a peak current velocity of 2.25 m/sec. The MCT rotors are between 15 and 20 m diameter and can pitch 180° degrees for use in bidirectional flows. The world's first commercial-scale and grid-connected tidal stream generator – SeaGen – in Strangford Lough is presented in Figure 2.13.

Regarding the numerical analysis of tidal barrages, there are many similarities with the methods that are used in hydropower technology, and they will not be further described here. To estimate TECs' response, performance and functionality, appropriate numerical

Figure 2.13 SeaGen, in Strangford Lough. *Source:* Courtesy of Fundy 2003; this file is licensed under the Creative Commons Attribution-Share Alike 3.0 Unported license.

models have to be developed. These models can be divided into global models and local models. The global models are capable of estimating the power performance and for the structural assessment of different parts that compose the tidal turbines. Local flow field models are capable of detailed estimation of the interaction between incoming fluid and TECs. For both types of models, hydrodynamic loads should be estimated and applied (with the use of different approximations and different levels of robustness). For the case of the global models, momentum theory, blade element method and blade element momentum theory are used to estimate the produced power of the tidal current turbine. These methods have the advantages of low computational cost, fast prediction of produced power and capability of design parametric analysis in a reasonable timeframe. Usually, balancing of the momentum and the energy of the water flow through the TECs provides the produced power.

Alternatively, inviscid flow models and viscous flow models can be used to estimate the hydrodynamic loads on blades or foils. With inviscid flow models, the inviscid and irrotational flow approximation is adopted. Usually, the lifting line method, vortex lattice method and boundary integral method are used for the estimation of the hydrodynamic loads. To update the inviscid flow models, usually viscous flow correction models are used, especially for when the turbine blades are positioned at high attack angles (blade stall and post-stall conditions). It should be mentioned that when the tidal current turbine operates at a high angle of current attack, the viscous effects, namely static/dynamic stall and boundary layer flow separation, dominate the response and behaviour of the blades and should be accounted for in the numerical analysis. High-fidelity numerical analysis of tidal current turbines can be performed with the use of viscous flow models based on the solution of Navier–Stokes equations. For tidal current turbines, usually the Reynolds averaged Navier–Stokes equation method that accounts for the flow vortices and turbulence is used (Pinon, 2012). For achieving more detailed levels of analysis, detached eddy simulation and large eddy simulation models that account for local flow perturbations and blade-generated wakes are used. The main drawback of this type of analysis is the huge computational time, the high level of required computer capacity and the simplification of the supporting substructure with the use of boundary conditions. Apart from the inviscid flow and viscous flow models, unsteady flow models can be used to estimate transient loads on the blades induced by the non-homogeneous unsteady incoming flow (high turbulence levels and velocity shear) during their rotation.

Physical model testing of tidal current turbines is, for the most part, in the early stages of development. Derived data from physical model tests are used to validate the developed numerical methods and to assess the range of their validity. Moreover, physical model tests can be performed to assess the effects of unsteady hydrodynamic loading on tidal turbines' components. Drawbacks and limitations of physical model testing of tidal current turbines are the: (a) blockage ratio, (b) scaling inequalities, (c) carriage shake and vibrations and (d) carriage speed tolerance and uniformity.

A comprehensive study especially for the blockage ratio should be performed prior to the tests, since large values of the blockage ratio will result in the fluid not expanding around the device (as it does in the real physical problem) and will result in completely different measured (larger) produced power. The blockage ratio is defined as the ratio of the rotor's area to the basin's cross-section area.

As for the case of the other offshore renewable energy systems (WECs and OWTs), the physical modelling of the tidal current turbines is dominated by the modelling of the PTO mechanism of the turbine and the applicability of the similitude laws. The PTO parameters that should be physically modelled are the rotation per minute of the shaft and the applied torque at the rotor. Usually, the PTOs can be physically modelled in the lab with mechanical, hydraulic or magnetic configurations or, alternatively, with the use of electrical torque control (hybrid testing). Also, the rotor can be forced by a speed-controlled motor to a specific rotation per minute in cases where the focus of the experimental tests is to estimate the hydrodynamic loads in unsteady conditions. Finally, the blades are usually scaled, redesigned physical models of reference blades.

2.6 Combined Offshore Energy Systems

The LCOE of offshore renewable energy technologies has to be decreased for these structures to be commercialized further, especially for the case of TECs and WECs. The exploitation of these energy resources has a direct positive effect on the general growth and enhancement of the energy security of different countries. In any case, the exploitation of the vast offshore wind, tidal and ocean wave energy potential should be realized in a sustainable manner, considering energy effectiveness, cost efficiency, safety, adequate resistance and durability in harsh sea environmental conditions, and environmental impact issues. It may be beneficial to combine offshore renewable energy systems of different technologies into one structure or, even better, into a farm array operating in shallow, intermediate or deep waters. This in turn will create new design, research and technological challenges. Novel and reliable combined offshore energy systems (COES) should be developed, satisfying functionality (high energy effectiveness), survivability and integrity requirements, but at the same time corresponding to cost-efficient solutions. Moreover, computational tools for the integrated dynamic analysis of COES should be developed, while appropriate laboratory experiments for demonstrating the potential of these solutions should be conducted.

Possible advantages as a result of the use of COES are: (a) increase of the energy production per unit area of ocean space; (b) efficient use of the ocean space; (c) decrease of the system balancing cost, since the wave and tidal resources are more predictable than wind resources; (d) decrease of the cost related to the required electric grid infrastructure, since the different technology energy systems will share the electric grid, especially with regard to the transportation of the produced power to onshore stations; (e) for some types of COES, the sharing of the cost related to the foundation substructure (e.g. semisubmersible or spar platform) will lead to the decrease of the cost and if possible of the LCOE; (f) decrease of the costs related to operation (e.g. installation) and maintenance (e.g. inspection) compared to the costs that similar pure offshore energy systems have operating as OWTs, TECs or WECs; (g) contribution to the requirement by national institutions for addressing energy demands using renewable energy sources; (h) ensuring uniform distribution and decentralization of the electrical energy production over even remote areas (islands); (i) for the case that COES are operating in farm configuration, the WECs or TECs can protect OWTs by creating shadow effects in the local wave climate by decreasing the wave height induced by their appropriate operation; and (j) TECs and WECs can be installed in the ocean after the installation and during the operation of OWTs, mainly for the case where the OWTs are in an array farm configuration.

COES can be categorized into the following types: (a) hybrid combined energy substructures (HCES), farm combined configurations (FCCs) and (c) multipurpose substructures (MPS). HCES are offshore structures that produce power by two or three different resources (e.g. wind and wave) using the same foundation substructure. HCES can be further categorized as fixed-bottom and floating-type HCES. Moreover, HCES can be classified depending on the number and technology type of offshore energy converters that they use; they can be classified as HCES that use one wind turbine and multiple WECs or TECs, or as HCES that use two or more wind turbines in one platform with large dimensions and a small number of WECs or TECs. For the FCCs, two major categories can be defined. The first category concerns independent arrays of OWTs, TECs or WECs that operate in ocean areas that are different but in close proximity, and are sharing the electric grid infrastructure as well as operational and maintenance costs. The second category concerns combined arrays of OWTs, TECs or WECs that, additionally to the above, are sharing the ocean area. For the latter category, the WECs or TECs can be positioned at the gap space between OWTs in an existing farm in a uniform or non-uniform distribution depending on the environmental characteristics and the type of operation of WECs or TECs. MPS are usually large ocean structures with a completely different basic operation (e.g. floating airport or floating bridge) that provide the foundation substructure space to the possible energy system configurations for deployment.

Recently, EU research projects have been introduced to accelerate the development of combined offshore energy systems; these projects are the MARINA (2016), ORECCA (2016), TROPOS (2016), H2Ocean (2016) and MERMAID (2016). Several researchers have studied combined concepts utilizing different types of floating support platforms, OWTs, WECs and TECs (Aubult, 2011; Soulard, 2013; Michailides, 2014).

The only combined wind and wave concept that was tested as a prototype in real sea conditions (off the coast of Lolland in Denmark) is the Poseidon P37 concept, which completed over 20 months of grid-connected tests. The Poseidon P37 (Plant, 2016) consists of a foundation support platform with dimensions of 37 m breadth, 25 m length and 6 m height; it weighs approximately 320 tons. It consists of three two-bladed wind turbines, 11 flap-type WECs and a mooring system.

In the EU project (MARINA, 2016), three combined concepts have been selected and studied both numerically and experimentally under operational and extreme conditions. The selection was based considering five criteria: the cost of energy, constructability, install ability, operation and maintenance, and survivability (Gao, 2016). These combined concepts are the semisubmersible with rotating flaps combination (SFC) (Michailides, 2014), spar torus combination (STC) (Muliawan, 2013) and array of oscillating water columns combination (AOWCC) (O'Sullivan, 2013). The combined-concept SFC (Michailides, 2016) consists of a braceless semisubmersible floating platform with four cylindrical-shaped columns (one central column and three side columns) and three rectangular-shaped pontoons with large dimensions, a 5 MW wind turbine placed on the central column of the semisubmersible platform, three rotating flap-type WECs hinged at the pontoons of the semisubmersible through two rigid structural arms and linear PTO mechanisms, and three catenary mooring lines positioned at the three side columns of the semisubmersible. STCs are characterized by their simplicity and consist of a spar platform, a 5 MW wind turbine, a torus-type WEC that is coaxially located with the spar and connected with the spar with PTO mechanisms, and a delta-type mooring line system (Michailides, 2014). AOWCCs consist of a V-shaped concrete large floating platform, a 5 MW wind turbine placed at the head of the V shape of the platform and 32 oscillating water columns operating as WECs and placed in the large concrete platform. To evaluate the behaviour of the aforementioned combined concepts, numerical models for the coupled dynamic analysis have been developed, while laboratory experiments in controlled environmental conditions for demonstrating the functionality and survivability of these concepts have been successfully conducted. In Figure 2.14, a physical model of an SFC placed into the ocean basin is presented.

With regard to the required numerical analysis of combined concepts, the same approximations as for the OWTs, WECs and TECs (as presented in the previous sections of this chapter) should be considered. Additionally, and in order to estimate the functionality and survivability of the combined concepts, the following issues should be addressed: (a) hydrodynamic interaction between different bodies that might be used for the combined concepts, (b) mechanical interaction between the different energy converters, (c) control strategies for maximizing the produced energy, (d) establishment of survival modes, (e) joint probability distributions for the estimation of the

Figure 2.14 Physical model of an SFC (semisubmersible with rotating flaps combination) placed into ocean basin.

environmental loads (e.g. wind, wave and current), (f) use of high-fidelity CFD tools for the estimation of wake effects and shadow effects of the offshore energy systems and (g) optimization techniques using genetic algorithms or, even better, neural networks for the overall design of the combined systems.

2.7 Multipurpose Offshore Structures and Systems

In order to share ocean space, infrastructures and costs in diverse offshore activities such as energy, transport, aquaculture, protection or leisure, multi-purpose offshore structures and systems (MPOSSs) have been developed. MPOSSs have intense fluid–structure interaction phenomena and special inherent features (e.g. flexibility, nonlinear behaviour, hydrodynamic interaction, control and multicriteria optimization) that must be appropriately addressed during their design and analysis phase. The design and realization of MPOSSs present increasing complexity and require extensive sophistication and technological advances; robust computational models and numerical analysis tools must be used. Furthermore, suitable laboratory physical model tests are required to ensure structural integrity and efficient performance of these complicated systems.

To investigate the possibility of the development of MPOSSs, research programs have been raised during the last two decades. Very recently, EU research projects have been introduced to accelerate the development as well as the establishment of regulations and rules of MPOSSs. Some projects are the MARINA (2016), ORECCA (2016), TROPOS (2016), H2Ocean (2016), MERMAID (2016) and PLOCAN (2016). The future use of the MPOSSs will result in: (a) new business opportunities, (b) new related technology, (c) cost reduction of the use of offshore structures, (d) efficient use of the ocean space, (e) low-carbon offshore structures that are environmentally friendly and meet the requirements of EU or other nations' policies, (f) increases of local industries in the sites where the MPOSSs are applied and of their economy, (g) new employment opportunities and new research and (h) greater educational horizons.

MPOSSs can be categorized, based on the main functionality use of the floating structure, as: (a) industrial MPOSSs, (b) energy MPOSSs and (c) leisure MPOSSs. The main function of the industrial MPOSSs is transport or energy-related issues (oil and gas). This type of MPOSS can integrate offshore wind turbines, wave energy converters, tidal energy converters, aquaculture systems and/or leisure functions related to maintenance and repair of offshore structures. Energy MPOSSs deploy large platforms that involve offshore renewable energy systems. Usually, energy MPOSSs involve offshore wind farms and other renewable energies (e.g. wave, tides and solar) and can integrate aquaculture installations (e.g. fish cages) and leisure components (e.g. diving, equipment testing and validation laboratory). Leisure MPOSSs are usually deployed in shallow waters, and the main functionalities include entertainment and living actions. Usually, offshore renewable energy systems and aquaculture can be integrated into leisure MPOSSs.

With regard to the required numerical analysis of MPOSSs, the same approximations as for the different types of offshore structures (e.g. OWTs, WECs, TECs and SFTs) as presented in other sections of this chapter should be taken into account. Additionally, and in order to estimate the functionality of the MPOSSs, the following issues should be addressed: (a) hydrodynamic interaction between different bodies, (b) mechanical interaction between different mechanical equipment, (c) control strategies, (d) flexibility and hydroelasticity, (e) joint probability distributions for the estimation of the environmental loads (e.g. wind, wave and current), (f) use of high-fidelity CFD tools for the estimation of nonlinear effects at the local model level, (g) multicriteria optimization techniques, (h) feasibility studies and (i) synergies studies between the different technologies to estimate the effects of one technology's operation on the functionality of the second one.

2.8 Submerged Floating Tunnels

Tunnels that are placed in water for transportation purposes have been constructed since 1900. The most commonly used structural types of tunnels in the oceans are (a) bridges in which the deck is above water, and (b) immersed tunnels that lay at the seabed. An alternative solution for crossing sea straits, fjords or inland waters is the submerged floating tunnel (SFT) (Larssen, 2010). SFTs (also called Archimedes bridges) are concepts going back at least 150 years. Historic records show that a rather complete understanding of this idea was brought forward by Sir James Reed in the UK in 1886; he proposed an SFT for railway purposes supported on caissons, but this idea was rejected by the UK Parliament. Later, in 1924, Trygve Olsen Dale of Norway proposed a new type of SFT. Since SFTs can be placed at a specific depth below the water surface, they do not need the long and/or steep approaching roadways that are necessary for conventional underground tunnels or traditional immersed tunnels resting on the seabed, and thus they can be more economic. Also, their application is enhanced by the existence of poor foundation and/or deep-water conditions (Jakobsen, 2010). Basic advantages of the SFTs compared to the fixed ones are mainly: (a) reduced environmental impact, (b) their transportation and reallocation ability, (c) the relatively short installation periods, (d) the flexibility for future extensions, (e) reduced requirements for the foundation, (f) their application in seismically active offshore and coastal areas and (g) their lower construction and installation costs. As a result of all these advantages, the application of SFTs in the coming years should be considered as highly visible (Tveit, 2010). On the other hand, complicated numerical models are required for the implementation of appropriate numerical analyses and assessment of their static and dynamic responses and behaviour.

The main component of SFTs is the traffic tube or tubes where the transportation (e.g. cars, trains, bicycles and/or people) takes place. The tube(s) floats at a certain depth and is balanced in place by all the induced loads and mainly by the buoyancy, weight and anchoring system. Usually, SFT tubes are designed to be installed at a water depth between 20 and 50 m, so that the hydrostatic pressure will be easy to deal with during the design, and moreover for ships to pass easily over the SFT. A very important factor in the design of SFTs is the relation between self-weight and buoyancy; the ratio between buoyancy and weight load must be less than 1.0, and usually it is between 0.5 and 0.8. SFTs have to be designed to withstand all possible environmental loadings as well as accidental loads (e.g. ship collision) without losing part of their structural integrity. Environmental loadings that should be addressed during the numerical analysis of SFTs are incident waves, currents, tidal variations, earthquakes, pretension losses of concrete's steel bars, corrosion of structural steel bars and braces, temperature differences, hydrostatic pressure, ice, material degradation and marine growth.

Usually, SFTs consist of: (a) tubes, (b) station-keeping systems and (c) soil support connections. Based on the station-keeping system, SFTs can be divided into those that are kept in place with either the use of floating pontoons or the use of any anchoring system that is restrained to the seabed; the former solution is suitable for seas with very deep water (Walter, 2010). In Figure 2.15, a sketch of the numerical model for the case of an SFT station kept with pontoons is presented.

The SFT tubes should be designed to provide enough space for the traffic lanes and for all the mechanical equipment. Depending on the SFT's design, the length of the tubes can be some hundred meters in length with a cross section that usually is circular (primarily from a hydrodynamic loading point of view) or alternatively can be elliptical, rectangular or polygonal. Sole and dual tubes can be constructed; for the case of dual tubes, these are connected with tubular steel braces. The material of the tubes is usually concrete or steel. The tubes carry all the loading types of SFTs, namely permanent, variable and accidental loads, and should be designed for an ultimate limit state, serviceability limit state and accidental limit state. With regard to the station-keeping system of SFTs, the possible systems can be categorized as: (a) a pontoon system, (b) a tether or column system and (c) unanchored. Pontoons can be used for the station keeping of SFTs in a specific location along the length of the tube (e.g. 200 m); usually, the pontoons are barge-shaped floating structures that are connected with the tubes through truss steel structures. Alternatively, tethers that are in pretension or fixed-bottom columns can be used for the station keeping of SFTs. The tethers are designed to always be in tension without any possible slack condition, and the tethers usually are oriented in

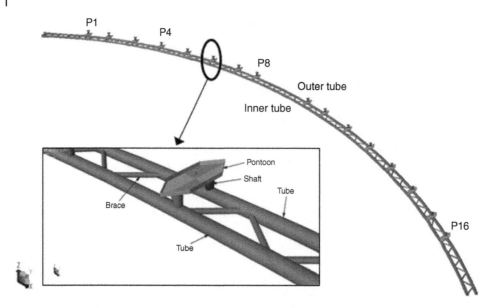

Figure 2.15 A numerical model of a submerged floating tunnel (SFT) with pontoons.

the vertical direction. Unanchored systems can be used when the SFT is short (e.g. 100 m). The soil support connections of SFTs are usually constructed by concrete; special attention should be paid during the soil–SFT interaction's numerical analysis for its reliable design.

With regard to the required numerical analysis, initially all the possible loads should be accounted: wave, current, self-weight, traffic, tide, wind, snow and ice, ship impact, passing ship and deformations. Usually, for a description of a sea state, a 50-year return period is considered, and the corresponding maximum significant wave height and wave period are calculated and used. Also, swells may be significant; they should be defined and afterwards included in the numerical analysis. Usually, directionality functions are used to obtain two-dimensional descriptions of the sea states. Second-order wave loads and the corresponding slowly varying forces are included in the numerical analysis (especially for the design of tubes). With regard to the current loads, drag forces should be included in the analysis as well as vortex-shedding loads due to pressure changing, since the tubes have large diameters (e.g. 13 m) when they have the shape of a cylinder. The self-weight consists of two parts, the permanent self-weight and the variable one; the permanent self-weight includes the weight of the structural parts of tubes, the weight of the material of the tubes, the required permanent ballast for station keeping, the weight of the pavement and the weight of the mechanical equipment. Meanwhile, the variable weight includes the weight of marine growth and the weight that the concrete absorbs from surrounding water. For the traffic loads, design standards and recommendations are used for their estimation. Tidal loads can play a governing role in the design of SFTs; the highest and lowest astronomical tides over a 100-year return period should be used (Tveit, 2000). The accidental case where a ship impacts the pontoon of SFTs should be considered during the design of SFTs. To reduce the effect of ship impact on the global response of SFTs, usually nonlinear weak-link mechanisms are studied, and sophisticated nonlinear analysis is required for this type of analysis. Time-varying flows induced by a ship passing above the SFTs should be studied and taken into account as well. Finally, the effect of deformations due to creep, shrinkage, post tensioning and temperature should also be included in the numerical analysis.

Usually, there are two different types of numerical models for SFTs, global models and local models. Local models are usually developed with the use of panels and the finite element method, taking as input load time series from the different types of loads. Local models are characterized by their high-fidelity simulation of the physical problem that they handle. For the development of global models, usually the total length of tubes of SFTs and tethers are modelled through beam elements with appropriate mechanical and loading characteristics. For wave loads, the boundary element method and Morison's equation are often used. For the type of SFTs with pontoons, the pontoons are modelled with rigid bodies, and hydrodynamic analysis is carried out; this type of analysis is multibody and characterized by high computational cost. Since tubes are slender structures, elasticity dominates the response of SFTs; hydroelasticity should be analysed and examined for all the cases of SFTs separately (Loukogeorgaki, 2012; Michailides, 2013).

2.9 Floating Bridges

Floating bridges have been used for crossing seas, rivers and fjords, especially when the water is very deep. In ancient years, floating bridges were constructed mainly for military operations, with the use of small vessels placed side by side and used as a roadway for soldiers and equipment; King Xerxes successfully constructed a floating bridge consisting of small interconnected vessels for crossing the Hellespont in 480 BC during the second Persian invasion to Greece. Military operations can be achieved nowadays with floating bridges, too (Fu, 2012).

Floating bridges take advantage of the natural law of buoyancy of the structure's volume that is submerged in the water to resist the environmental and structural loads. Conventional piers or seabed foundations are not required. Usually, an anchoring system (e.g. mooring lines or tethers) is used for keeping the bridge in place. However, floating bridges are an obstacle for the crossing ships, and a navigational opening near the two edges of the floating bridge or at the bridge's central span is required. The stochastic nature of all the possible loads and the resulting floating bridge's dynamic behaviour present a very complicated dynamic system that is understood to have inherent nonlinear features (Seif, 1998). The design of a floating bridge is dominated by the environmental loadings.

Floating bridges can be constructed in deep waters where the construction of piers is very expensive, in places with poor seabed foundation properties, in areas with intense astronomical tidal phenomena, in places where it is impossible due to geological issues to build a structure, in military operations where the available time is limited, in places with active intense seismicity and in cases where the bridge serves temporary needs and then, after a period of time, is disassembled (Watanabe, 2000).

Floating bridges can be divided into: (a) continuous pontoon-type floating bridges and (b) separated floating-foundation-type floating bridges. Continuous pontoon-type floating bridges are acting as beams on elastic foundations in both vertical and transverse directions. The pontoons can be constructed by reinforced or post-tensioned concrete (Watanabe, 2003). Usually, a continuous-type bridge consists of pontoons that are connected with special joints and form a continuous structure where the top level of pontoons can be used for the traffic operation.

With regard to the separated floating-foundation-type floating bridges, different sub-types can be met, namely, the separated pontoon type, the semisubmerged foundation type and the long-spanned separated foundation type. Usually, separated floating bridges consist of individual pontoons that are placed transversely to the bridge and spanned by a superstructure of steel or concrete that is completely out of water. Figure 2.16 presents sketches of different possible types of floating bridges (Watanabe, 2003).

Different types of floating bridges have been constructed in different countries. Figure 2.17 shows different floating bridges that will be briefly described here. Figure 2.17a presents the Homer Hadley Bridge (left floating bridge) and the Lacey V. Murrow Memorial Bridge (right floating bridge). The Lacey V. Murrow Memorial Bridge was constructed in 1940 in Seattle, USA; has a length equal to 2018 m; and is a continuous pontoon-type floating bridge. Figure 2.17b shows the Hood Canal Bridge, which was constructed in 1961 in Washington State, USA, and has a floating length equal to 1972 m. The Hood Canal Bridge is also a continuous pontoon-type bridge. In Figure 2.17c, the Kelowna Floating Bridge (left floating bridge) and the William R. Bennett Floating Bridge

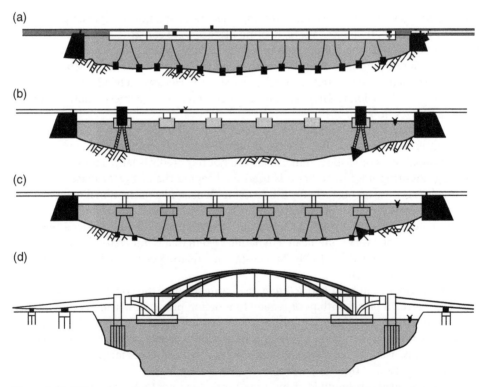

Figure 2.16 Different possible types of floating bridges: continuous pontoon type, separated pontoon type, semi-submerged foundation type and long-spanned separated foundation type. *Source:* Watanabe (2003).

(a)

(b)

(c)

Figure 2.17 (a) The Homer Hadley Bridge (left) and the Lacey V. Murrow Memorial Bridge (right); (b) the Hood Canal Bridge; (c) the William R. Bennett Bridge. *Source:* (a) Courtesy of Tradnor 2005; this file is licensed under the Creative Commons Attribution-Share Alike 3.0 Unported license. (b) Courtesy of Mabel 2012; this file is licensed under the Creative Commons Attribution-Share Alike 3.0 Unported license. (c) Courtesy of Kirby 2009; this file is licensed under the Creative Commons Attribution-Share Alike 2.0 Generic license.

(right floating bridge) that were built on Lake Okanagan in British Columbia, Canada, are presented. In 2008, the William R. Bennett Floating Bridge replaced the Kelowna Floating Bridge, which was constructed in 1958. The newer floating bridge is also a continuous pontoon-type bridge but with the use of a fixed-bottom elevated approach structure on one edge of the bridge.

Other types of bridges are described here. The Nordhordland Bridge was constructed in 1994 in Hordaland, Norway, and has a floating length equal to 1246 m. This bridge forms an arch with a curvature equal to 1700 m and is a separated pontoon-type floating bridge. The pontoons are made by concrete, and the superstructure is constructed by steel box girders. The Yumenai bridge is a very modern, long-spanned separated foundation-type bridge. The floating length of the bridge is 410 m, and the main span has a length equal to 280 m. Moreover, floating bridges can be constructed to serve as a foot-bridge. The West India Quay floating footbridge, which has a length equal to 90 m in London's Docklands and was constructed in 1966, is one of this type. The West India Quay footbridge is a separated pontoon-type bridge in which the pontoons are filled with foam.

For the design and analysis of floating bridges, mainly the wave loading dominates their response and behaviour, as well as the possible moving loads (Fu, 2005). Moreover, the following loads should be appropriately estimated and afterwards accounted during the numerical analysis: (a) wave loads; (b) wind loads; (c) tidal vertical and horizontal loads; (d) hydrostatic pressure loads; (e) internal wave loads due to ship passing near the floating bridge; (f) marine growth; (g) water absorption by concrete; (h) deformations due to temperature, creep and settlements; (i) tsunami loads; (j) ice loads; (k) snow loads; (l) earthquake loads; (m) stable and variable loads; (n) traffic loads; and (o) ship collision loads. Usually, a large number of combinations with appropriate safety factors are required during the design stage of floating bridges according to regulations and codes. With regard to the required numerical analysis, usually and as in the case of the SFTs, global and local models of the floating bridge or different parts should be developed with the inclusion of all possible loading types.

2.10 Aquaculture and Fish Farms

Aquaculture is the farming of aquatic organisms such as fish and plants under controlled conditions. Farming includes procedures to enhance production via regular feeding, protection from predators, monitoring and checking the health condition of the organisms. Aquaculture includes farming of fish, shrimp, oyster, seaweed and similar types of marine organisms that are normally consumed by humans, directly or indirectly (White *et al.*, 2004; Swann, 1992).

Aquaculture is performed both for freshwater and saltwater fish and for other aquatic organisms. The focus of this section of the book is mainly fish farming and fish cages. This is different from commercial fishing and harvesting wild fish. The aquatic organisms are cultivated in aquaculture with regular operation and maintenance, which shows the importance of designing such structures to be capable of handling environmental loads while maintaining proper conditions for aquatic organisms (Rubino, 2008).

In Figure 2.18, the world's fisheries and aquaculture productions (both marine and inland) are plotted for the year 2012. As it is clear, 50% of the total production comes from fisheries (capture-marine). Marine aquaculture is about 16% of the total

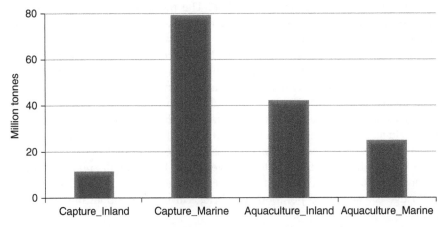

Figure 2.18 World fisheries and aquaculture productions in 2012, both inland and marine. *Source:* FAO (2014).

production, and it is booming very rapidly to respond to international food demand. For more information, refer to the Food and Agriculture Organization of the United Nations (FAO), "The State of World Fisheries and Aquaculture" (2014).

Caged fish farming in ocean is like an operating factory in which different processes occur. This includes but is not limited to transportation of fish, feeding, monitoring and maintenance of cages and fish, cleaning nets and removing dead fish, examining the health and quality of products, and so on. Some of these tasks are performed manually, and some of them are automatic by using a controlling system and autonomous mechanisms (Naylor *et al.*, 2001).

Figure 2.19 shows a schematic layout of a small fish farm consisting of four cages. Shared mooring lines and an anchoring system can be used for cost optimization, taking into account the required safety and structural integrity. The following are important items in ocean fish farms (Vázques Olivares, 2003; Mathisen, 2012; Adoff, 2013; Kankainen *et al.*, 2014; AKVA Group, 2015).

- Feed platform: In advanced fish farms, floating platforms, for example a moored barge, are used for feeding fish. The ship/barge can provide a platform for personnel as well as control units.
- Feed system: This system is arranged for regular and planned feeding of fish. The food is pumped through pipes to different cages from the feed platform.
- Structural components: Cage (plastic or steel), net and mooring system
- Software: A central controller gathers the information from the monitoring system to control the feeding and ensure integrated operation. The current and wave data from environmental sensors will be processed by the software system as well. This is needed for marine operation and checking the environmental windows for different activities in the farm.

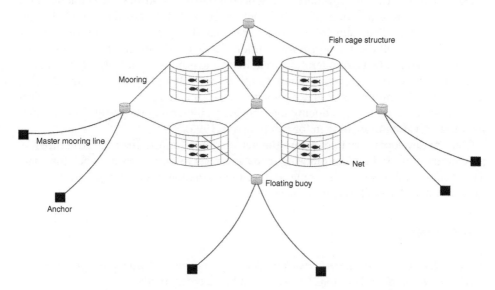

Figure 2.19 Schematic layout of a small fish farm consisting of four fish cages; the shared mooring system and anchoring are shown as well.

- Monitoring system: To ensure optimum operations and healthy fish production, a monitoring system is used. This includes an ROV (remotely operated underwater vehicle), AUV (autonomous underwater vehicle), underwater lights, video camera, sensor systems and similar monitoring systems for checking and observing cages, a mooring system, fish and the feeding process.
- Service boats: Monitoring and maintenance of the fish farm and fish cages are performed using small service boats.
- Cleaning system: This includes a mort (dead fish) collection system and net cleaning.

With fish cage design, one of the most important issues is to ensure survivability of the system under extreme ocean loads. The current, wave and interaction between them are the key environmental actions. Moreover, elasticity of the floating pipe will be important, and the construction is considered as slender marine structures (comparing the pipe diameter and the cage diameter) (Strand, 2013). A fish cage has a floating part that is subjected to wave loads (Thomassen, 2008). This is normally a floating pipe (steel or plastic) with a diameter of 0.2–0.6 m. The normal cage diameter is in the range of 10–70 m. Fatigue and ultimate load checks are essential for such construction (Ágústsson, 2004).

Among important parameters that affect the fish cage design are net shape under environmental conditions and strength of net to reduce fish escaping. Fish escaping through holes is a major concern for concept developers and fish farm owners. Different materials such as polyester, nylon and non-fibre (e.g. steel) are being used for net construction. The net deformations under current loads should not exceed the fish's required water volume. Reduced net volume stresses fish, and decreased oxygen can endanger fish. The stressed fish (due to lack of space) and those that experienced decreased oxygen are subjected to diseases. All these factors reduce the quality of the fish and affect the production of the fish farm. Hence, several studies are being done to maximize water flow through nets, increase the available oxygen and optimize net volume under current loads (Grue, 2014).

Deformation of the net is primarily influenced by current load and net characteristics. The drag load is a function of the current speed and drag coefficients of the net (Fredheim, 2014). The main net properties are strength, rigidity, weight in water, water flow and drag coefficients. Non-fibre solutions have an advantage because the semi-rigid net remains intact in moderate current loads and is not significantly affected under extreme current loads. Several research projects are attempting to optimize the fish farm concept by minimizing the net deformation, maintaining the net shape in survival environmental conditions, reducing the drag loads, increasing the water flow through nets and providing more oxygen for fish (Lawson, 1995).

Concept developers try to decrease the net weight, minimize the cage structure (i.e. pipe dimension), reduce the mooring loads and reserve more buoyancy for the cage. These efforts will result in less drag, reduced deformation, less damage, more fatigue life and reduced overall weight of the system.

References

ABS, 2014. *Dynamic Loading Approach for Floating Production, Storage and Offloading (FPSO) Installations*, Houston, TX: American Bureau of Shipping.

Adoff, G., 2013. *A Guide to Marine Aquaculture: An Introduction to the Main Challenges When Establishing and managing marine aquaculture plants*, Leezen, Germany: www.aquafima.eu.

Ágústsson, G., 2004. *Design Considerations and loads on Open Ocean Fish Cages South of Iceland*, Reykjavík: Faculty of Engineering, University of Iceland.

AKVA Group, 2015. *Cage Farming Aquaculture*, Bryne, Norway: www.akvagroup.com.

Anonymous, 2005. LNG FPSO concept announced. *Naval Architect*, 15(1):20–27.

Anonymous, 2014. Aerodynamic thrust modelling in wave tank tests of offshore floating wind turbines using a ducted fan. *Journal of Physics: Conference Series*, 012089.

Aubult, A. A., 2011. Modeling of an oscillating water column on the floating foundation WindFloat. In: *Proceedings of the 30th International Conference on Ocean, Offshore and Arctic Engineering*, Rotterdam, the Netherlands: s.n., pp. OMAE2011-49015.

Ayala-Uraga, E., 2009. *Reliability-based Assessment of Deteriorating Ship-shaped Offshore Structures*, Trondheim, Norway: NTNU.

Bachynski, E. M. T., 2012. Design considerations for tension leg platform wind turbines. *Marine Structures*, 29(1):89–114.

Bachynski, E. M. T., 2014. Wind-wave misalignment effects on floating wind turbines: Motions and tower load effects. *Journal of Offshore Mechanics and Arctic Engineering*, 136(4):041902.

Bagbanci, H., 2012. Review of offshore floating wind turbine concepts, In *Maritime Engineering and Technology*. London: Taylor & Francis Group.

Baker, C., 1991. Tidal power. *Energy Policy*, 19(8):792–797.

Biasotto, P., Bonniol, V. and Cambos, P., 2005. *Selection of Trading Tankers for FPSO Conversion Projects*, Houston, USA: Offshore Technology Conference (OTC).

Breton, S. F., 2009. Status, plans and technologies for offshore wind turbines in Europe and North America. *Renewable Energy*, 34(3):646–654.

Bryden, I. E. A., 1998. Matching tidal current plants to local flow conditions. *Energy*, 23(9):699–709.

Bryden, I. G., 2004. Assessing the potential of a simple tidal channel to deliver useful energy. *Applied Ocean Research*, 26(5):198–204.

Byrne, B. H. G., 2003. Foundations for offshore wind turbines. *Philosophical Transactions of the Royal Society of London*, 2909–2930.

Chakrabarti, S., 2005. *Handbook Of Offshore Engineering*, New York: Elsevier.

Charlier, R., 2007. Forty candles for the Rance River TPP tides provide renewable. *Renewable & Sustainable Energy Reviews*, 11(9):2032–2057.

Clark, R., 2007. *Elements of Tidal-Electric Engineering*. Hoboken, USA: John Wiley & Sons.

Dallyn, P. E., 2015. Experimental testing of grouted connections for offshore substructures: A critical review. *Structures*, 90–108.

Denny, E., 2009. The economics of tidal energy. *Energy Policy*, 1914–1924.

Devold, H., 2013. *Oil and Gas Production Handbook*. Oslo, Norway: ABB.

EWEA, 2015. EWEA. [Online] Available at: http://www.ewea.org/fileadmin/files/library/publications/statistics/EWEA-European-Offshore-Statistics-H1-2015.pdf [Accessed 30 October 2016].

EWEA, 2016. EWEA. [Online] Available at: http://www.ewea.org/fileadmin/files/library/publications/statistics/EWEA-European-Offshore-Statistics-2015.pdf [Accessed 30 October 2016].

Falcão, A., 2008. Phase control through load control of oscillating-body wave energy converters with hydraulic PTO system. *Ocean Engineering*, 35:358–366.

Falcão, A., 2010. Wave energy utilization: A review of the technologies. *Renewable and Sustainable Energy Reviews*, 14:899–918.

Falnes, J., 2007. A review of wave-energy extraction. *Marine Structures*, 20:185–201.

Finn, L. D., 1976. *A New Deepwater Offshore Platform: The Guyed Tower*. Houston, USA: OTC.

Food and Agriculture Organization of the United Nations (FAO), 2014. *The State of World Fisheries and Aquaculture*, Rome: FAO.

Fredheim, A., 2014. *From Oil Exploration to Fish Farming: Synergies in Research and Collaboration across the Blue Sectors*, Trondheim, Norway: SINTEF Fisheries and Aquaculture.

Fu, S. C. W., 2005. Hydroelastic analysis of a nonlinearly connected floating bridge subjected to moving loads. *Marine Structures*, 85–107.

Fu, S. C. W., 2012. Dynamic responses of a ribbon floating bridge under. *Marine Structures*, 246–256.

Gao, Z. M. T., 2016. Comparative numerical and experimental study of two combined wind and wave energy concepts. *Journal of Ocean Engineering and Science*, 1:36–51.

Grue, I. H., 2014. *Loads on the Gravity-Net-Cage from Waves and Currents*, Trondheim, Norway: NTNU.

H2Ocean, 2016. *H2Ocean*. [Online] Available at: www.h2ocean-project.eu [Accessed 30 October 2016].

Hansen, A., 1993. Aerodynamics of horizontal-axis wind turbines. *Annual Review of Fluid Mechanics*, 25:115–149.

Huijs, F. R. E., 2014. Comparison of model tests and coupled simulations for a semi-submersible floating wind turbine. In: *Proceedings of the 33rd International Conference on Ocean, Offshore and Arctic Engineering*, pp. OMAE2014-23217.

Hwang, J.-H., Roh, M.-I. and Lee, K.-Y., 2012. Integrated engineering environment for the process FEED of offshore oil and gas production plants. *Ocean Systems Engineering*, 2(1):49–68.

Jakobsen, B., 2010. Design of the submerged floating tunnel operating under various. *Procedia Engineering*, 71–79.

Johnson, K. L. 1980. *The Guyed Tower Offshore Platform: Preliminary Design Considerations*. Cambridge, MA: Massachusetts Institute of Technology.

Jonkman, J. M. D., 2011. Dynamics of offshore floating wind turbines: Analysis of three concepts. *Wind Energy*, 14(4):557–569.

Kankainen, M. and Mikalsen, R., 2014. *Offshore Fish Farm Investment and Competitiveness in the Baltic Sea*, Helsinki, Finland: Finnish Game and Fisheries Research Institute.

Karimirad, M., 2013. Modeling aspects of a floating wind turbine for coupled wave-wind-induced dynamic analyses. *Renewable Energy*, 53:299–305.

Karimirad, M. and Mazaheri. S., 2007. *Design of SBMs in Persian Gulf*. Iran: Second Offshore Conference.

Karimirad, M. and Michailides, C., 2015a. Dynamic analysis of a braceless semisubmersible offshore wind turbine in operational conditions. *Energy Procedia*, 80:21–29.

Karimirad, M. and Michailides, C., 2015b. V-shaped semisubmersible offshore wind turbine: an alternative concept for offshore wind technology. *Renewable Energy*, 83:126–143.

Karimirad, M. and Michailides, C., 2016. V-shaped semisubmersible offshore wind turbine subjected to misaligned wave and wind. *Journal of Renewable and Sustainable Energy*, 8:023305.

Karimirad, M. and Moan, T., 2012a. A simplified method for coupled analysis of floating offshore wind turbines. *Marine Structures*, 27:45–63.

Karimirad, M. and Moan, T., 2012b. Feasibility of the application of a spar-type wind turbine at a moderate water depth. *Energy Procedia*, 24:340–350.

Karimirad, M. and Moan, T., 2012c. Stochastic dynamic response analysis of a tension leg spar-type offshore wind turbine. *Wind Energy*, 16(6):953–973.

Kofoed, J. F. P., 2006. Prototype testing of the wave energy converter Wave Dragon. *Renewable Energy*, 31:181–189.

Larssen, R. S. S., 2010. Submerged floating tunnels for crossing of wide and deep fjords. *Procedia Engineering*, 171–178.

Lawson, T., 1995. *Fundamentals of Aquacultural Engineering*, New York: Springer.

Lemonis, G. C. J., 2004. Wave and tidal energy conversion. In: *Encyclopedia of Energy*. New York: Elsevier, 385–396.

LondonArray, 2016. *LondonArray*. [Online] Available at: www.londonarray.com [Accessed 30 October 2016].

Lopez-Cortijo, J., Fene, I., Duggal, A., Van Dijk, R. and Matos, S., 2003. *Case Study: An FPSO with DP for Deep Water Applications*, Marseille, France: DOT.

Loukogeorgaki, E. M. C., 2012. Hydroelastic analysis of flexible mat-shaped floating breakwaters under oblique wave action. *Journal of Fluids and Structures*, 31:103–124.

Magagna, D. U. A., 2015. Ocean energy decelopment in Europe: Current status and future perspectives. *International Journal of Marine Energy*, 11:84–104.

MARINA, 2016. *MARINA Platform*. [Online] Available at: http://www.marina-platform. info/index.aspx [Accessed 30 October 2016].

Martin, H. K. R., 2012. Methodology for wind/wave basin testing of floating offshore wind turbines. In: *Proceedings of the ASME 2012 31st International Conference on Ocean, Offshore and Arctic Engineering*, pp. OMAE2012-83627.

Mathisen, S., 2012. *Design Criteria for Offshore Feed Barges*, Trondheim, Norway: NTNU.

Mazumder, R., 2005. Tidal rhythmites and their implications. *Earth–Science*, 79–95.

McCormick, M., 1981. *Ocean Wave Energy Conversion*. New York: John Wiley & Sons.

MERMAID, 2016. *MERMAID*. [Online] Available at: www.mermaidproject.eu [Accessed 30 October 2016].

Michailides, C., 2015. Power production of the novel WLC wave energy converter in deep and intermediate water depths. *Recent Patents on Engineering*, 9:42–51.

Michailides, C. A. D., 2012. Modeling of energy extraction and behavior of a flexible floating breakwater. *Applied Ocean Research*, 35:77–94.

Michailides, C. A. D., 2014. Optimization of a flexible floating structure for wave energy production and protection effectiveness. *Engineering Structures*, 85:249–263.

Michailides, C. G. Z. M. T., 2014. Response analysis of the combined wind/wave energy concept SFC in harsh environmental conditions. In: *1st International Conference on Renewable Energies Offshore, RENEW 2014*, pp. 877–884.

Michailides, C. G. Z., 2016a. Experimental and numerical study of the response of the offshore combined wind/wave energy concept SFC in extreme environmental conditions. *Marine Structures*, 50:35–54.

Michailides, C. G. Z., 2016b. Experimental study of the functionality of a semisubmersible wind turbine combined with flap-type wave energy converters. *Renewable Energy*, 93:675–690.

Michailides, C. K. M., 2015. Mooring system design and classification of an innovative offshore wind turbine in different water depth. *Recent Patents on Engineering*, 9(2):104–112.

Michailides, C. L. C., 2014. Effect of flap type wave energy converters on the response of a semi submersible wind turbine in operational conditions. In: *33rd International Conference on Ocean, Offshore and Arctic Engineering*, pp. OMAE2014-24065, pp. V09BT09A014.

Michailides, C. L. E., 2013. Response analysis and optimum configuration of a modular floating structure with flexible connectors. *Applied Ocean Research*, 43:112–130.

Mierendorff, R., 2011. *Critical Success Factors for the Efficient Conversion of Oil Tankers to Floating Production Storage Offloading Facilities [FPSOs]*. Lismore, Australia: Southern Cross University.

Mueller, M., 2002. Electrical generators for direct drive wave energy converters. *IEE Proceedings – Generation, Transmission and Distribution*, 149:446–456.

Muliawan, M. K. M., 2013. Dynamic response and power performance of a combined spar-type floating wind turbine and coaxial floating wave energy converter. *Renewable Energy*, 50:47–57.

Müller, K. S. F., 2014. Improve tank test procedures for scaled floating offshore wind turbines. In: *The International Wind Engineering Conference e Support Structures & Electrical Systems*.

Myhr, A., 2014. Levelised cost of energy for offshore floating wind turbines in a life cycle perspective. *Renewable Energy*, 66:714–728.

Naylor, R. J. G., Primavera, J., Kautsky, N., Beveridge, M. C. M., Clay, C., Folke, C., Lubchenco, J., Mooney, H. and Troell, M., 2001. *Effects of Aquaculture on World Fish Supplies*, Issues in Ecology series, Washington, DC: Ecological Society of America.

Nematbakhsh, A. B. E., 2015. Comparison of wave load effects on a TLP wind turbine by using computational fluid dynamics and potential flow theory approaches. *Applied Ocean Research*, 53:142–154.

ORECCA, 2016. *ORECCA*. [Online] Available at: www.orecca.eu [Accessed 30 October 2016].

O'Sullivan, K. M. J., 2013. Techno-economic optimisation of an oscillating water column array wave energy converter. In: *Proceedings of the 10th European Wave and Tidal Energy Conference*. Aalborg, Denmark: s.n.

Paik, J. K. and Thayamballi, A. K., 2007. *Ship-Shaped Offshore Installations, Design, Building, and Operation*. New York: Cambridge University Press.

Panicker, N., 1976. Power resource potential of ocean surface waves. In: *Proceedings of the Wave and Salinity Gradient Workshop*. Newark, DE: s.n., pp. 1–48.

Pinon, G., 2012. Numerical simulation of the wake of marine current turbines with a particle method. *Renewable Energy*, 46:111–126.

Plant, F. P., 2016. *Floating Power Plant*. [Online] Available at: http://www.floatingpowerplant.com/company/ [Accessed 30 October 2016].

PLOCAN, 2016. *PLOCAN*. [Online] Available at: http://www.plocan.eu/index.php/es/ [Accessed 30 October 2016].

Power, L. D., Hayes, D. A. and Brown, C. P., 1984. Design of guylines for the Lena guyed tower. *Journal of Energy Resources Technology*, 106(4):489–495.

Price, T., 2005. James Blyth – Britain's first modern wind power pioneer. *Wind Engineering*, 191–200.

Rhinefrank, K. A. E., 2006. Novel ocean energy permanent magnet linear generator buoy. *Renewable Energy*, 31:1279–1298.

Robertson, A., 2014. *Definition of the Semisubmersible Floating System for Phase II of OC4*, NREL/TP-5000-60601, Golden, CO: National Renewable Energy Laboratory.

Ross, D., 1995. *Power from Sea Waves*, Oxford: Oxford University Press.

Rourke, F. B. F., 2010. Tidal energy update 2009. *Applied Energy*, 398–409.

Rubino, M. (ed.), 2008. *Offshore Aquaculture Offshore Aquaculture: Economic Considerations, Implications & Opportunities*, Silver Spring, MD: US Department of Commerce, National Oceanic & Atmospheric Administration.

Salter, S. T. J., 2002. Power conversion mechanisms for wave energy. *Proceedings of the Institution of Mechanical Engineers, Part M: Journal of Engineering for the Maritime Environment*, 216:1–27.

Seif, M. I. Y., 1998. Dynamic analysis of floating bridges. *Marine Structures*, 29–46.

Shell, 2010. *Perdido*. [Online] Available at: http://www.shell.com/about-us/major-projects/perdido.html [Accessed 31 August 2017].

Shell, 2015. *Floating LNG*. [Online] Available at: http://www.shell.com/energy-and-innovation/natural-gas/floating-lng.html [Accessed 31 August 2017].

Soulard, T. B. A., 2013. C-HYP: A combined wave and wind energy platform with balanced contributions. In: *Proceedings of the 32nd International Conference on Ocean, Offshore and Arctic Engineering*. Nantes, France: n.p., pp. OMAE2013-10778.

Strand, I. M., 2013. *Modeling of Hydroelastic Response of Closed Flexible Fish Cages Due to Sea Loads*, Trondheim, Norway: NTNU.

Swann, L., 1992. *A Basic Overview of Aquaculture*, Lafayette, IN: Purdue University.

Terpstra, T., d'Hautefeuille B. B. and MacMillan, A. A. 2001. *FPSO Design and Conversion: A Designer's Approach*. Paper presented at the Offshore Technology Conference, Houston, TX, 30 April–3 May 2001.

Texaco Press, 2000. *Texaco's Petronius Platform Installation Completed in Gulf of Mexico*. [Online] Available at: https://www.chevron.com/stories/texaco-press-release-texacos-petronius-platform-installation-completed-in-gulf-of-mexico [Accessed 22 November 2015].

Thomassen, P. E., 2008. *Methods for Dynamic Response Analysis and Fatigue Life Estimation of Floating Fish Cages*, Trondheim, Norway: NTNU.

Thorpe, T., 1999. *A Brief Review of Wave Energy*, technical report no. R120, Energy Technology Support Unit, Harwell, UK: AEA Technology.

TOTAL, 2015. *Overcoming the Harsh Conditions of the Ocean Deeps*. [Online]. Available at: www.total.com [Accessed 31 August 2017].

TROPOS, 2016. *TROPOS*. [Online] Available at: www.troposplatform.eu [Accessed 30 October 2016].

Tveit, P., 2000. Ideas on downward arched and other underwater concrete tunnels. *Tunneling and Underground Space Technology*, 69–78.

Tveit, P., 2010. Submerged floating tunnels (SFTs) for Norwegian fjords. *Procedia Engineering*, 135–143.

Vázques Olivares, A. E., 2003. *Design of a Cage Culture System for Farming in Mexico*, Reykjavik, Iceland: United Nations University.

Walter, C., 2010. Conceptual study for a deep water, long span, submerged floating. *Procedia Engineering*, 61–70.

Washio, Y. O. H., 2000. The offshore floating type wave power device "Mighty Whale": open sea tests. In: *Proceedings of 10th International Offshore Polar Engineering Conference*. Seattle, WA: n.p., pp. 373–380.

Watanabe, E. M. T., 2000. Design and construction of a floating swing bridge. *Marine Structures*, 437–458.

Watanabe, E. U. T., 2003. Analysis and design of floating bridges. *Progress in Structural Engineering and Materials*, 127–144.

White, K., O'Neill, B. and Tzankova, Z. 2004. *At a Crossroads: Will Aquaculture Fulfill the Promise of the Blue Revolution?* SeaWeb. www.AquacultureClearinghouse.org.

Whittaker, T. C. D., 2007. The development of Oyster – a shallow water surging wave energy converter. In: *Proceedings of 7th European Wave Tidal Energy Conference*. Porto, Portugal: n.p.

Yan, G. and Yonglin, J., 2008. LNG-FPSO: Offshore LNG solution. *Frontiers of Energy and Power Engineering in China*, 249–255.

Yemm, R. H. R., 2000. The OPD Pelamis WEC: Current status and onward programme. In: *Fourth European Wave Energy Conference*. Aalborg, Denmark: Iben Ø, Søren I, pp. 104–109.

3

Offshore Environmental Conditions

3.1 Introduction

This chapter focuses on the environmental conditions that dominate the design and operation of offshore structures. The proper understanding of these environmental conditions is essential for a subsequent successful design of offshore structures. Initially, basic information about wave processes is presented; the wave is the most important and design-driven environmental element of offshore structures. Next, we discuss wind environmental parameters that are important for offshore structures, especially offshore wind turbines. Then, we deal with tidal currents and their different types. Afterwards, joint distribution of wind and wave conditions, scour and erosion, and extreme environmental conditions are discussed. Finally, we close this chapter by explaining the relevant application of the discussed environmental effects.

3.2 Wave Conditions

Waves are surface waves that occur on the free surface of oceans, seas, lakes, rivers, and canals. Waves are the result of wind blowing over an area of a fluid surface. Waves range in size from small ripples, to waves over 30 m high that travel thousands of miles in the oceans. The period of waves in the oceans can be less than 1 sec (e.g. capillary waves) and has little application in offshore engineering. There exist free surface waves with periods from a few seconds up to a few minutes. Basic characteristics of normal free surface waves with a period between 1 sec and 30 sec will be discussed in this chapter, since it has more application in offshore engineering. Figure 3.1 shows the classification of the spectrum of ocean waves according to wave period.

The inertia that leads to the motion of the returning water surface produces a surface oscillation. When the free water surface oscillates in the vertical direction, the gravity force will act to return the water surface to its equilibrium position. As a wave propagates, a continuous interaction between gravity and inertia exists, resulting in the oscillatory water motion. The water particles are accelerating and decelerating continuously. For the case of two-dimensional (2D) freely propagating periodic gravity waves, a small-amplitude theory can be developed with the use of some basic assumptions (e.g. irrotational flow; stationary, impermeable, and horizontal bottom; pressure along the air–sea interface is constant; and

Offshore Mechanics: Structural and Fluid Dynamics for Recent Applications, First Edition.
Madjid Karimirad, Constantine Michailides and Ali Nematbakhsh.
© 2018 John Wiley & Sons Ltd. Published 2018 by John Wiley & Sons Ltd.

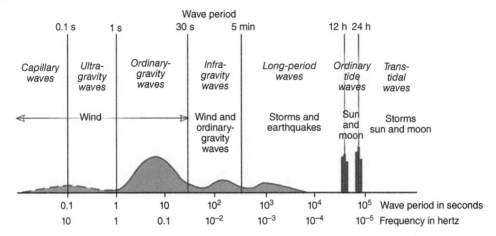

Figure 3.1 Classification of the spectrum of ocean waves according to wave period. *Source:* Munk (1950). This file is licensed under the Creative Commons Attribution 3.0 Unported license (Ryazanov, 2015).

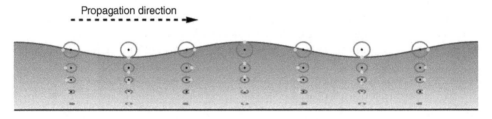

Figure 3.2 Particle motion for deep waters. *Source:* Courtesy of Kraaiennest, 2008. This file is licensed under the Creative Commons Attribution-Share Alike 3.0 Unported license.

wave height is small compared to the wavelength and water depth). Based on mathematical formulations and as the wave propagates, the water particles move in a clockwise orbit. The relevant water particle velocities and orbit dimensions decrease in size with increasing depth below the still-water line. Based on the ratio of water depth d to wavelength L, a very basic classification of the wave propagation properties can be achieved in shallow, intermediate, and deep waters. For the case of deep waters, the water particles are moving in a circular path (Figure 3.2), while moving in an elliptical path for the case of shallow waters. A very good description of wave propagation is in Sorenson (2006).

3.2.1 Basic Characteristics of Free Surface Normal Waves

Free surface waves generated by winds represent one of the main environmental effects on offshore structures. The description of the sea surface can be achieved with the use of irregular waves as the sum of regular-type waves and can be classified in two different types: (a) long-crested waves and (b) short-crested waves. A basic assumption that is used for the case of long-crested waves is that all the waves have the same direction (2D case). The real sea surface is assumed, that is, simulated as the sum of long-crested regular waves all with the same direction. Moreover, it assumes that the wave process is stationary, the wave elevation is normally distributed, and the wave process is ergodic. For a specific location in space, the surface elevation ζ is approximated with:

$$\zeta(t) = \sum_{n=1}^{N} \zeta_n \cos(\omega_n t + \varepsilon_n)$$

Equation 3.1

where n is the number of considered different regular waves and ε_n is the phase angle. For a specific regular component n, the energy of linear waves is:

$$E_n = \frac{1}{2}\rho g \zeta_n^2$$

Equation 3.2

where ρ is the density of water and g the acceleration of gravity. The total amount of energy for the case of long-crested waves and for a sea state can be estimated as follows:

$$E_n = \rho g \sum_{n=1}^{N} \frac{1}{2}\zeta_n^2(\omega_n)$$

Equation 3.3

where $\zeta_n(\omega_n)$ is the wave amplitude with frequency ω_n. With the use of the wave spectrum $S(\omega)$ of the $\zeta(t)$, and for the case that $N \rightarrow \infty$ and consequently $\Delta\omega \rightarrow 0$, the total amount of energy and $\zeta(t)$ are given as:

$$E_n = \rho g \int_{0}^{\infty} S(\omega) d\omega$$

Equation 3.4

$$\zeta(t) = \sum_{n=1}^{N} \sqrt{2S(\omega_n)\Delta\omega} \cos(\omega_n t + \varepsilon_n)$$

Equation 3.5

For the case that the different components n propagate with different angles θ (short-crested waves assumption), and by adding the different wave components (both direction and frequency), the wave elevation is:

$$\zeta(x,y,t) = \sum_{i=1}^{I}\sum_{j=1}^{J} \zeta_{Aij} \cos(\omega_i t - \kappa_i x \cos\theta_j - \kappa_i y \sin\theta_j + \varepsilon_{ij})$$

Equation 3.6

where $\kappa_i = \dfrac{\omega_i^2}{g}$. The total amount of energy for the case of short-crested waves can be estimated as:

$$E = \rho g \sum_{i=1}^{I}\sum_{j=1}^{J} \frac{1}{2}\zeta_{Aij}^2 = \sum_{i=1}^{I}\sum_{j=1}^{J} S(\omega,\theta) d\omega d\theta$$

Equation 3.7

where $S(\omega,\theta)$ is the directional spectrum. With the use of the directional spectrum, and for the case that $\Delta\omega \rightarrow 0$ and $\Delta\theta \rightarrow 0$, the total amount of energy and $\zeta(x,y,t)$ are given as:

$$E_n = \rho g \int_{0}^{2\pi}\int_{0}^{\infty} S(\omega,\theta) d\omega d\theta$$

Equation 3.8

$$\zeta(x,y,t) = \sum_{i=1}^{I}\sum_{j=1}^{J} \sqrt{2S(\omega_i,\theta_j)\Delta\omega\Delta\theta} \cos(\omega_i t - \kappa_i x \cos\theta_j - \kappa_i y \sin\theta_j + \varepsilon_{ij})$$ Equation 3.9

3.2.2 Swells

Swells are the waves that are not generated or affected by locally generated waves, but instead generated by winds elsewhere (Wright, 1999). These waves travel from the origin of generation, usually have a very long wavelength of 300–600 m, and their wave height is a few centimeters. Often, a local storm generates local waves composed of a range of frequencies. The components of the waves with lower frequency travel with faster speed away from the storm. Therefore, soon after, the waves with longer wavelength separate themselves from shorter waves and travel away from the local storm. Usually, the orientation of traveling of these long-period waves is 45° with respect to the wind direction (mainly due to the Coriolis effect). Figure 3.3 presents swell waves in New Zealand.

3.2.3 Wave Propagation in Space

As the wave propagates and moves in the space close to the coasts, different phenomena are taking place mainly due to the variation of water depth. This section presents critical information about two basic processes of waves in space, wave refraction and shoaling.

Wave refraction can be considered as the bending in the wave's crest due to the variation of water depth. As the waves are moving in shallow waters, they tend to travel slower than when in deeper waves. The variation of wave propagation speed at different locations of the wave creates a curve pattern, which is the refraction effect. As the wave moves toward the shore, even a straight-line shore, the water depth usually decreases (Figure 3.4). The decrease in water depth corresponds to decrease in wavelength of waves. Conservation of energy requires that the energy coming in and out of any section

Figure 3.3 Swell at Lyttelton Harbour, New Zealand. *Source:* Courtesy of Phillip Capper, 2008. This file is licensed under the Creative Commons Attribution 2.0 Generic license.

Figure 3.4 Shoaling effect: increase in the amplitude of waves due to decrease in the speed of wave motion as it travels toward shallow waters. *Source:* Courtesy of Régis Lachaume, 2005. This file is licensed under the Creative Commons Attribution-Share Alike 3.0 Unported license.

of the sea has to remain the same (Fox, 1985). Therefore, the waves in shallow waters possess higher amplitude for compensating their lower speed. So, typically as the waves enter the shore, the wavelength decreases while the wave height increases. This effect is called the shoaling effect.

3.2.4 Wave Measurement

Either mechanical- or electric-based gauges are often used to measure wave heights and wavelength. The most typical mechanical-based gauge is the floating buoy that is anchored to the seabed with a tether. The buoy is oscillating on the water free surface. The amplitude and frequency of motion of the buoy can be used for measuring the amplitude and frequency of the waves. The measured data are recorded and are sent to the end user for further post-processing. A popular alternative electrical-based method for measuring wave height and amplitude is called the *radar altimetry method* (Wright, 1999). In this approach, a radar sends pulses (e.g. 1000 pulses per second) to the sea surface. These pulses will pass the air and will reach the water surface, then part of them reflects back to the satellite or another receptor. The time taken for the pulse to reach the surface and come back is an indicator of the amplitude of the waves. The accuracy of such measurement is on the order of 10 cm.

3.3 Wind

3.3.1 Global Wind Pattern

The global pattern of winds is dominated by the difference in the amount of energy received from the sun at different latitudes of the earth; also, the earth's rotation plays an important role. The earth receives energy with higher intensity from regions close to the equator (around $1 \, \text{kW/m}^2$), while at the latitude of 60°, the intensity falls down. As we move from equator to the poles of the earth, the temperature is decreased and the air pressure is increased. We may expect that globally the wind has a circular motion from the poles (high pressure) to equator (low pressure), warms up, and move back to the poles. However, the wind pattern is more complicated than that due to the effects of earth rotation (Coriolis effect).

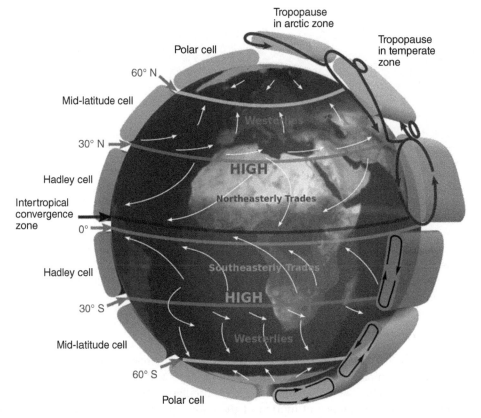

Figure 3.5 World's dominant winds. *Source:* Courtesy of Kaidor, 2013. This file is licensed under the Creative Commons Attribution-Share Alike 3.0 Unported license.

The earth rotates around its axis with a constant angular velocity of 1400 km/s at the equator and zero at the poles. Therefore, as the air moves from the poles to the lower latitude regions, it also moves in from east to west. This creates a global wind pattern that is called a *polar front* (Figure 3.5). Polar fronts affect considerably the weather of the Northern Hemisphere. As the air reaches the latitude of 60°, it warms up, rises up, and recirculates back to the poles. This circular path of the air is called a *polar cell*. The polar front winds in the Northern Hemisphere blow from north east toward south west, and in the Southern Hemisphere blow from south east toward north west. The same mechanism results in motion of air from the latitude of 30° toward the equator. The wind blows from high-pressure higher latitude regions to low-pressure and lower latitude regions. At the same time, the wind moves west due to Coriolis effects and reaches the equator. The global wind pattern due to this air motion is called the *trade winds* due to the importance that they have for the trading ships. In the equator, the air becomes warm, rises up, and recirculates back to the latitude of 30°. The circular path between the equator and latitude of 30° is called the Hudley cell. Between the latitude of 30° and 60°, the air has a circular motion that moves up and toward east, and then recirculates back; the global winds in this region are called Westerlies. In total, there are six circular patterns of air around the globe.

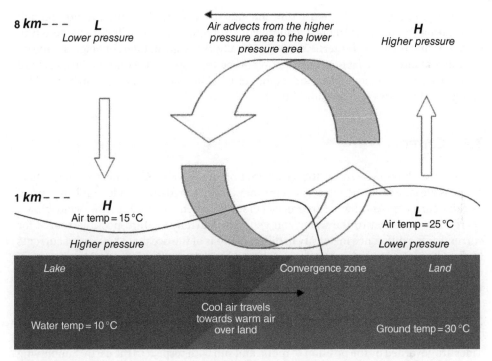

8 km- - - **L**
Lower pressure

Air advects from the higher pressure area to the lower pressure area

H
Higher pressure

1 km- - -
H
Air temp = 15 °C

Higher pressure

L
Air temp = 25 °C

Lower pressure

Lake

Convergence zone

Land

Cool air travels towards warm air over land

Water temp = 10 °C

Ground temp = 30 °C

Figure 3.6 Sea breeze. *Source:* Courtesy of Vaughan Weather, 2006. This file is licensed under the Creative Commons Attribution-ShareAlike 3.0 License.

In addition to the global patterns of wind, there are more local patterns that affect the wind patterns. For example, as the sun heats up the earth in the day, due to the large heat capacity of water, the land usually becomes warmer than the sea. Therefore, on land, the hot air will rise up, and cold air from the sea will breeze from the sea to the shore (Figure 3.6). At night, the land loses its heat while again due to high capacity of water it still remains warm; therefore, at night usually there is a breeze from the land to the ocean.

3.3.2 Wind Measurement

For measuring the wind speed, anemometers are very often used. There are different types of anemometers; the one that is used widely consists of several cups connected to a shaft. Based on the rotational speed of the shaft, the wind speed is measured (Michailides, 2013). Additionally, there are other types of anemometers like the horizontal-axis wind speed gauge. The speed on the horizontal blades needs to be converted appropriately to obtain the wind speed.

Alternatively, another type of anemometer that is used is based on electrical current principles; it is called the hot-wire anemometer and is composed of very thin-diameter wire (e.g. on the order of micrometers). As the air passes the wire, it cools down, affecting the resistance of the wire. Based on measuring the variation of the resistance of the wire, the wind speed passing the wire is then measured.

Usually, and prior to the construction and operation of any offshore structure, enough data during the time for the installation site are required, especially for the design of offshore wind turbines (Butterfield, 2005). Usually, the wind at different heights is measured for obtaining a reliable approximation of the wind speed. Very often, the measured wind speed data are incomplete, and computational intelligence models can be used for filling the data series (Panapakidis, 2016).

3.4   Currents

Currents can be considered as the steady motion of the water. Currents can have different possible generation sources. Currents may be generated due to tides, and are named as tidal currents. Another source of generating current is wind; as the wind blows, energy is transferred to the water and generates current. A third possible source for currents is the difference in density in different areas of the oceans. This type of currents is often developed in a global scale, and their velocity is smaller compared to the case of tidal and wind-generated currents. Therefore, for engineering purposes, the tidal and wind-generated currents are more important and will be discussed in this section.

3.4.1   Tidal Currents

Tides are governed by the position of moon and sun with respect to the earth. Simplifying the planetary motion, the earth is rotating around the sun, the moon is rotating around the earth, and the earth rotates around its own axis. Slightly increasing the complexity of planetary motion, in addition to the aforementioned motions, the earth also has an eccentric rotation around the center of mass of the moon and the earth (Wright, 1999). The interaction of gravity forces between earth and moon is the main source for the generation of tidal currents. The locations on earth closer to the moon experience higher gravitational forces. Furthermore, the direction of forces between earth and moon will change as the moon and earth are rotating, which results in variation of gravitational force directions. The centrifugal forces due to eccentric motion of the earth will partly compensate the gravitational forces of the moon to the earth, while the rest will result in the rising and falling of the seas, called the lunar tides. The lunar tides depend not only on how close or far a certain location is from the moon, but also on the direction of this force. If the force is exactly opposite to the gravity of the earth, the gravity will highly damp the tidal force motion. If the force has an angle with respect to the gravitational force of earth, it may result in higher lunar tides. Since the period of the eccentric motion of the earth and rotation of earth around itself is one day, the period of the currents is in this range. Solar tides also exist; although the mass of sun is incomparable with the mass of moon, the distance between the earth and sun is 360 times greater than between the moon and earth. The gravitational force is proportional to the power of two of the distance. When the sun and moon are aligned, the lunar and solar tides are summed together, while if they are perpendicular they partly cancel each other.

Tides can be diurnal with frequency once per day or semi-diurnal with frequency twice per day. The tides in different locations of the earth can be classified based on the relative amplitude (F) of diurnal and of semi-diurnal tides. If F is greater than 3, the type of tide is mainly diurnal, and it is dominated by lunar tides. If F is less than 0.25, it indicates that the relative position of the moon and sun will affect the tides. If F is between these two threshold values, both effects have relative importance.

Different devices have been developed for measuring the tidal currents. They are usually based on either mechanical or acoustic principles. One of the mechanical-based devices utilizes the rotation of a propeller and is called the rotor current meter. As the current passes the propeller, the equipment starts to rotate. The number of rotations of the device per minute can be used to calculate the current speed. Alternatively, there are other types of mechanical current meters that are based on the measuring of a tilting angle of a hinged bottom flap. It is composed of an anchor embedded in the seabed and a small mooring line connected to the flap. Based on the intensity of the current, the flap is tilted. The measured data are transmitted with radio devices to the end user.

3.4.2 Wind-driven Currents

Wind-driven currents, local or global, are generated by the wind blow. In this subsection, global and local (e.g. longshore currents, rip currents, and upwelling currents) wind-driven currents are briefly described.

3.4.2.1 Global Wind-driven Currents

Global wind-driven currents are dominated by the global wind patterns, the Coriolis effects, and the geomorphology of the continents. The wind-driven currents are usually surface currents since the energy is transported by wind to the surface of the ocean, and typically not more than 100 m below the water's surface is affected by the current (Ekman, 1905). The most important currents on the earth are driven by gyres; gyres are very large-scale circular motions (five major ones on earth) of the water. Figure 3.7 shows the world's largest gyres. One example of these five gyres is North Atlantic gyres, which highly contribute to temperature in western Europe and the east coast of the United States (Figure 3.7). In this gyre, due to the Northern Hemisphere trade winds that are affected by the Coriolis effects, the water moves from west to the east.

3.4.2.2 Longshore Currents

Longshore currents happen close to the shorelines. The direction of the wave propagation is often not normal to the shore; also, the shape of the beach cannot be considered as a straight line. As the waves break on the shore, energy is released and a current parallel to the shore, which is named the *longshore current*, is generated. Longshore currents contribute to sediment transportation along the beaches. This happens because the waves that reach the beach by an angle transport some sediments to the shore. Longshore currents result in significant erosion of the beach.

3.4.2.3 Rip Currents

Rip currents are narrow, fast currents moving away from the beach. A narrow long current may be created that is called a *rip current*. Rip current velocities can be greater than 2 m/s. The width of rip currents can be started from a few meters up to 30 m wide, while the length of the rip current can reach 100 m. The rip current is composed of three main sections: feeder, neck, and head. After the rip current is formed by the feeders, it travels back in the neck regime with the highest speed, and after passing the breaking wave zone of the sea it damps out in the head of the rip.

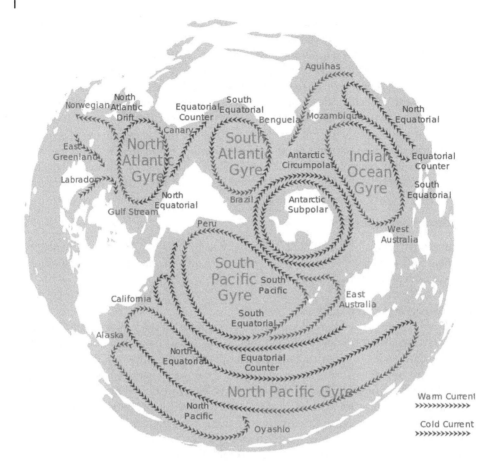

Figure 3.7 World's largest gyres. *Source:* Courtesy of Avsa, 2009. This file is licensed under the Creative Commons Attribution-Share Alike 3.0 Unported license.

3.4.2.4 Upwelling Currents

Upwelling current is a type of wind-driven current created by surface wind, Coriolis effect, and Ekman transport (Ekman, 1905). As the wind blows on the surface of the water, it pushes the surface of the water in the direction of the wind. This results in the water near the surface being dragged. So, a net upward motion of the water exists. At the same time, due to the Coriolis effect, the water is not moving in the same direction with the wind, and it has about 45° deviation. The deviation angle varies in different levels of depth of the water. Both the net upward motion and Coriolis effect result in spiral motion of the water (Ekman transport). As a result, if the wind blows parallel to the ocean, Ekman transport leads to upward motion of the water from the seabed. Consequently, a net current is created below the water's free surface to replace the water transported by the Ekman effect. This creates a current in the lower level toward the shore and a net current away from the shore at the free surface defined as the upwelling current.

3.5 Joint Distribution of Waves and Winds

In design of offshore structures, investigating the structural integrity and considering the environmental conditions, including wind, wave, and current loads, are essential. Also, an assessment of the service life of offshore structures needs reliable and realistic methods accounting for combined wind and wave loads. For example, fatigue and ultimate limit state checks include combinations of loads occurring during the design life. This is quite a complex analysis of simultaneous wind and wave conditions. Moreover, proper numerical simulations depend on accurate models representing the environmental states and related characteristics, such as wave and wind kinematics.

Normally, metocean variables have some relationships. Storm surges, strong winds, and large waves are likely to show some dependencies, as they are usually generated from the same storm conditions. Wave height, wave period, and direction are dependent as they are representing properties of the same physical process. Tides and storm surges are partially dependent due to interactions between these processes. Water levels and waves are independent in deep water, with increasing dependency in shallow water. Waves and currents are independent in deep water, with increasing dependency in shallow water. Water levels and currents are dependent, representing properties of the same driving physical process (Cooper *et al.*, 2008).

Wave, wind, and current are normally linked and correlated to each other; hence, coupled or integrated simulations accounting for simultaneous action of environmental conditions are needed to accurately predict the dynamic responses of ocean and offshore structures. Among different correlations between environmental states, wave and wind have a significant relationship, in particular wind-generated waves. Recent development of offshore wind applications has increased the demand for having better representation of such joint wind and wave distributions.

One of the first attempts to study the correlation of wind and wave was made by Sir Francis Beaufort (1774–1857). The Beaufort scale (Beaufort wind force scale) is an empirical measure relating wind speed to wave conditions; see Table 3.1. The Beaufort scale extends from Force level 0 (calm) to Force 12 (Hurricane), with Force 12 defined as a sustained wind of 32.7 m/s (64 knots) or more (Met Office, 2016). The Beaufort scale roughly shows what may be the ordinary gravity waves (1–30 sec) under specific wind conditions. The Beaufort scale applies only when the sea is fully developed; this means that the wind blows enough times so the waves have reached their maximum height for a particular wind speed. When the fetch and duration of the wind are limited, special care is needed. The *fetch* is the distance over which the wind has blown, and the *duration* is the time it has been blowing. Remember that the sea's surface is influenced not only by wind but also by other factors such as swell (waves from far away), precipitation, tidal streams, and currents.

Developing optimized and robust offshore wind energy converters (OWECs) with a long lifespan is necessary to make offshore wind energy economically viable. Design of offshore wind turbines is highly dependent on accurate joint distributions of wave and wind. This is because coupled simulations accounting for aero-hydro numerical modelling are required. The resulting action and action effects are relatively sensitive to the correlation model for wind and waves. For floating wind turbines and combined wave–wind platforms, this is more significant as wind and wave loads are equally important.

Table 3.1 Beaufort scale, and wind and wave conditions: this guide shows roughly what may be expected in the open sea.

| Beaufort number | Description | Specification for use at sea | Wind speed at 10 m above sea level (m/s) | | | Height of waves (m) | |
			Mean	Limits	Probable	Maximum
0	Calm	Sea like a mirror	0.0	0.0–0.2	0.0	0.0
1	Light air	Ripples are formed, without foam crests	0.8	0.3–1.5	0.1	0.1
2	Light breeze	Small wavelets, still short but more pronounced. Crests have a glassy appearance and do not break.	2.4	1.6–3.3	0.2	0.3
3	Gentle breeze	Large wavelets. Begin to break. Glassy appearance. Perhaps scattered white horses	4.3	3.4–5.4	0.6	1.0
4	Moderate breeze	Small waves, becoming longer, fairly frequent horses	6.7	5.5–7.9	1.0	1.5
5	Fresh breeze	Moderate waves, taking a more pronounced form; many white horses are formed. Chance of some spray	9.3	8.0–10.7	2.0	2.5
6	Strong breeze	Large waves begin to form; white foam crests are more extensive everywhere. Probably some spray	12.3	10.8–13.8	3.0	4.0
7	Near gale	Sea heaps up; white foam from breaking waves begins to be blown in streaks along direction of the wind.	15.5	13.9–17.1	4.0	5.5
8	Gale	Moderately high waves of greater length; edges of crests break into spindrift. The foam is blown in well-marked streaks along the direction of the wind.	18.9	17.2–20.7	5.5	7.5
9	Strong gale	High waves. Dense streaks of foam along the direction of the wind. Crests of waves begin to topple, tumble, and roll over. Spray may affect visibility.	22.6	20.8–24.4	7.0	10.0

10	Storm	Very high waves with long, overhanging crests. The resulting foam, in great patches, is blown in dense white streaks along the direction of the wind. On the whole, the surface of the sea takes on a white appearance. The "tumbling" of the sea becomes heavy and shock-like. Visibility affected.	26.4	24.5–28.4	9.0	12.5
11	Violent storm	Exceptionally high waves (small and medium-sized ships might be lost behind the waves for a time). The sea is completely covered with long white patches of foam lying along the direction of the wind. Everywhere, the edges of the wave crests are blown into froth. Visibility affected.	30.5	28.5–32.6	11.5	16.0
12	Hurricane	The air is filled with foam and spray. Sea completely white, with driving spray; visibility very seriously affected.	N/A	32.7 and over	14.0	N/A

Note: In enclosed waters (or when near land with an offshore wind), wave heights will be smaller and the waves steeper (Met Office, 2016). The maximum wave height is the height of the highest wave expected in a period of 10 min.

In Figure 3.8, an example of an offshore wind-measuring mast is shown. The recording of meteorological figures such as wind speed and direction will be carried out using anemometers and wind vanes, which can be post-processed to find wind distribution. Other meteorological parameters such as temperature, humidity, pressure, precipitation, and solar effects will also be recorded. Wave measurements can be taken with buoys, ship-based equipment, or satellites. Several approaches for obtaining wind–wave correlation have been developed. Some of these models are briefly discussed in this chapter. To relate the wave and wind (i.e. determining sea state characteristics for a certain wind condition), the following methods may be applied (Mittendorf, 2009):

- Statistical correlation methods with jointly measured wind and wave data
- Wave-forecasting methods
- Numerical sea state hindcast
- Joint probability modelling.

The mean wind speed and significant wave height are correlated. One of the simplest approaches to correlate mean wind speeds and significant wave heights is to assign values with equal cumulative probabilities. Then, the relation between mean wind speed and significant wave height with equal probability is derived, for example significant wave heights ($H_S(V_m)$) as a third- or fifth-degree polynomial function of mean wind speed. If the measured metocean data consist of pairs of measured wave height and wind speed, it is pretty simple to directly derive a joint description of the data. For example, models for expected significant wave height H_S are calibrated to the data for a given mean wind speed V_m. Due to a limited amount of data and larger scatter for high wind velocities (i.e. 25 m/s, 10 min averaged at 100 m), the deviations between the measurement and fitted model to the data are normally significant. So, results for higher wind or wave values should be used with care. Advanced methods take into account the observed scattering in the data. An analysis of the wind and wave data with respect to

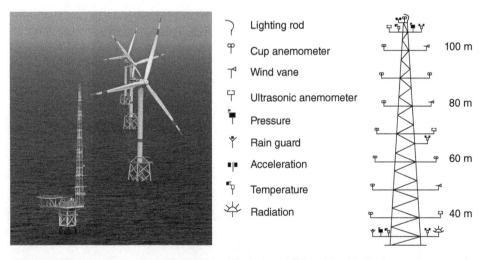

Figure 3.8 Offshore Meteorological Met Mast (Alpha Ventus Offshore Wind Park). *Source:* Courtesy of Martina Nolte (Nolte, 2012). This file is licensed under the Creative Commons Attribution-Share Alike 3.0 Germany license. The file is modified to show the sensors normally used for measuring wind and other parameters (fino3.de, 2012).

certain direction sectors and single storm events results in a more detailed view with less scatter. Also, considering the fetch length as well as the duration of the wind blowing may reduce the variability of the wind and wave data (Mittendorf, 2009).

In the past years, many empirical formulas describing the significant wave height from known properties of the wind field have been defined using ship observations and/or site measurements. Although these simple formulas are very practical for engineering purposes, their application range is limited. An estimate of the entire wave spectrum from known values such as the significant wave height and wave period can be performed based on models of the spectrum.

Among the first attempts to predict fetch-limited wave heights, Sverdrup and Munk (1947) and Bretschneider (1952) were successful in relating the wave energy and frequency to the fetch. The non-dimensional growth curves became known as the Sverdrup–Munk–Bretschneider (SMB) curves. With the advance of spectrum analysis and increasing use of computers, along with new instrumentation and measurement techniques, new approaches have subsequently evolved. However, the basic principles implemented in SMB have been continuously practiced by most of the later wind–wave analysis and model developments (Liu, 1996).

For fully developed wind-generated waves in deep water, the Neumann and Pierson model (Neumann and Pierson, 1966) relates the mean wind speed (V_m) to the significant wave height (H_S) and mean wave period (T_Z):

$$H_S = 0.21 V_m^2 / g$$

$$T_Z = 1.62 \pi V_m / g$$

Equation 3.10

If the fetch is large enough, the wave growth will cease and the sea state is called *fully developed*. The Pierson–Moskowitz (PM) wave spectrum is used for fully developed sea. The PM spectrum considers the wind speed as an input to predict the wave energy (mean wind speed and significant wave height have a relation). The Joint North Sea Wave Project (JONSWAP) wave spectrum is normally applied for fetch-limited sea states in which the relationship between wave energy, wind speed, fetch length, and duration is considered (Ochi, 1998). The fetch duration and length are relatively challenging to estimate.

In regions with limited water depth, the TMA spectrum (Bouws *et al.*, 1985) relating the mean wind speed and significant wave height may be used; this is a modified JONSWAP spectrum where the JONSWAP spectrum is multiplied by a depth- and frequency-dependent function (Hughes, 1984). The TMA spectrum takes into account the shallow-water effects such as shoaling and wave breaking.

It is possible to forecast the propagation of wave energy when performing a wave forecast. However, the growth of the wave energy is dependent on the wind, as the winds cause the waves. Hence, the wave growth is diagnosed from the forecast wind. *Hindcasting* refers to the diagnosis of waves based on historical wind data. A computation based on present wind data, a so-called *nowcast*, is performed based on wave analysis (World Meteorological Organization, 1998).

The measurements of waves at suitable points provide data for verification of wave forecasts or validation of models for hindcasting. Hindcasting using wind fields derived from historical weather charts or archived air-pressure data from atmospheric models is the only way to obtain a satisfactory dataset of the waves for a sufficiently long time

period in many applications. The availability of satellite data allows wave height clima-
tology to be precisely defined, at least in regions not affected by tropical storms, considering the spatial resolution of the satellite data. The generation of a reliable historical
wind field is very important for hindcasting. Manual analysis of marine variables to
construct the wind fields can be performed if a large computer system with a real-time
observational database is not available. Evaluation of the wind–wave model is performed using the sea-state measurements and related weather data (i.e. buoy data for
wind and wave measurements).

The joint probability model for mean wind speed and significant wave data considering the variability within the data is needed to derive the design conditions from simultaneously measured wind and wave data. The wind and wave loads are based on
simultaneous processes using the design considerations; the resulting parameters are
sensitive to wind and wave correlation. The main environmental design parameters are
mean wind speed, significant wave height, and wave period. It is practical to model one
environmental variable as independent and the other environmental variables as
dependent. The joint density distribution of the characteristic parameters – mean wind
speed, significant wave height, and wave period – is presented by Equation 3.11
(Johannessen *et al.*, 2001):

$$f(V_m, H_S, T) = f(V_m) f(H_S | V_m) f(T | H_S, V_m) \qquad \text{Equation 3.11}$$

where $f(V_m)$ is the marginal distribution of mean wind speed, $f(H_S | V_m)$ is the condi-
tional distribution of significant wave height for a given mean wind speed, and
$f(T | (H_S, V_m))$ is the conditional distribution of a wave period given significant wave
height and mean wind speed.

The marginal distribution for the mean wind speed can be presented by a two-parameter
Weibull distribution. The Weibull shape (α) and scale parameter (β) are obtained by
fitting mean wind speed measurement data to Weibull distribution (Johannessen
et al., 2002).

$$F(V_m) = 1 - \exp\left[-(V_m / \beta)^\alpha\right] \qquad \text{Equation 3.12}$$

The shape and scale parameters are dependent on the offshore site and averaging period.
For the Statfjord site (an oil and gas field in the Norwegian sector of the North Sea operated
by Statoil, located at 59.7°N and 4.0°E, and 70 km from the shore), based on simultaneous
wind and wave measurements taken between 1973 and 1999 from the northern North
Sea, shape and scale parameters recommended for one hour of mean wind speed at 10 m
are, respectively, 1.708 and 8.426; refer to Karimirad and Moan (2012).

The marginal distribution of 10-min mean values of the wind speed at 100 m height
by fitting to a two-parameter Weibull distribution, based on FINO measurement
for the time period from November 2003 to May 2005, is determined by shape and
scale parameters of 11.789 and 2.310, respectively; for more information, refer to
Mittendorf (2009).

Normally, the wind data are divided into classes of mean wind speed, that is, each 1 or
2 m/s, and the corresponding wave heights to every wind speed class are also fitted to
proper distributions (i.e. Weibull distribution). The Weibull parameters of the wave
heights are derived as a function of the wind speed. Therefore, a continuous conditional

distribution of significant wave height is attained. The shape and scale parameters for Weibull distribution of significant wave height as a function of mean wind speed based on FINO measurement are presented by Mittendorf (2009):

$$\alpha = 1.535 + 0.01304 V_m$$

$$\beta = 0.7704 + 0.01304 V_m^{1.7696}$$

Equation 3.13

The corresponding values of the shape and scale parameters for Weibull distribution of significant wave height as a function of mean wind speed based on the Statfjord site are given by Johannessen *et al.* (2001):

$$\alpha = 2.0 + 0.135 V_m$$

$$\beta = 1.8 + 0.10 V_m^{1.322}$$

Equation 3.14

Having the shape and scale parameters for Weibull distribution of significant wave height as a function of mean wind speed, the conditional mean and standard deviation of the significant wave height given the mean wind speed are presented by:

$$E(H_S) = \beta \Gamma(1 + 1/\alpha)$$

$$\sigma(H_S) = \beta \left[\Gamma(1 + 2/\alpha) - \Gamma(1 + 1/\alpha) \right]^{0.5}$$

Equation 3.15

The conditional distribution of mean wave periods for given wave heights and mean wind speeds, considering all wind–wave classes, may be presented as a log-normal distribution:

$$f(T|H_S, V_m) = \frac{1}{T_Z \sigma_{\ln(T_Z)} \sqrt{2\pi}} \exp \left(\frac{1}{2} \left[\frac{\ln(T_Z) - \mu_{\ln(T_Z)}}{\sigma_{\ln(T_Z)}} \right]^2 \right)$$

Equation 3.16

where $\sigma_{\ln(T_Z)}$ is the standard deviation of $\ln(T_Z)$ and $\mu_{\ln(T_Z)}$ is the expectation value of $\ln(T_Z)$; the standard deviation and expectation value of $\ln(T_Z)$ are given by the following expressions:

$$\mu_{\ln(T_Z)} = \ln \left[\frac{\mu_{T_Z}}{\sqrt{1 + \upsilon_{T_Z}^2}} \right]$$

$$\sigma_{\ln(T_Z)} = \ln \left[1 + \upsilon_{T_Z}^2 \right]$$

$$\upsilon_{T_Z} = \frac{\sigma_{T_Z}}{\mu_{T_Z}}$$

Equation 3.17

where σ_{T_Z} is the standard deviation of T_Z from measurement, and α_{T_Z} is the mean value of T_Z from measurement; the mean value and the standard deviation of T_Z are calculated for every combination of significant wave height and mean wind speed.

By inserting these parameters, see Equations 3.12 to 3.17; the joint density distribution for wave and wind is derived. This allows a simultaneous description of all considered sea state parameters considering their distribution. The accuracy of this representation depends on the measured database. Hence, synchronous wind and wave data are required, and for a reliable long-term prediction, a longer data series is needed (Mittendorf, 2009). This model can be used for long-term predictions using the contour surface approaches; see, for example, Karimirad, *Stochastic Dynamic Response Analysis of Spar-type Wind Turbines with Catenary or Taut Mooring Systems* (2011; see also Nerzic *et al.*, 2007).

3.6 Oceanographic and Bathymetric Aspects

The ocean and seas cover 70.8% of the earth's surface. By international agreement, the ocean is divided into three named parts: (a) Pacific Ocean ($181.34E + 06\,km^2$), (b) Atlantic Ocean ($106.57E + 06\,km^2$), and (c) Indian Ocean ($74.12E + 06\,km^2$). The minimum width of the Atlantic is around 1500 km, and the north–south extent of the Atlantic and the width of the Pacific are more than 13,000 km. However, the representative depth is around 3.5 km. This means the horizontal dimension of the ocean is much larger than the vertical dimension (i.e. 1000 times greater).

Due to the small ratio of depth to width, the vertical velocities are much smaller than horizontal velocities and the velocity field in the ocean is nearly 2D. This assumption has a great influence on turbulence. 3D turbulence is different than 2D turbulence; in three dimensions, vortex stretching has an essential role in turbulence, while in two dimensions, vortex lines are vertical, and there is little vortex stretching (Stewart, 2008).

The surface of earth has two types, oceanic and continental. The ocean has a mean depth of 3400 m, while the continents have a mean elevation of 1100 m. The ocean surpluses onto the continents and creates continental shelves as the ocean water volume is more than the capacity of the ocean basins. The continental shelves are shallow seas with typical depths of 50–100 m, and normally they are part of adjacent countries. Among important shelves is the North Sea, shared between Germany, the UK, the Netherlands, Belgium, France, Sweden, Denmark, and Norway. The crust plates move relative to each other, making distinctive features of the sea floor, including trenches, mid-ocean ridges, basins, and island arcs; see Figure 3.9.

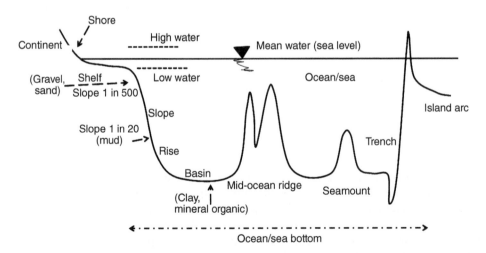

Figure 3.9 Features of the sea floor, including trenches, mid-ocean ridges, basins, and island arcs. For more information, refer to Stewart (2008).

Ocean subsea features affect the ocean circulation. Stewart (2008) explained some of the ocean subsea features: "Trenches are long, narrow, and deep depressions of the sea floor, with relatively steep sides. Ridges are long, narrow elevations of the sea floor with steep sides and rough topography. Basins are deep depressions of the sea floor of more or less circular or oval form."

Offshore construction projects on the ocean seafloors need wide data about deformability; strength; and hydraulic, acoustic, thermal, and seismic properties for selecting stable environments and ocean seafloor properties to ensure those structures, pipelines, and other installations on the surface and buried into the marine sediments are properly functioning. The core samplers and sediment grab devices with subsequent lab analysis and *in situ* probes are used to measure sediment properties. However, these point measurement techniques are expensive due to ship, time, and limited coverage area. For near-surface sediment information with improved coverage rates, the sub-bottom acoustic and electromagnetic sensors can provide proper profiles. Also, the fusion techniques are being developed to provide an areal extent of sediment information from multiple sensors (Harris *et al.*, 2008).

The design of offshore structures is highly dependent on proper understanding of oceanographic and bathymetric features. Hence, geotechnical and geophysical investigations are essential prerequisites for selection and characterization of a site for moored floating platforms and bottom-resting structures. Several issues affect the site survey, among them: (a) availability of personnel and equipment; (b) cost of survey support platforms; (c) platform value, including the costs of engineering, procurement, construction, and installation; (d) platform functionality; (e) risk associated with possible failure of the platform; (f) environmental loadings and operational loads; (g) anchoring and support-structure type; (h) seabed soil conditions and topography; and (i) pre-survey requirements.

The accuracy of the data needed for an offshore project is based on the type, cost, and extent of the project or project phases. Sometimes, general information may be enough; however, in some cases, precise information is needed. As an example, to find the best place for a living-quarter platform, low-precision data from a regional survey over a large area is enough (regional surveys compare sites or cover large distances, and detailed data generally are neither possible nor needed). However, design of a living-quarter platform requires high-precision data from the particular site (site-specific studies need more detailed data that are used in design). Normally, regional surveys contain geophysical information collection with limited soil sampling, for example gravity coring. However, deep soil borings and *in situ* tests are not normally used for site selection, while site-specific data are usually collected by soil sampling and by *in situ* testing. Depending on the project, extra geophysical and geological data (beyond those collected during the regional survey) may also be required (Thompson and Beasley, 2012).

A geophysical survey provides understanding of the conditions surrounding the site, and helps to identify the potentially hazardous geological features. There are several geotechnical hazards, including earthquake loading, liquefaction, faulting, gas hydrates, submarine landslides, erosion, and the presence of underconsolidated sediments. These conditions should be considered for assessments of offshore sites based on the acceptable level of risk of the project. Thorough investigation of geotechnical hazards from earthquakes, winds, waves, and currents is needed. The historical environmental information using examination of existing maps, charts, and bottom environmental data is useful for investigation of environmental factors and hazardous features at the start of the project.

The installation of a platform on the seabed requires a precise determination of horizontal and vertical position. This is a critical aspect of geophysical and geotechnical investigation for an installation that requires defining the location of the platform with respect to the surface vessel and location of the vessel with respect to geographical coordinates. Global Positioning Systems (GPS) accuracy is around 3 m, and this can be improved to 1 m by using a differential GPS (DGPS) system (based on a network of ground-based reference stations). Sonar transponders (with accuracies on the order of 0.1% of the distance being measured) are used to measure the relative seafloor-to-surface positions (Thompson and Beasley, 2012).

The geophysical surveys provide valuable information for understanding the geological setting that may not be easily detected by data from borings. Also, it detects the 3D features over a large area and provides the correlations with soil-boring data. Hence, the required number of borings will be limited, and costs will be reduced. The available modern geophysical instruments are numerous with a variety of capabilities. These geophysical systems may be categorized as following:

- High-resolution reflection systems: Measure reflected acoustic (sound) energy from the seabed.
- Seismic refraction systems: Measure refracted sound energy from the seabed.
- Electrical resistivity systems: Measure electrical energy/resistance of near-seabed soil.

The high-resolution reflection systems use sound or laser light to measure the seabed and the sub-seabed. These sensors are in three main groups, (1) seabed measuring systems (e.g. echosounders and multi-beam sounders), (2) imaging sensors (e.g. side-scan sonar, laser-scan, and acoustic scanning systems), and (3) sub-bottom profilers (e.g. pingers, boomers, etc.). Sound (acoustic energy) is the most practical source for measuring and sensing underwater. An acoustic energy source makes a pulse of sound traveling in water and penetrates the seabed. The reflection of this sound, received by a receiver system, is used to define the seabed and sub-seabed characteristics. The acoustic signal is affected by the material properties through which it passes. The signal is faster in denser material, and hence the rate of progress is affected by material densities. The acoustic geophysical survey systems use fundamental characteristics of sound such as amplitude and frequency. It is possible to investigate different aspects of the physical environment by using different amplitude and frequency ranges. To get high-resolution information, high-frequency, low-amplitude signals are needed. A low-frequency, high-amplitude signal travels more into the seabed; however, it has lower resolution (ISSMGE, 2005).

There are two types of seabed sediments: (1) terrigenous (land derived) and (2) pelagic (ocean derived). Pelagic sediments (formed in the sea) include abyssal clays, siliceous oozes, and calcareous oozes. These soils are composed of clays and of skeletal material from plants or animals, and roughly cover 75% of the seafloor. The terrigenous soils are made on or adjacent to land. These sediments (also known as neritic sediments) are mainly found on the continental shelves and slopes. Terrigenous soils include gravels, sands, silts, and clays. They are transported by currents, wind, or iceberg rafting to the deep ocean. The main terrigenous soils are (a) terrigenous silty clays, or muds; (b) turbidites; (c) slide deposits and volcanic ash; and (d) glacial marine soils.

Normally, soil properties at a site are calculated by sampling the soil and examining it in the laboratory. The techniques of drilling and sampling used on land may be used for shallow-water applications (i.e. by utilizing a jack-up barge or a fixed platform). However, in deep water due to greater depth, more complex tools and techniques than sampling in near shore are needed. For example, the sampling should be performed from a floating vessel. For limited soil depth, the gravity corers and vibracorers are usually used, while for large soil depth, drilling rigs and wireline sampling techniques are usually applied. The performance of these sampling techniques depends on the handling capability of the supporting vessel and the weather conditions (Thompson and Beasley, 2012).

3.7 Scour and Erosion

The soil erosion is called scour. Wave and current can cause scour and erosion around offshore and coastal structures. The scour is normally greater in sand than in clay. Also, the potential of scour usually increases as water depth decreases. Gravity-based offshore structures installed on sand might require a graded-gravel fill around the periphery for erosion and scour protection. Skirts along the periphery can help prevent the scour underneath the base. In the case of piles, scour or erosion is accommodated by accounting for a certain scour depth in the design considerations (ISSMGE, 2005).

There are two main types of scour: general and local. When the ocean water flows over the seabed, the general lowering of the seabed happens due to erosion of the soil. Due to the presence of offshore structures (e.g. pipeline, pier, pile, anchor, and similar) close to the seabed or in the seafloor, the water stream has to change its pattern and go around the structures, which changes the flow pattern (streamlines). Such interaction results in local scour of the soil supporting the structure. Due to excessive scour, the stability of the structure can be lost, and failure may happen (Briaud, 2008). So, scour is an important issue when a foundation is designed for marine structures. The estimation of the scour depth is essential to evaluate the structural integrity and stability of the design under maximum possible scour and environmental loads. Numerical and analytical methods are used to predict the scour. To mitigate the scour problem, scour protection based on soil type, water and environmental conditions, and the structure can be applied, for example by placing a filter and then an armor layer (made of rocks or riprap) image.

Normally, a threshold of erosion based on critical shear stress or critical velocity is defined. This threshold depends largely on soil properties, for example the grain size. For coarse-grained soils (i.e., sands), the relationship between erosion and erosion threshold is fairly linear. For fine-grained soils (i.e., clays), there is more scatter. Therefore, other parameters than the grain size affect the threshold values. The water and soil can affect the critical shear stress and consequently the erosion rate. For example, decreasing the salt concentration can decrease the critical shear stress and increase the scour. The following soil properties may affect the scour (Thompson and Beasley, 2012):

- Grain size
- Undrained shear strength
- Plasticity index
- Water content.

At the top of the water column, the horizontal velocity of water is largest and the shear stress is zero, while at the bottom of the water column, the shear stress is largest and the horizontal velocity of the water is zero due to the boundary layer effects. The shear stress results in shear strain. The shear strain (γ) is defined as the ratio of the change in horizontal displacement between two points (∂x) when the element is sheared to the vertical distance (∂z) separating the points (in Chapter 5, the definitions of stress, strain, and their relation are discussed in more detail).

$$\gamma = \partial x / \partial z \hspace{5cm} \text{Equation 3.18}$$

The shear stress is proportional to the rate of shear strain in fluid.

$$\tau = \mu \left(\partial \gamma / \partial t \right) \hspace{4cm} \text{Equation 3.19}$$

where μ is the dynamic viscosity. Inserting Equation 3.18 in Equation 3.19, it is possible to show that shear stress is proportional to the gradient of the horizontal velocity profile with depth.

$$\tau = \mu \left(\partial \left(\partial x / \partial z \right) / \partial t \right) = \mu \left(\partial V_x / \partial z \right) \hspace{2cm} \text{Equation 3.20}$$

Normally in scour evaluation, the resistance of a single particle (or cluster of soil particles) is studied. Hence, the magnitude of the erosive shear stresses is fairly small compared to values in other areas (in the other geotechnical engineering fields, the resistance of a larger soil mass is usually considered).

Scour is affected by ocean waves. Gravity waves, storm surge, and flood affect the scour depth more compared to scour under current-only conditions. Wave scour is significantly influenced by depth of water, size or diameter of the structure, and wave characteristics such as the wave height, wave length, wave period, phase angle, orbital velocity, and semi-orbital length. The velocity of the water particles is mainly horizontal in current-only conditions. However, the water particles have orbital velocity in wave conditions. The particle motions (the size of the orbit) reduce with water depth. In deep water the orbital motion is circular, while in shallow water the orbital motion is elliptical. The orbital motion is negligible below a depth of half of the wave length. The wake pattern produced by oscillatory flow, such as that seen in waves, is governed by the Keulegan–Carpenter number (KC), which is a key parameter in marine scour.

$$KC = V_m T / D \hspace{5cm} \text{Equation 3.21}$$

where V_m is the maximum velocity of the water particle (in scour, the maximum orbital velocity at the seabed), T is the wave period, and D is the structure diameter. For small KC numbers (KC < 6), the orbital motion is small compared to the structure size; while for large KC numbers (KC ≥ 6), the orbital motion is large compared to the structure size, and vortices form around the structure resulting in scour (Thompson and Beasley, 2012).

As the propeller of a ship or marine vessel produces a turbulent jet eroding the soil (if the velocities go beyond the critical velocity of the soil), the propeller-induced scour may happen and should be accounted for in the design of offshore structures.

An ocean structure in water results in disturbances in the fluid flow. The water should change its pattern when it reaches the obstacle. To have the same flow rate, the velocity will be increased and the water accelerates. Scour happens if the water velocity exceeds

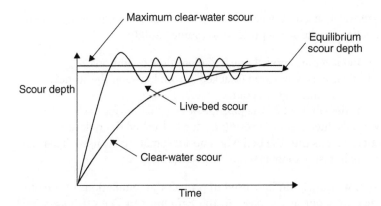

Figure 3.10 Clear-water scour and live-bed scour. For more information, refer to Arneson *et al.* (2012). Normally, clear-water scour produces higher scour depth compared to live-bed scour. As some of the soil particles in suspension during live-bed scour fall down on soil, this reduces the size of scour.

the critical velocity of the soil. Depending on the magnitude of the approach velocity and local velocity compared to the critical velocity of soil, a clear-water scour or live-bed scour may happen; see Figure 3.10.

- Clear-water scour occurs when the local velocity at the structure is higher than the critical velocity, while the approach velocity is lower than the critical velocity. In this case, water does not carry much soil particles, and the water particles remain in suspension.
- Live-bed scour happens when the approach velocity and the local velocity are higher than the critical velocity of the soil. In this case, water carries a significant amount of soil.

The foundation design is based on the equilibrium scour depth and considers the scour depths on the fatigue. Hence, evaluation of the maximum scour depth and time development of the scour hole are very important. Also, scour protection requires the knowledge of time development and eroded volume that could be filled in with rocks before the scour hole is totally developed (Margheritini *et al.*, 2006).

Several approaches can be used to prevent or mitigate scour around foundations, either planned-for approaches or as a remedial (if scour is more severe than was estimated). Scour protection options include geotextile containers/sandbags, concrete armor units, concrete block mattresses, grout bags/mattresses, gabions/gabion mattresses, and flow inhibitors. Rock armor is the most commonly used form of scour protection at offshore wind farms using gravel, quarry run stone, or blasted rock to cover a particular area of seabed to a specified thickness. The common practice using of rock armor is (Whitehouse *et al.*, 2011):

- Placement of small rock or gravel on the seabed as a preparatory layer and a filter layer on the seabed. This can be placed before installation of the foundation or before scour develops.
- Placement of larger rock as an armor layer on top of the preparatory filter layer. There are known criteria for ensuring stability of interface layers in these types of systems (but given the practicalities of offshore installation, it is not always possible to follow these strictly).
- Placement of graded rock in the scour hole around the structure/foundation.

Stone protection is extensively used in ocean, coastal, and offshore engineering. There are different failure mechanisms of stone protection (Sumer, 2008):

1) Stones of the protection layer are moved due to violent flow conditions.
2) The underlying bed material is winnowed ("sucked") from between the stones.
3) Edge of the protection layer undergoes scour.
4) The protection layer is destabilized by the passage of bed-form troughs.
5) The protection layer fails due to global scour (i.e., due to the bed degradation).
6) The protection material sinks into the bed when no filter material is used. This can be due to different mechanisms such as scour, or liquefaction.

Usually, scour protection with rock dump is used to protect the foundations of the off-shore wind turbines against scour at the base. In the Egmond aan Zee offshore wind farm, this type of scour protection system is used to prevent removal of the sediment base. The survey campaign (to evaluate the performance/stability of the scour protection and to quantify the edge scour development at the circumference of the scour protection) showed significant edge scour (Petersen *et al.*, 2014).

Scour depth in irregular waves can be estimated by the following formula (Thompson, 2006):

$$S_{ew} = 1.3D\left(1 - \exp\left(-0.03KC\right)\right) \qquad \text{Equation 3.22}$$

where the KC number is defined earlier, and the maximum orbital velocity of the water particle at the seabed is calculated using the following formula:

$$V_m = \sqrt{2\int_0^\infty S(f)\,df} \qquad \text{Equation 3.23}$$

where $S(f)$ is the power spectrum of the water particle velocity, and f is the wave frequency. The irregular wave series with $KC = 3.2$ generate scour depths in the range of 0.1–0.2 of the diameter of the pile. There is a potential for a significant reduction in the expected scour depth, when the effect of wave back filling is taken into account. The current-wave equilibrium scour depth may be predicted by:

$$S_{ew} = 1.3D\left(1 - \exp\left(-A\left(KC - B\right)\right)\right) \text{ for } KC^3\{3-4\} \qquad \text{Equation 3.24}$$

where the coefficients A and B are functions of the combined effect of the current velocity (V_c) and the orbital velocity under the wave at the seabed (V_m) (Thomsen, 2006):

$$A = 0.03 + 0.75V_{cw}^{2.6}$$

$$B = 6\exp\left(-4.7V_{cw}\right) \qquad \text{Equation 3.25}$$

$$V_{cw} = V_c/(V_c + V_m)$$

3.8   Extreme Environmental Conditions

Extreme events have a great impact on marine structures, although their probability is low. Hence, a precise estimation of extreme action and action effects considering extreme environmental conditions (i.e. at a return period of 50 or 100 years) is needed for proper design of offshore structures. Various combinations of metocean conditions (mainly operational conditions) considering issues such as wake effects are analyzed for checking the fatigue limit state to evaluate the cyclical loading on the structures. The cyclical loads operating at relatively high frequency are important for fatigue damage, in contrast to extreme loads. The extreme metocean conditions are important for ultimate limit state analysis. The harsh conditions, storms, and hurricanes result in extreme load and load effects, normally with very low frequency.

The design process should consider how the structure needs to be engineered to withstand harsh conditions. The offshore conditions (air, sea, and seabed, to a sufficient level of detail) are needed to develop a conceptual design. The design process needs to consider a range of environmental conditions presenting the statistically probable extreme events (normally, the design should survive a 100-year event). The main design variables are extreme environmental conditions considering:

- Winds (including gusts)
- Waves, both wind-generated and swell
- Currents (including wind-induced currents and storm surges)
- Water levels (including storm surges)
- Wake effects (i.e. in an array of offshore energy structures).

The design of offshore structures is traditionally based on extreme wave conditions' so-called design wave, which is an individual wave with a great height and specified return period; the height is so large that it is only exceeded on average once over a specified time. The return period depends on the service life of the system; for offshore wind, it is 25–50 years for foundation structures. Considering such service life, the design wave should have a return period that is longer than the service life (e.g. 50 or 100 years) to ensure survival of the construction under extreme environmental conditions (Cooper *et al.*, 2008).

If the offshore structure behaves quasi-statically, the design wave approach may be used. However, for new ocean structures like wind turbines or floating bridges, the response is essentially governed by dynamic behavior, and hence care is needed for using such simplified approaches. Still, it may be useful for feasibility studies and for comparing different designs to use simplified approaches (like the design wave), keeping in mind the accuracy and assumptions made.

To quantify the extreme conditions, the standard methodology is to consider a long-term record containing an adequate sample of extreme conditions, and to perform extreme-value analysis by fitting a proper probability distribution to the data and extrapolating the marginal extreme values. The extrapolation depends on fitting a probability distribution function (PDF) to the marginal extremes data collected from the offshore site. The common distributions widely used in engineering are:

- Weibull
- Gumbel

- Fisher–Tippett
- Generalized extreme variate
- Generalized Pareto distribution.

As an example, the Gumbel distribution for extreme values can be presented as follows:

$$F^{Gum}(x) = \exp\left(-\exp\left(-\alpha\left(x-u\right)\right)\right)$$

Equation 3.26

$$f^{Gum}(x) = \alpha\exp\left(-\alpha\left(x-u\right)-\exp\left(-\alpha\left(x-u\right)\right)\right)$$

The Gumbel parameters can be found using the initial distribution, that is, by knowing the peak value distribution, $F(x), f(x)$.

$$F(u) = 1 - \frac{1}{N}$$

Equation 3.27

$$\alpha = Nf(u)$$

The mean value of Gumbel distribution is defined by:

$$\mu^{Gum} = u + \frac{0.57722}{\alpha}$$

Equation 3.28

Starting with Rayleigh distribution, the mean value $F^{Gum}(x) = 0.5$ is found for the largest maximum among N maxima:

$$F(u) = 1 - \exp\left(-\frac{u^2}{2\sigma_x^2}\right) = 1 - \frac{1}{N}$$

$$u = \sigma_x\sqrt{2\ln N}$$

Equation 3.29

$$\alpha = \frac{\sqrt{2\ln N}}{\sigma_x}$$

$$\mu^{Gum} = u + \frac{0.57722}{\alpha} = \sigma_x\left(\sqrt{2\ln N} + \frac{0.57722}{\sqrt{2\ln N}}\right)$$

Equation 3.30

The largest, most probable extreme, when $f^{Gum}(x)$ is maximum, is $\sigma_x\sqrt{2\ln N}$.

Applying classical models for calculating the extreme values needs comprehensive effort to define the type of extreme-value distribution and its parameters, which can be uncertain. Several methods such as Monte Carlo methods, the Weibull tail, the Gumbel method, the Winterstein method, and the peaks-over-threshold (POT) method are presented to estimate the extreme value. The analytical models are used for determining the linear response, while the distribution of the nonlinear response generally needs to be treated in a semi-empirical manner by modeling the distribution of the response peaks or up-crossing rates (Karimirad, 2011).

Extreme-value statistics for a 1-h or 3-h period may be obtained by taking into account the regularity of the tail region of the mean up-crossing rate. The prediction of low-exceedance probabilities needs a large sample size, which results in time-demanding calculations as extreme values have a low probability of occurrence. The analyses of offshore structures subjected to stochastic wave and wind loading are time-consuming. For floating concepts, this computational cost is even worse as the total simulation time is higher. So, extrapolation methods can be used to estimate the extreme-value responses of these structures.

The mean up-crossing rate can be implemented for extreme-value prediction. For complicated marine structures subjected to wave and wind loads (i.e. offshore energy structures), the response is nonlinear and non-Gaussian. Consequently, the methods based on the up-crossing rate are more robust and accurate. The up-crossing rate is the frequency of passing a specified response level (the up-crossing rate is lower for higher response levels).

The Poisson distribution represents the extreme values since the occurrence of extreme values is rare. Also, the Poisson distribution can be defined based on up-crossing rate. For each response level, it is possible to count the number of up-crossings directly from the time histories. Long time-domain simulations are needed to obtain up-crossing rates for high response levels. For example, a 1-h simulation cannot provide any information about an up-crossing rate of 0.0001. Extrapolation methods are applied to extrapolate raw data and provide up-crossing rates for higher response levels (Naess *et al.*, 2008). The probability of extreme values using Poisson distribution can be written as:

$$P\big(Y(T) \le x\big) = \exp\left(-\int_0^T v_x^+(t)\,dt\right)$$

Equation 3.31

where T is the total time duration (i.e. 3 h), and $v_x^+(t)$ is the up-crossing rate of the level x. For more information, refer to Karimirad and Moan (2011).

The extreme events are independent, and their scarceness increases statistical uncertainty. For fitting probability distributions, different fitting techniques can be used, such as method of moments (MOM), least squares methods (LS), and maximum likelihood estimation (MLE). The choice of PDF is performed considering standard error and what fits the data best. The extreme wave height may be higher using the Weibull distribution compared to the Gumbel distribution. Using the best fit, the extreme-value extrapolations are derived for a set of return periods (i.e. 1-, 10-, 50-, and 100-year). As extreme values vary depending on how well the data fit the distribution, a sensitivity analysis should be performed.

Offshore energy structures are booming, in particular in shallow-water and intermediate-water areas. If the long-term record is not available for the site, appropriate methods may be employed to develop the data; for example, wave transformation models to transform deep-water wave into shallow water.

To perform a reliable extreme-value analysis, any secular trend in the data should be filtered out to remove possible bias in the distribution of the data around a long-term mean value (Cooper *et al.*, 2008). Nonrandom trends such as sea level rise should be filtered out. The statistical confidence in the extrapolation of the probability distribution decreases for return periods larger than 2–3 times the length of the input data recorded at the offshore site. So, to have a reliable extrapolation for a return period of 50 years, 20–25 years of data are needed. Otherwise, the error of extrapolation process increases for longer return periods.

There should be judgement on the extreme values predicted considering the resulting environmental parameters to remain physically realistic and not exceed any critical limitation. As an example, for large waves, the limiting conditions are mainly wave steepness and water depth beyond which waves may break.

The metocean conditions – namely, storm surge, current, water level, wave, and wind – are generated from either astronomical forcing or meteorological forcing, which are two independent physical mechanisms. Unless there is some form of interaction, the metocean conditions created by these separate mechanisms (astronomical or meteorological) remain independent of each other. As an example, in deep water, waves are practically independent of tides. However, in shallow water, waves are dependent on tide when they start to feel the seabed. In areas of strong current, the dependency of waves increases.

It is unlikely that all extreme conditions, winds, waves, water levels, and currents occur simultaneously (i.e. with a reduced probability). Joint distribution can be used to define the probability of such extreme events; joint distribution of wave and wind is explained in Section 3.4. Considering joint probabilities of variables and the degree of correlation of these extreme variables, some allowance may be applied in the design.

The extremes are typically provided for each month as well as for the complete year in metocean reports. The minimum data measured at an offshore site are normally extreme wind and waves for return periods of 1 year, 10 years, and 100 years. When measuring data, care is needed for reporting realistic data for design. For example, it is possible that the wind speeds estimated at the ocean surface are high due to low central pressure, which resulted in the surface boundary layer. So, the anemometers are subjected to strong low-level jet associated with the low-pressure system. So, the wind speeds measured at the anemometers are not related to the ocean surface winds, and modification factors to adjust wind speeds from the anemometers to the ocean surface are not usable.

The annual extreme values for winds and waves are calculated from a couple of hundred (e.g. 200–300 events) of storms with a given threshold (e.g. wind speed above 20 m/s and wave height above 9 m). The unstable conditions lead to extreme wind conditions like wind gusts, which are transient in speed and direction, and for which the assumption of stationarity does not hold. Extreme load calculation should be performed for such conditions to properly investigate the ultimate limit state checks.

The transient wind conditions occur when the wind speed or wind direction changes. Transient wind conditions are not stationary wind conditions, such as gusts, squalls, extremes of wind speed gradients, strong wind shears, extreme changes in wind direction, and simultaneous changes in wind (DNV, 2007). The short-term wind speed process will usually not be a narrow-banded Gaussian process, although it may be Gaussian for homogeneous terrain. This is important for prediction of extreme values of wind speed and their probability distributions' expression in terms of spectral moments.

The probability distributions for 1-h mean wind speed at 10 m above mean surface level (MSL) may be expressed in terms of the three-parameter Weibull distribution. The location, shape, and scale parameters are defined for this distribution. The non-Gaussian properties significant for crest height should be taken into account for phenomena that are sensitive to higher order wave characteristics. Proper distribution should be used in a long-term analysis to provide proper evaluations for extreme crests accounting for higher order wave properties.

The contour line approach has been used in offshore engineering to investigate the extreme-value responses. The contour lines have constant probability density and are usually obtained using 3-h sea states. In this method, the most critical sea state along the 100-year or 10,000-year contour line is defined for a given problem. Based on the joint distribution, contour lines for sea states with different return periods can be established. Using this sea state as the short-term design condition, the 90% percentile of the 3-h extreme response is suggested as an approximation for the 100-year response. For a non-linear problem, the 3-h extreme-value distribution can be obtained using time-domain analyses or model testing in the ocean basin.

Directional values of the significant wave height and corresponding values of the spectral peak period may be used for extreme-value analysis. If directional metocean values are used for design loads, one musts verify that the chosen combination of significant wave height and spectral peak period results in responses with acceptable return period. This is usually performed by a full long-term analysis accounting for the exceedance probabilities of each directional sector. Also, it is recommended to use monthly extremes rather than seasonal ones.

Extreme wave heights can be obtained from a long-term analysis. The Næss distribution (Næss, 1985) may be used as a short-term distribution for individual waves. The ratio between the 10,000-year design wave and the 100-year design wave recommended by the NORSOK may be used (NORSOK, 2007). For load calculation using the design wave approach, a fifth-order Stokes profile is recommended with respect to wave profile and wave kinematics in accordance with the NORSOK recommendations. For calibration of a design wave to match the result of a long-term analysis, a first-order Stokes wave is likely acceptable (Eik and Nygaard, 2003).

3.9 Environmental Impact of Offshore Structures' Application

The environmental issues of offshore industry, both oil/gas and new ocean industries, have been increased. In particular, new developments have raised new concerns as impacts of recent activities are not yet fully understandable in some cases. This requires more investigations to clearly classify the environmental impacts of offshore structures' application and document possible remedies to avoid permanent problems.

Since 1950, about 7000 platforms have been constructed and installed around the world; most of these structures (65% of the total) are located along the American coast of the Gulf of Mexico (Trabucco *et al.*, 2012). Various offshore oil and gas applications have a wide range of emissions and discharges, some of which require specific management measures. For example, the key issues are oil and chemicals in water, impacts from cuttings piles and atmospheric emissions, noise, and light. Toxic components in crude oil include polycyclic aromatic hydrocarbons (PAHs), phenols, naphthalene, phenanthrene, and pyrenes. Also, offshore oil and gas activities may disturb seabed habitats during installation and decommissioning. Accidental events may lead to the release of oil and chemicals to the environment.

The exploration and production of oil and gas have the potential for different impacts on the environment, based on the stage of the process, the nature and sensitivity of the surrounding environment, and pollution prevention, mitigation, and control techniques (Trabucco *et al.*, 2012).

Environmental impacts of offshore oil and gas application can happen in all phases, including exploration, operation, and decommissioning (OSPAR Commission, 2009):

- Oil discharges from normal operations
- Chemical discharges
- Accidental (oil) spills
- Drill cuttings
- Atmospheric emissions and air pollution
- Low-level, naturally occurring radioactive material
- Noise
- Soil pollution (due to placement of installations and pipelines)
- Operational and accidental discharges of water-containing substances, such as oil components, PAHs, alkyl phenols, and heavy metals.

During the planning stages of a major development project, one must perform an Environmental Impact Assessment (EIA) in a systematic manner. This ensures determining what kind of effects the project will cause for the environment and what should be done to control negative impacts when the effects are significant and pass the defined acceptable limits (Jones *et al.*, 2007).

A key part of modern structural engineering is public acceptance. It is necessary to introduce the environmental, social, and economic impacts of proposed projects properly to the public. This can be done via community meetings, public hearings, news releases, tours, exhibits, and so on. During the early phase of a project, it is important to gain public support to ensure success (Lwin, 2000).

Impact significance determination is extensively considered as a vital and critical activity (Lawrence, 2007). An Environmental Risk Assessment (ERA) is a qualitative and quantitative valuation of environmental status. An ERA considers both human health risk assessment and ecological risk assessment. Human health risk assessment includes (a) hazard identification, (b) dose–response assessment, and (c) exposure assessment and risk characterization. However, ecological risk assessment determines the likelihood of the occurrence of adverse ecological effects due to exposure to stressors (Shearer, 2013).

It is important to (a) monitor offshore cuttings piles (diminishing oil leaching rates from cuttings piles); (b) implement environmental management systems; (c) reduce the areas affected by contaminated drill cuttings; (d) investigate the biological effects from produced water in wild fish; and (e) reduce the discharge of hazardous chemicals, diesel-based muds and cuttings, and untreated oil-based muds and cuttings, as well as the volume of hydrocarbons discharged in produced water. As an example, it is good to mention OSPAR attempts to protecting and conserving the Northeast Atlantic and its resources (OSPAR Commission, 2009). Among their achievements, a Harmonised Mandatory Control System for the Use and Reduction of Discharge of Offshore Chemicals and a Harmonised Offshore Chemical Notification Format (HOCNF) are important (discharges of oil in produced water have been reduced).

The deep-sea corals and sponges are vulnerable spawning and nursing areas for fish. Oil and chemical discharges affect marine mammals, seabirds, and fish. Although discharge of oil-based and synthetic-based drilling fluids is extensively reduced, still the discharge of water-based fluids and drill cuttings is a concern for the sea environment. The development of offshore oil and gas activities will expand into deeper waters and into arctic regions. More extreme climatic conditions and seasonal ice cover in these areas will increase the risk of accidental release of oil.

Through exploration and production activities, oil is mainly released from such activities in produced water; however, deck and machinery drainage also contain minor quantities of oil. Another potential source is dropout of oil when flaring during well testing and well work-overs. During drilling, the operation of offshore structures and from shipping accidental oil release may happen. Oil affects the marine ecosystems, including: (a) feathering of seabirds and fur of some marine mammals; (b) mammals, birds, and turtles may ingest oil; (c) fish eggs and larvae are susceptible to toxic effects; (d) invertebrates, corals, barnacles, limpets, finfish, and shellfish may accumulate oil; and (5) algae can become tainted.

During the past years, in the areas with intense extraction and production activity, monitoring programs have developed. Also, national and international sea protection policies and legislations have been introduced, for example the Barcelona Convention with the Offshore Protocol representing a regional regulatory framework for the Mediterranean basin (Barcelona Convention, 1979).

The renewable energy technologies have been extensively developed to reduce the reliance on fossil fuels and to reduce global warming. The European Council backed Commission proposals on energy and climate change in 2007, agreeing on a binding target to reduce EU emissions by 20% by the year 2020, increased to 30% should other industrialized nations take similar steps (OSPAR, 2008).

Development of renewable energies has some impacts, for example underwater noise during installation of structures; electromagnetic fields; bird displacement; public perception; suspended sediment concentrations from foundations installation and cable laying; scour pit development; and seabed morphological effects within arrays of foundations, species composition, and rates of organisms colonizing the subsea structures. There can be potential environmental impacts associated with the location, construction, operation, and removal phases of offshore energy structures. To assess properly such impacts, enough site-specific data on the biological environment (biological communities, population dynamics, distribution, and abundance), habitat types and characteristics, and physical and chemical features (morphology, waves, currents, temperature, and salinity) are needed.

The environmental issues and potential impacts should be critically reviewed to highlight the concerns and to result in more targeted methods for assessment, monitoring, and management of such impacts. As an example, concerning the offshore wind application, the following items need more investigation (OSPAR, 2008):

- Impacts of underwater noise from construction activities and operation
- Bird displacement and collision risk
- Seabed morphology (gravity base and multi-pile foundations)
- Public perceptions and acceptance
- Cumulative impacts.

To minimize the environmental impact of offshore structures, ideally speaking, all the structures and their attachments should be removed after design life. However, decommissioning and removal of offshore installations are extremely costly, and some of the national and international regulations set about 40 years ago need to be revised, such as the requirement set by the Convention on the Continental Shelf (1985) and the United

Nations Convention on the Law of the Sea (1982) for removing abandoned offshore installations totally. Currently, a more flexible and phased approach is used (Patin, 2016):

- In a deep-water offshore site, it is allowed to just remove the upper parts from above the sea surface to 55 m deep and leave the remaining parts. From the technical-economic perspective, it is more reasonable to leave very large structures in deep water totally or partially intact.
- In intermediate water (site with depths less than 100 m), immediate and total removal of offshore structures (mainly platforms) is in practice. In shallow waters, total structure removal makes more sense.

The removed fragments can be transported to the shore, buried in the sea, or reused for some other purposes. The innovative and practical secondary reuse of abandoned offshore platforms for other purposes can help extensively lower the cost while helping to reduce the environmental impact of offshore installations from a long-term perspective.

The structures or their fragments left on the sea bottom may cause physical interference with fishing activities. So the chance of fishing vessel and gear damages and resultant losses can remain even after termination of production activities in the offshore site (even for some decades after the oil and gas operators leave the site). Moreover, the pipelines left on the seabed are particularly hazardous, and their degradation and uncontrolled dissipation may result in unexpected circumstances during bottom trawling. The fate of underwater pipelines is still not affected by clear regulations, as most of the national and international rules and regulations are about the decommissioning and abandonment of large offshore structures (e.g. drilling platforms).

References

Arneson, L.A., L.W. Zevenbergen, P.F. Lagasse and P.E. Clopper. (2012). *Evaluating Scour at Bridges*. Publication No. FHWA-HIF-12-003. Washington, DC: US Department of Transportation, Federal Highway Administration.

Barcelona Convention. (1979). *Convention for the Protection of the Mediterranean Sea against Pollution.*

Bouws, E., H. Günther and C.L. Vincent. (1985). Similarity of the wind wave spectrum in finite depth water. *J. Geophys*, 975–986.

Bretschneider, C. (1952). Revised wave forecasting relationships. In *Proceedings of the 2nd International Conference on Coastal Engineering.*

Briaud, J.-L. (2008). Case histories in soil and rock erosion: Woodrow Wilson Bridge, Brazos River Meander, Normandy Cliffs, and New Orleans Levees. In *Proceedings of the Fourth International Conference on Scour and Erosion (ICSE-4)*, Tokyo (pp. 1–27).

Butterfield, S.M. (2005). Engineering challenges for floating offshore wind turbines. In *Proceedings of the Copenhagen Offshore Wind Conference*, Copenhagen, Denmark.

Cooper, W., A. Saulter and P. Hodgetts. (2008). *Guidelines for the use of Metocean Data through the Life Cycle of a Marine Renewable Energy Development*. London: CIRIA.

DNV. (2007). *Environmental Conditions and Environmental Loads*. Norway: DNV.

Eik, K.J. and E. Nygaard. (2003). *Statfjord Late Life Metocean Design Basis*. Norway: STATOIL.

Ekman, V.W. (1905). On the influence of the earth's rotation on ocean currents. *Ark. Mat. Astron. Fys.*, 2, 1–52.

Faltinsen, O. (1993). *Sea Loads on Ships and Offshore Structures*. Cambridge: Cambridge University Press.

fino3.de. (2012). *FINO – Research Platforms in the North Sea and the Baltic*. http://www.fino3.de/en/fino3

Fox, R.W. (1985). *Introduction to Fluid Mechanics*. New York: John Wiley & Sons.

Harris, M.M., W.E. Avera, A. Abelev, F.W. Bentrem and L.D. Bibee. (2008). Sensing shallow seafloor and sediment properties: recent history. In *OCEANS* (pp. 1–11). Quebec, Canada: IEEE.

Hughes, S.A. (December 1984). *The TMA Shallow-Water Spectrum Description and Applications*. Vicksburg, MS: Coastal Engineering Research Center, Department of the Army.

International Society for Soil Mechanics and Geotechnical Engineering (ISSMGE). (2005). *Geotechnical & Geophysical Investigations for Offshore and Nearshore Developments*. London: ISSMGE.

Jay, S., C. Jones, P. Slinn and C. Wood. (2007). Environmental impact assessment: retrospect and prospect. *Environ. Imp. Assess. Rev.*, 27, 287–300.

Johannessen, K., T.S. Meling and S. Haver. (2001). Joint distribution for wind and waves in the northern North Sea. In *ISOPE*. Cupertino, CA: International Society of Offshore and Polar Engineers.

Johannessen, K., T.S. Meling and S. Haver. (2002). Joint distribution for wind and waves in the northern North Sea. *International Journal of Offshore and Polar Engineering*, 12, 1.

Karimirad, M. (2011). *Stochastic Dynamic Response Analysis of Spar-Type Wind Turbines with Catenary or Taut Mooring Systems*. PhD thesis. Trondheim, Norway: NTNU.

Karimirad, M. and T. Moan. (2011). Extreme dynamic structural response analysis of catenary moored spar wind turbine in harsh environmental conditions. *J. Offshore Mech. Arctic Engin.*, 133, 041103-1.

Karimirad, M. and T. Moan. (2012). Wave- and wind-induced dynamic response of a spar-type offshore wind turbine. *J. Waterway, Port, Coastal, Ocean Engin.*, 9–20.

Lawrence, D. (2007). Impact significance determination – back to basics. *Environ. Impact Assess. Rev.*, 27, 755–769.

Liu, P.C. (1996). Fifty years of wave growth curves. In *Coastal Engineering 1996: Proceedings of the 25th International Conference on Coastal Engineering, Orlando, Florida*.

Lwin, M.M. (2000). *Floating Bridges: Bridge Engineering Handbook*, ed. W.-F. Chen and L. Duan. CRC Press.

Margheritini, L., P. Frigaard, L. Martinelli, and A. Lamberti. (2006). Scour around monopile foundations for offshore wind turbines. In *The First International Conference on the Application of Physical Modelling to Port and Coastal Protection*. Porto: Faculty of Engineering, University of Porto.

Met Office. (2016, 2 13). *Beaufort*. Fact sheet 6 — The Beaufort Scale. National Meteorological Library and Archive. http://www.metoffice.gov.uk/media/pdf/b/7/Fact_sheet_No._6.pdf

Michailides, C.L. (2013). Monitoring the response of connected moored floating modules. In *23rd International Offshore and Polar Engineering Conference*.

Mittendorf, K.E. (2009). Joint description methods of wind and waves for the design of offshore wind turbines. *Marine Tech. Soc. J.*, 43, 23–33.

Munk, W.H. (1950). Origin and generation of waves. In *Proceedings of the 1st International Conference on Coastal Engineering, Long Beach, California* (pp. 1–4).

Næss, A. (1985). The joint crossing frequency of stochastic processes and its application to wave theory. *J. Appl. Ocean Res.*, 7, 1.

Naess, A., O. Gaidai, and P.S. Teigen. (2008). Extreme response prediction for nonlinear floating offshore structures by Monte Carlo simulation. *Appl. Ocean Res.*, 29, 221–230.

Nerzic, R., C. Frelin, M. Prevosto, and V. Quiniou-Ramus. (2007). Joint distributions of wind/waves/current in West Africa and derivation of multivariate extreme I-FORM contours. In *Proceedings of the Sixteenth (2007) International Offshore and Polar Engineering Conference* (pp. 36–42). Lisbon, Portugal: International Society of Offshore and Polar Engineers (ISOPE).

Neumann, G. and W.J. Pierson. (1966). *Principles of Physical Oceanography.* Englewood Cliffs, NJ: Prentice-Hall.

Newman, J.N. (1977). *Marine Hydrodynamics.* Cambridge, MA: MIT Press.

Nolte, M. (2012, May 13). Nordsee-Luftbilder. DSCF8888.jpg. https://commons. wikimedia.org

NORSOK. (2007). *Actions and Action Effects, N-003.* Norway: NORSOK Standard.

Ochi, M.K. (1998). *Ocean Waves: The Stochastic Approach.* Cambridge: Cambridge University Press.

OSPAR. (2008). *Assessment of the Environmental Impact of Offshore Wind-Farms.* OSPAR Commission. www.ospar.org

OSPAR Commission. (2009). *Assessment of impacts of offshore oil and gas activities in the North-East Atlantic.* OSPAR Commission, Offshore Industry Series. www.ospar.org

Panapakidis, I.M. (2016). Missing wind speed data: clustering techniques for completion and computational intelligence models for forecasting. In *26th International Offshore and Polar Engineering Conference* (pp. 503–510).

Patin, S. (2016, May 16). *Decommissioning, abandonment and removal off obsolete offshore installations.* http://www.offshore-environment.com/abandonment.html

Petersen, T.U., Sumer, B. Mutlu, J. Fredsøe, D.R. Fuhrman and E.D. Christensen. (2014). *Scour around Offshore Wind Turbine Foundations.* Denmark: Department of Mechanical Engineering, Technical University of Denmark.

Ryazanov, M. (2015). *Munk ICCE 1950 Fig1.svg.* https://commons.wikimedia.org/wiki/File:Munk_ICCE_1950_Fig1.svg: http://journals.tdl.org/ICCE/article/view/904

Shearer, K. (2013). *Assessment of Cumulative Impacts in Offshore Wind Developments.* Strathclyde, UK: Department of Mechanical and Aerospace Engineering, University of Strathclyde.

Sorensen, R. (2006). *Basic Coastal Engineering*, 3rd ed. New York: Springer.

Stewart, R.H. (2008). *Introduction to Physical Oceanography.* College Station: Texas A&M University.

Sumer, B.M. (2008). Coastal and offshore scour / erosion issues – recent advances. In *Proceedings of the Fourth International Conference on Scour and Erosion* (pp. 85–94), Japan.

Sverdrup, H. and W. Munk. (1947). *Wind Sea and Swell: Theory of Relation for Forecasting.* Washington, DC: Hydrographic Office, US Navy.

Thompson, D. and D.J. Beasley. (2012). *Handbook for Marine Geotechnical Engineering.* Washington, DC: NAVFAC.

Thomsen, J.M. (2006). *Scour in a Marine Environment Characterized by Currents and Waves.* Aalborg, Denmark: Aalborg Universitet.

Trabucco, B., C. Maggi, L. Manfra, O. Nonnis, R. Di Mento, M. Mannozzi, C.V. Lamberti, A.M. Cicero and M. Gabellini. (2012). Monitoring of impacts of offshore platforms in the Adriatic Sea. In *Advances in Natural Gas Technology*, ed. Hamid Al-Megren (pp. 285–300). Rijeka, Croatia: InTech.

Wright, J., A. Colling and D. Park. (1999). *Waves, Tides and Shallow-Water Processes.* Houston: Gulf Professional Publishing.

Whitehouse, R.J.S., J.M. Harris, J. Sutherland and J. Rees. (2011). The nature of scour development and scour protection at offshore windfarm foundations. *Marine Poll. Bull.,* 62, 73–88.

World Meteorological Organization. (1998). *Guide to Wave Analysis and Forecasting.* Geneva: Secretariat of the World Meteorological Organization.

4

Hydrodynamic and Aerodynamic Analyses of Offshore Structures

4.1 Introduction

Offshore structures are imposed on different environmental loads. These loads can be due to wind, wave and currents, or they can be because of an offshore structure's inter-action with the sea floor, or even because of earthquake, icing or lighting effects (Figure 4.1). Although all of these loads need to be considered in the design of offshore structures, the most important and usually design-driven loads are hydrodynamic and aerodynamic loads. The hydrodynamic loads can be due to very small-scale effects such as a small puddle on the water's free surface, or it can be due to very large-scale effects such as currents around the earth. From the time-scale point of view, it can happen very frequently such as waves and currents, every few years such as hurricanes or once every hundred years such as tsunamis. However, from an engineering point of view, our focus in this book is on normal-scale offshore structures that have applications in oil and gas industry, energy-harvesting systems, offshore transportation and aquaculture.

4.2 Wave Kinematics

In many offshore structures, the most important design-driven environmental condi-tion is the wave condition. Therefore, in this chapter, the wave environmental condition will be discussed in more detail; then, we discuss tide and currents. Since wind load is an important environmental load in offshore wind turbines, special attention is devoted to study wind kinematics and loads.

4.2.1 Regular Waves

The free surface waves are periodic motions of water that occur due to perturbation on the free surface and propagate due to gravity force. Initial perturbation may be as simple as a stone thrown in the water, or as big as an earthquake. The most common distur-bance source for ocean waves is wind. The time period of waves generated by wind varies from the range of a couple of seconds up to less than an hour.

The equation of motion for the free surface waves in deep waters can be obtained with a few assumptions. These assumptions are not far from practical applications, and

Offshore Mechanics: Structural and Fluid Dynamics for Recent Applications, First Edition.
Madjid Karimirad, Constantine Michailides and Ali Nematbakhsh.
© 2018 John Wiley & Sons Ltd. Published 2018 by John Wiley & Sons Ltd.

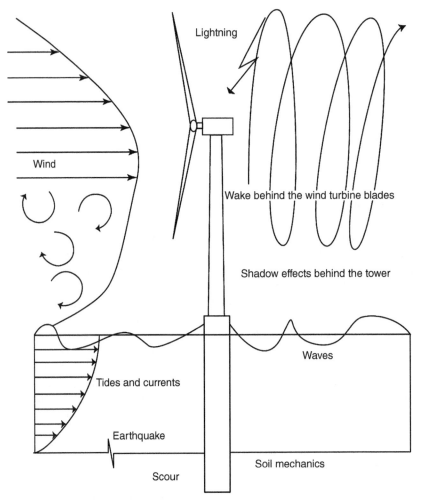

Figure 4.1 Different environmental loads on an offshore wind turbine. *Source:* Figure concept adopted from Musial and Ram (2010).

most industrial computational tools, developed for offshore structures analyses, are mainly based on these assumptions.

For developing the equation of free surface waves, we assume that the flow is incompressible and irrotational. If the velocity of flow in $x - z$ coordinate system is defined as V, a potential function ϕ is defined such that:

$$V = \nabla \phi = \frac{\partial \phi}{\partial x} i + \frac{\partial \phi}{\partial z} k \qquad \text{Equation 4.1}$$

where ϕ is a useful scalar variable for developing an equation of motion for irrotational, incompressible flows.

In the irrotational flow, the vorticity vector is equal to zero. Vorticity is introduced in most of the basic fluid dynamics textbooks (Fox and Mcdonald 1994). However, roughly speaking, it can be considered the rotation of a fluid element around itself. Formally speaking, vorticity can be defined as the curl of the velocity vector. Since the flow is assumed to be irrotational, it can be stated that:

$$\nabla \times V = 0 \qquad\qquad \text{Equation 4.2}$$

Combining Equations 4.1 and 4.2 leads to:

$$\frac{\partial^2 \phi}{\partial x^2} + \frac{\partial^2 \phi}{\partial z^2} = 0 \qquad\qquad \text{Equation 4.3}$$

Equation 4.3 is called a Laplace equation. It is a well-known equation in the field of mathematics and engineering, and can be seen in lots of physical phenomena. A Laplace equation is a second-order partial differential equation with respect to x and z coordinates. Proper boundary conditions are required to solve the Laplace equation. The first boundary condition can be obtained by noticing that at the sea floor, the velocity in the z direction is equal to zero. Assuming the sea is very deep:

$$\frac{\partial \phi}{\partial z} = 0 \quad \text{at } z = -\infty \qquad\qquad \text{Equation 4.4}$$

This boundary condition also implies that the disturbance given in the z direction will vanish as we move toward the sea floor.

Now we need to find a proper boundary condition on the free surface. As the wave propagates, it disturbs the free surface. Let's say that the function that describes this disturbance is called ξ. ξ only explains the water profile at the free surface, not inside the fluid. Therefore, ξ is not a function of z but only is a function of x and t ($\xi(x, t)$). For each x coordinate, free surface elevation with respect to mean undisturbed free surface ($z = 0$) can be found by ξ. In general, any disturbance can be described by Fourier transformation. If we assume that the disturbance on the water's free surface is small, a simple form of Fourier transform should be accurate enough to describe the disturbance, in the following form:

$$\xi(x,t) = A\cos(\omega t - kx + \varepsilon) \qquad\qquad \text{Equation 4.5}$$

Partial derivative of ξ with respect to t is the velocity of free surface in the z direction. On the other hand, by definition, $\dfrac{\partial \phi}{\partial z}$ gives the velocity of fluid. Therefore, at the free surface, the following equation holds:

$$\frac{\partial \phi}{\partial z} = \frac{\partial \xi}{\partial t} \quad \text{at } z = 0 \qquad\qquad \text{Equation 4.6}$$

If we assume that the given disturbance at the water free surface already propagated all along the water free surface in x direction and no barrier limits the wave propagation, the boundary condition in the x direction is not required to solve the Laplace equation. Figure 4.2 shows the boundary conditions used for deriving the equation of motion for free surface waves.

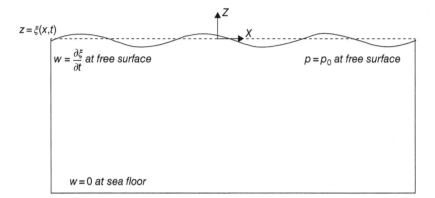

Figure 4.2 Boundary conditions imposed on the Laplace equation to derive an equation of motion for the free surface waves.

Boundary conditions in Equations 4.4 and 4.6 can be used to solve Equation 4.3, which result in the following solution:

$$\phi = \frac{gA}{\omega} e^{\gamma z} \sin\left(kx - \omega t\right),$$

Equation 4.7

and the velocities of fluid flow are the following:

$$u = \frac{\partial \phi}{\partial x} = \frac{gAk}{\omega} e^{\gamma z} \cos\left(kx - \omega t\right)$$

Equation 4.8

$$w = \frac{\partial \phi}{\partial z} = \frac{gAk}{\omega} e^{\gamma z} \sin\left(kx - \omega t\right)$$

Equation 4.9

We can impose an additional boundary condition on the Laplace equation, which results in relation between wave number and wave frequency. The pressure at the water free surface is atmospheric pressure. By using an unsteady Bernoulli equation in the form of:

$$\rho g z + p + \rho \frac{\partial \phi}{\partial t} + \rho V^2 = C,$$

Equation 4.10

we can relate the known atmospheric pressure to the potential field at the free surface. In the above equation, the velocity term is a nonlinear term and can be neglected in linear analysis. Also, the pressure on the free surface is constant, equal to p_0. Therefore, the equation on the free surface can be reduced to:

$$\rho g z + \rho \frac{\partial \phi}{\partial t} = C - p_0 = C_{new}$$

Equation 4.11

Since only derivatives of the ϕ function are the physical quantities of interest, adding or subtracting constant values does not affect them. Therefore, Equation 4.11 can be simplified in the form of:

$$gz + \frac{\partial \phi}{\partial t} = 0 \text{ at the free surface} \qquad \text{Equation 4.12}$$

And z at the free surface is $z = \xi(x,t)$; therefore,

$$g\xi + \frac{\partial \phi}{\partial t} = 0 \text{ at the free surface} \qquad \text{Equation 4.13}$$

By using boundary conditions described in Equations 4.13 and 4.6, it can be shown that k and ω are not independent parameters, and the following relation between them is valid in infinite-depth water:

$$\omega^2 = gk \qquad \text{Equation 4.14}$$

So far, the equation of motion is derived for a wave propagating in infinite water depth. If the water depth is finite, the same Laplace equation and boundary condition at the free surface is valid; however, the boundary condition at the sea floor (Equation 4.4) is replaced with:

$$\frac{\partial \phi}{\partial z} = 0 \text{ at } z = -h \qquad \text{Equation 4.15}$$

Solving the Laplace equation with boundary conditions described in Equation 4.15 and Equation 4.13, the velocities in x and z directions will be as follows:

$$u = \frac{\partial \phi}{\partial x} = \frac{gAk}{\omega} e^{\gamma z} \frac{\cosh(k(z+h))}{\cosh(kh)} \cos(kx - \omega t) \qquad \text{Equation 4.16}$$

$$w = \frac{\partial \phi}{\partial z} = \frac{gAk}{\omega} e^{\gamma z} \frac{\sinh(k(z+h))}{\cosh(kh)} \sin(kx - \omega t) \qquad \text{Equation 4.17}$$

Furthermore, using the pressure constraint at the water's free surface and writing a Bernoulli equation with a new potential function lead to the following relation between the frequency and wave number:

$$\omega^2 = gk \tanh(kh) \qquad \text{Equation 4.18}$$

Figure 4.3 shows the relation between the wave number and frequency for different values of *kh*, which is an indication of the ratio of wave depth to wavelength. Note that for *kh* >2, the finite and infinite water depth formulas nearly converge.

All of the equations in this subsection are derived based on linear theory of free surface waves. This model is widely used as an engineering tool to analyze different offshore structures and usually leads to fairly good estimates of the wave load analysis of offshore structures. The model, however, can be extended to second-order and or sometimes even third-order wave theory (Faltinsen *et al.*, 1995; Sclavounos, 2012) to obtain higher accuracy. Considering the velocity square in the Bernoulli equation, and taking into account the free surface wave effects above the mean water level, are among

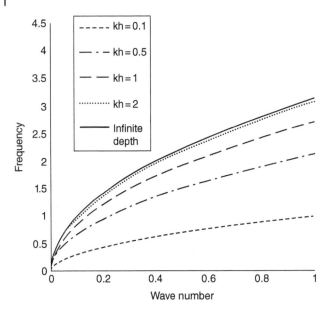

Figure 4.3 Relation of wave number and wave frequency for different **kh** values and infinite water depth.

the factors that can be studied for nonlinear analysis of free surface waves. For further study about such effects, the reader is referred to Faltinsen (1993) and Newman (1977).

4.2.2 Ocean Waves

If we look at ocean waves, especially from the shore, it looks like an extremely complicated physical phenomenon. Breaking waves, rapid variation of water depth near the shore, and waves traveling in random heights and directions are among sources of complications. Although considering all these factors together is extremely difficult, the mathematical model can be simplified by assuming nonbreaking, unidirectional, infinite-depth ocean waves and gradually adding more complication to the problem. By measuring the ocean waves with experimental tools such as floating buoys, researchers have noticed that the ocean waves are composed of a nearly specific range of frequencies. Therefore, if we assume the waves are linear and unidirectional (see Section 4.2.1), the ocean free surface wave profile can be described as a summation of linear wave profiles with different frequencies:

$$\xi(x,t) = \sum_{i=1}^{N} A_i \cos(\omega_i t - k_i x + \varepsilon_i) \qquad \text{Equation 4.19}$$

For example, Figure 4.4 shows the summation and components of four different waves with different frequencies and amplitudes. For each wave frequency, there is a wave amplitude, wave number and phase angle. Wave frequency and number are related by Equation 4.18, and phase angle, ε_i, has a uniform random distribution between 0 and 2π. The wave amplitude can be expressed by the wave spectrum as follows (Faltinsen, 1993):

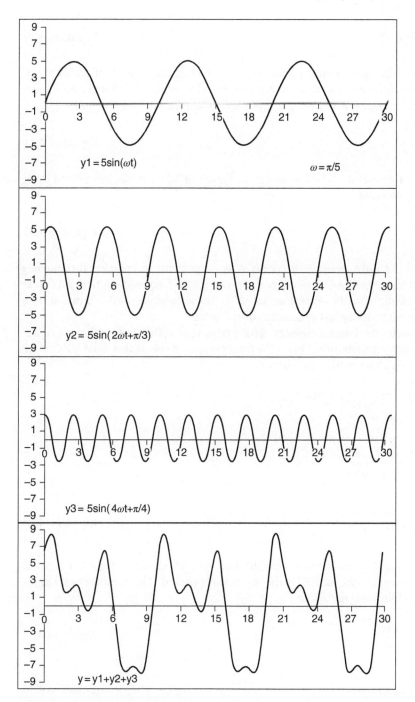

Figure 4.4 Summation and components of four different free surface waves with different amplitudes, frequencies and phases.

$$\frac{1}{2}A_i^2 = S(\omega_i)\Delta\omega_i \qquad\qquad \text{Equation 4.20}$$

This wave spectrum can be better understood by referring the concept of wave energy. Based on kinetic and potential energy, it can be shown (Newman, 1977) that total energy of a wave averaged over one period, per unit mean water free surface, is equal to:

$$E = \frac{1}{2}\rho g A^2 \qquad\qquad \text{Equation 4.21}$$

Since ρ and g are constant variables, energy in Equation 4.21 can be normalized by ρg and can be rewritten as:

$$\bar{E} = \frac{1}{2}A^2 \qquad\qquad \text{Equation 4.22}$$

We can simply call $\bar{E}$ normalized wave energy. A wave spectrum in a particular frequency can be defined as normalized wave energy per unit gap of frequency in that particular frequency. According to the experimental data, instantaneous wave elevation as a function of wave frequency has a Gaussian distribution.

Based on experimental data, researchers have tried to fit different functions to represent this ocean wave distribution. One of the frequently used spectrums is the JONSWAP spectrum, which can be written as follows:

$$S(\omega) = 155\frac{H_{1/3}^2}{T_1^4 \omega^5}\exp\left(\frac{-944}{T_1^4 \omega^4}\right)3.3^Y$$

$$Y = \exp\left(-\left(\frac{0.191\omega T_1 - 1}{\sqrt{2}\sigma}\right)^2\right)$$

$$\sigma = \begin{cases} 0.07 \ \omega \le 5.24/T_1 \\ 0.09 \ \omega > 5.24/T_1 \end{cases} \qquad\qquad \text{Equation 4.23}$$

where $H_{1/3}$ is defined as the mean of one-third of the highest wave heights; it is called the significant wave height, and T_1 is the mean wave period. The mean period can be obtained by the following equation:

$$\omega_1 = \frac{2\pi}{T_1} = \frac{m_0}{m_1} \qquad\qquad \text{Equation 4.24}$$

$$m_k = \int_0^\infty \omega^k s(\omega)d\omega \qquad\qquad \text{Equation 4.25}$$

Figure 4.5 shows the wave spectrum according to a significant wave height equal to 10 m and a mean wave period of 12 s. Note that constant values for mean wave period and significant wave height are usually a valid assumption for the short-term wave condition

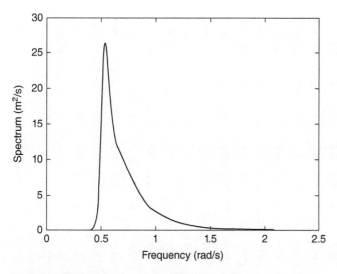

Figure 4.5 JONSWAP spectrum for short-term representation of wave condition with 10 m significant wave height and 12 second mean wave period.

of ocean waves, which is limited between half to ten hours. Therefore, this is called a short-term analysis of ocean waves.

In a long-term analysis of ocean waves, the mean period and significant wave height are not constant anymore. Therefore, it can be stated that long-term ocean wave conditions are composed of a number of short-term ocean wave conditions. Table 4.1 shows the long-term ocean waves registered for the North Atlantic sea based on experimental measurements (Hasselmann *et al.*, 1973). The characteristic parameter for the wave period in Table 4.1 is defined based on the second mean period (T_2), defined as:

$$\omega_2 = \frac{2\pi}{T_2} = \left(\frac{m_0}{m_2} \right)^{1/2}$$

Equation 4.26

For the given JONSWAP spectrum, the following relation between T_1 and T_2 can be derived (Faltinsen, 1993):

$$T_1 = 1.073 T_2$$

Equation 4.27

Usually, for hydrodynamic analysis of offshore structures, the short-term wave conditions that have the highest significant wave height and the conditions in which the mean or peak natural frequency coincides with the offshore structures' natural frequencies are more important, and the response of the structure needs to be carefully investigated for such conditions.

4.3 Wave Loads on Offshore Structures

Wave loads on offshore structures deal with the loads that are due to the harmonic water forces on the structure. Therefore, we are dealing with the forces due to the fluids on the structure. The fluid's force in general can be due to gravitational force, viscous

Table 4.1 Long-term North Atlantic wave conditions based on significant wave height and second mean period.

Tz(s)	3.5	4.5	5.5	6.5	7.5	8.5	9.5	10.5	11.5	12.5	13.5	14.5	15.5	16.5	17.5	18.5	Sum
Hs(m)																	
0.5	1.3	133.7	865.6	1186.0	634.2	186.3	36.9	5.6	0.7	0.1	0.0	0.0	0.0	0.0	0.0	0.0	3050.0
1.5	0.0	29.3	986.0	4976.0	7738.0	5569.7	2375.7	703.5	160.7	30.5	5.1	0.8	0.1	0.0	0.0	0.0	22575.0
2.5	0.0	2.2	197.5	2158.8	6230.0	7449.5	4860.4	2066.0	644.5	160.2	33.7	6.3	1.1	0.2	0.0	0.0	23810
3.5	0.0	0.0	34.9	695.5	3226.5	5675.0	5099.1	2838.0	1114.1	337.7	84.3	18.2	3.5	0.6	0.1	0.0	19128
4.5	0.0	0.0	6.0	196.1	1354.3	3288.5	3857.5	2685.5	1275.2	455.1	130.9	31.9	6.9	1.3	0.2	0.0	13289
5.5	0.0	0.0	1.0	51.0	498.4	1602.9	2372.7	2008.5	1126.0	463.6	150.9	41.0	9.7	2.1	0.4	0.1	8328
6.5	0.0	0.0	0.2	12.6	167.0	690.3	1257.9	1268.6	825.9	386.8	140.8	42.2	10.9	2.5	0.5	0.1	4806
7.5	0.0	0.0	0.0	3.0	52.1	270.1	594.4	703.2	524.9	276.7	111.7	36.7	10.2	2.5	0.6	0.1	2584
8.5	0.0	0.0	0.0	0.7	15.4	97.9	255.9	350.6	296.9	174.6	77.6	27.7	8.4	2.2	0.5	0.1	1309
9.5	0.0	0.0	0.0	0.2	4.3	33.2	101.9	159.9	152.2	99.2	48.3	18.7	6.1	1.7	0.4	0.1	626
10.5	0.0	0.0	0.0	0.0	1.2	10.7	37.9	67.5	71.7	51.5	27.3	11.4	4.0	1.2	0.3	0.1	285
11.5	0.0	0.0	0.0	0.0	0.3	3.3	13.3	26.6	31.4	24.7	14.2	6.4	2.4	0.7	0.2	0.1	124
12.5	0.0	0.0	0.0	0.0	0.1	1.0	4.4	9.9	12.8	11.0	6.8	3.3	1.3	0.4	0.1	0.0	51
13.5	0.0	0.0	0.0	0.0	0.0	0.3	1.4	3.5	5.0	4.6	3.1	1.6	0.7	0.2	0.1	0.0	21
14.5	0.0	0.0	0.0	0.0	0.0	0.1	0.4	1.2	1.8	1.8	1.3	0.7	0.3	0.1	0.0	0.0	8.0
15.5	0.0	0.0	0.0	0.0	0.0	0.0	0.1	0.4	0.6	0.7	0.5	0.3	0.1	0.1	0.0	0.0	3.0
16.5	0.0	0.0	0.0	0.0	0.0	0.0	0.0	0.1	0.2	0.2	0.2	0.1	0.1	0.0	0.0	0.0	1.0
Sum	1	165	2091	9280	19922	24879	20870	12898	6245	2479	837	247	66	16	3	1	100000

Source: Data from Det Norske Veritas (2000).

force and inertial force. Also, surface tension is another factor that is much less important in studying typical offshore structures. To estimate the importance of these different forces, nondimensional numbers can be used. These nondimensional numbers will help us understand the relative importance of these forces. If we neglect the surface tension, for our study the following two nondimensional numbers can give us an estimate of the importance of different effects (Newman, 1977). First is the Reynolds number, which is the ratio of inertial forces to the viscous force and can be written as follows:

$$\text{Re} = \frac{\rho U^2 L^2}{\mu U L} = \frac{\rho U L}{\mu} \qquad \text{Equation 4.28}$$

Second is the Froude number, which is the ratio of inertial force to gravity force and can be written as follows:

$$Fr = \frac{\rho U^2 L^2}{\rho g L^3} = \frac{U^2}{gL} \qquad \text{Equation 4.29}$$

The last ratio, which is the ratio of viscous to gravity force, will not make a new nondimensional number. In the remainder of this section, we are going to talk about wave force, where the importance of viscous effects is minor. In most offshore structures, the ratio of inertial and gravity forces is much larger than that of viscous forces; therefore, there are plenty of cases where the inviscid forces dominate over the hydrodynamic forces of offshore structures.

4.3.1 Wave Loads Induced by Inviscid Flows

The wave equation of motion in the linear condition is derived in Section 4.2.1. Two main parameters affecting wave forces on offshore structures are the wave motion and structure motion. To simplify the problem, let's first focus on fixed offshore structures, like fixed offshore wind turbines. Suppose a wave is propagating in the ocean. When an offshore structure blocks the wave's propagating direction, the structure will force the wave to move around the structure; in other words, the wave is diffracted. The wave forces on an offshore structure can be considered as a composition of forces due to unaffected wave, plus the forces that are induced by diffraction effects. The first term is called the Froude–Krylov force, which is usually a major component of wave force. The diffraction term also plays an important role, especially for situations when the characteristic length of the offshore structure is large compared to the wavelength.

Let's assume a fixed cylindrical-shaped structure, like the fixed offshore wind turbine shown in Figure 4.6. We want to compute the horizontal component of the Froude–Krylov force on this structure. Suppose that the wavelength is much larger than the diameter of an offshore wind turbine monopile, and the effect of limited water depth can be neglected for simplicity.

The wave force can be written as follows:

$$\Delta_z F_x = \int_A P dA_x = \int_A P dA_x = \int_0^{2\pi} P R d\theta \cos\theta = R \int_0^{2\pi} \rho g A e^{kz} \sin(\omega t - kx)\cos\theta d\theta$$

$$\text{Equation 4.30}$$

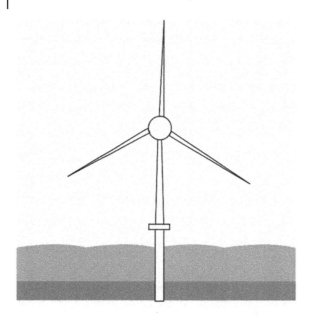

Figure 4.6 A monopile offshore wind turbine as an example of a fixed offshore structure.

If the center of the cylinder is considered as $x = 0$, the x location of the perimeter of the cylinder is $x = -R\cos\theta$. Therefore, Equation 4.30 can be written as follows:

$$\Delta_z F_x = \Delta z R \rho g A e^{kz} \int_0^{2\pi} \sin(\omega t + kR\cos\theta)\cos\theta d\theta \qquad \text{Equation 4.31}$$

or:

$$\Delta_z F_x = \Delta z R \rho g A e^{kz} \int_0^{2\pi} \left(\sin(\omega t)\cos(kR\cos\theta) + \cos(\omega t)\sin(kR\cos\theta)\right)\cos\theta d\theta$$

Equation 4.32

We assume the wavelength is much larger than the radius of the monopile, hence the term $kR\cos\theta$ is very close to zero. Therefore, the first term of the Taylor expansion can be used to estimate these two terms:

$$\cos(kR\cos\theta) \approx 1$$
$$\sin(kR\cos\theta) \approx kR\cos\theta \qquad\qquad \text{Equation 4.33}$$

$$\Delta_z F_x = \Delta z R \rho g A e^{kz} \int_0^{2\pi} \left(\sin(\omega t) + \cos(\omega t)kR\cos\theta\right)\cos\theta d\theta$$

$$\Delta_z F_x = \Delta z R \rho g A e^{kz} \int_0^{2\pi} \left(\sin(\omega t) + \cos(\omega t)kR\cos\theta\right)\cos\theta d\theta$$

Equation 4.34

$$\Delta_z F_x = \Delta z R \rho g A e^{kz} \left(\sin(\omega t)\int_0^{2\pi}\cos\theta d\theta + kR\cos(\omega t)\int_0^{2\pi}\cos^2\theta d\theta\right)$$

$$\Delta_z F_x = \Delta z K \pi R^2 \rho g A e^{kz} \cos(\omega t) = \rho \Delta z \pi R^2 a_{x,x=0}$$

It can be seen that the Froude–Krylov force can be written as the amount of water that is displaced by the cylinder, without considering any disturbance that occurs because of the presence of the cylinder.

The force calculated so far is only for one strip of the cylinder. We may do the integration along the length of the monopile to obtain the total force by Froude–Krylov force on the structure:

$$\Delta_z F_x = \Delta z K \pi R^2 \rho g A e^{kz} \cos(\omega t) = \rho \Delta z \pi R^2 a_{x,x=0}$$

$$\lim_{\Delta_z \to 0} \Delta_z F_x = dF_x$$

$$\lim_{\Delta_z \to 0} \Delta z = dz$$

$$F_x = \int_{-\infty}^{0} dF_x = \int_{-\infty}^{0} K\pi R^2 \rho g A e^{kz} \cos(\omega t) dz = K\pi R^2 \rho g A \cos(\omega t) \int_{-\infty}^{0} e^{kz} dz = \pi R^2 \rho g A \cos(\omega t)$$

Equation 4.35

The Froude–Krylov force is usually one of the most dominant hydrodynamic loads on offshore structures, and it is quite useful to get a first estimate of the order of wave forces on an offshore structure.

4.3.1.1 Inviscid Loads Due to Forced Oscillation of an Offshore Structure (Concept of Added Mass and Damping Coefficients)

Another set of problems are the ones that occur when the structure is forced to oscillate with a frequency in the water. The hydrodynamic forces in this case can be written as a function of acceleration and velocity of the structure. For example, if the structure is forced to oscillate in the surge direction, the hydrodynamic forces can be written as follows:

$$F_{x-hydordynamic} = \lambda \ddot{x} + \beta \dot{x}$$
Equation 4.36

where λ is called added mass, and β is called the damping coefficient of the structure. Note that also there is an additional term proportion to the displacement of the structure which is "hydrostatic" term and comes from buoyancy effect and is relatively straight forward for calculation. To better understand the concept of added mass and damping coefficient, let's start with an example in unbounded fluid:

Consider a sphere in an unbounded fluid that is forced to oscillate in the surge direction. We are interested in calculating the added mass and damping coefficient of the sphere (Figure 4.7). First, we need to calculate the potential field around the sphere, which has an acceleration of du/dt. The solution should satisfy the Laplace equation and boundary conditions, which can be written as:

$$\nabla^2 \phi = 0$$
$$u_r = U(t)\cos\theta \quad @r = R$$
$$\mathbf{u} = 0 \quad @r \to \pm\infty$$
Equation 4.37

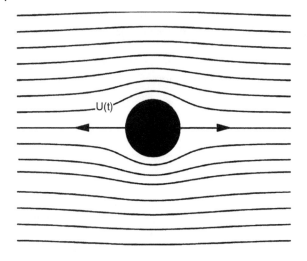

Figure 4.7 Forced oscillation of a sphere in unbounded fluid.

A three-dimensional (3D) doublet can be a good candidate to generate such kind of flow, and can be written as follows:

$$\phi = \frac{\mu}{4\pi} \frac{x}{r^3} = \frac{\mu}{4\pi} \frac{\cos\theta}{r^2}$$

$$u_r = \frac{\partial\phi}{\partial r} = -\frac{\mu\cos\theta}{2\pi r^3}$$

Equation 4.38

Letting the velocity be equal to U(t) at r = R results in:

$$u_r\big|_{r=R} = \frac{\partial\phi}{\partial r}\bigg|_{r=R} = \frac{\mu\cos\theta}{2\pi R^3} = U(t)\cos\theta$$

$$\phi = -\frac{UR^3}{2} \frac{x}{r^3}$$

Equation 4.39

Also, it can be verified that as we go toward infinity, the velocity reaches toward zero as well. So, the potential field can resemble an accelerating sphere in the X direction. The pressure on the cylinder from the fluid can be calculated using a Bernoulli relation: for two points, one on the surface of the sphere and the other one on the infinity, the unsteady Bernoulli equation can be written in the following form:

$$\frac{p}{\rho} + u^2 + \frac{\partial\phi}{\partial t} = \frac{p_\infty}{\rho} + u_\infty^2$$

Equation 4.40

The velocity at infinity is equal to zero, and let's assume that the reference pressure at infinity is equal to zero.

Neglecting the nonlinear term u square, the pressure can be written in the form of:

$$p = -\rho \frac{\partial\phi}{\partial t}$$

Equation 4.41

The force on the sphere can be calculated by integration of pressure on the sphere, which can be written as follows:

$$F_x = \int_A P \, dA_x = -\rho \int_A \frac{\partial \phi}{\partial t} \, dA_x \qquad \text{Equation 4.42}$$

The differential size of the sphere can be written in a spherical coordinate system by using circular stripes as follows:

$$dA_x = \cos\gamma \, dA = \cos\gamma \, (2\pi r \, ds)$$
$$ds = r \, d\gamma \qquad \text{Equation 4.43}$$
$$r = R \sin\gamma$$

Therefore, the force on the sphere can be written as follows:

$$F_x = -\int_0^\pi \rho \frac{\partial \phi}{\partial t} \left(2\pi R^2 \sin\gamma \cos\gamma \, d\gamma \right)$$
$$F_x = -\int_0^\pi \rho \frac{\dot{U}(t) R \cos\gamma}{2} \left(2\pi R^2 \sin\gamma \cos\gamma \, d\gamma \right)$$
$$F_x = -\rho \dot{U}(t) \pi R^3 \int_0^\pi \sin\gamma \cos^2\gamma \, d\gamma \qquad \text{Equation 4.44}$$
$$F_x = -\frac{2}{3} \rho \dot{U}(t) \pi R^3$$
$$F_x = -\frac{1}{2} \rho \forall \dot{U}(t)$$

Since the fluid is applying this force to the sphere, the sphere needs to apply the opposite force to overcome the fluid's force and move in the inviscid fluid. So this formula shows that if we move a sphere with radius R in the inviscid fluid, in addition to moving the mass of sphere, we need to apply a force equal to half of the imaginary sphere filled with water. We call this the *added mass of the sphere*. Referring back to Equation 4.36, the alpha coefficient and beta are equal to:

$$\lambda = \frac{1}{2} \rho \forall \qquad \text{Equation 4.45}$$
$$\beta = 0$$

If we had a term proportional to the velocity of the sphere, then the damping coefficient would not be zero anymore. The added mass is only a function of geometry of the structure. Since the sphere is symmetric, added mass in surge, heave and sway direction will get the same value in the unbounded fluid domain. Similar procedures can be used to calculate the added mass for ellipsoids with different geometric values, as is calculated by Kochin *et al.* (1966). An analytical solution for the added mass of a more complicated solution underwater is extremely hard to obtain, and usually numerical models are used for estimation (see Chapter 7).

4.3.1.2   Added Mass and Damping Coefficients in the Presence of a Free Surface

The above-mentioned problem was for the case where the structure was in unbounded fluid. In other words, there is no free surface in the domain. If a free surface is present, the solution for assessing hydrodynamic loads needs to satisfy the boundary condition not only on the structure, but also on the water's free surface.

Therefore, to assess the added mass and damping coefficient, the Laplace equation along with the following boundary conditions need to be satisfied (assuming infinite water depth):

$$\nabla^2 \phi = 0$$

$$\frac{\partial^2 \phi}{\partial t^2} + g\frac{\partial \phi}{\partial z} = 0 \ at \ z = 0 \qquad\qquad \text{Equation 4.46}$$

$$\frac{\partial \phi}{\partial n} = V.n$$

Also, we should make sure that the gradient of potential goes to zero as it moves far from the body. Satisfying the boundary condition on the free surface makes the solution slightly more difficult than that for unbounded fluid. For simplicity, let's start with two special cases where the frequency of a forced oscillating body goes toward zero or infinity.

If the structure has a forced harmonic motion in the surge direction, the motion of the structure can be written in the following form:

$$u_1 = \text{Re}\left(\xi_1 e^{iwt}\right) \qquad\qquad \text{Equation 4.47}$$

Since we are studying the linear potential field, the potential field in the domain will have the same frequency in time and can be written in the following form:

$$\phi = \text{Re}\left(\xi_1 \phi_1(x,y,z) e^{i\omega t}\right) \qquad\qquad \text{Equation 4.48}$$

The $\phi_1(x,y,z)$ can be considered the flow field due to unit surge motion of the structure. Applying the potential field in Equation 4.48, a set of Equation 4.46 can be rewritten in the following form:

$$\nabla^2 \phi_1 = 0$$

$$\frac{-\omega^2}{g}\phi_1 + \frac{\partial \phi_1}{\partial z} = 0 \ at \ z = 0 \qquad\qquad \text{Equation 4.49}$$

$$\frac{\partial \phi_1}{\partial n} = i\omega n_1$$

where n is the normal vector at the body surface, and n_1 is the component of that vector in the surge direction.

If we consider the surge motion in very low frequency, then the first term on the left-hand side of the free surface boundary condition in Equation 4.49 becomes equal to zero, and the boundary condition that remains to be satisfied is as follows (Newman, 1977):

$$\frac{\partial \phi_1}{\partial z} = 0 \ at \ z = 0 \ for \omega \to 0 \qquad\qquad \text{Equation 4.50}$$

If the structure moves with very high frequency, then to keep the first term limited, the potential needs to be zero at the free surface. Hence, we can write a free surface boundary condition in Equation 4.49 in the following form:

$$\phi_1 = 0 \ at \ z = 0 \ if \ \omega \to \infty \qquad\qquad \text{Equation 4.51}$$

The first case (oscillation frequency close to zero) is similar to studying the flow field near a plane boundary. The potential field for a 2D cylinder near the boundary can be obtained analytically. The basic idea of a solution is by using the method of images. In this method, the same object (here, a cylinder) is placed on the opposite side of the plane boundary, akin to a mirror. Then the velocities normal to the plane boundary cancel each other, and the plane boundary (Equation 4.50) will be satisfied. In other words, the summation of a potential field for the object and the mirror of it is a new potential field that satisfies the Laplace equation, the boundary condition on the plane boundary and the boundary condition on the surface of the object. So, all of our boundary conditions are satisfied. If we compute the hydrodynamic forces on the object, the added mass of cylinder in the limit of zero frequency on the presence of water free surface can be estimated. The only issue in this regard is that when the imaged object is added to satisfy the plane wall boundary condition, it also creates a potential field that slightly violates the boundary condition on the cylinder surface of the opposite side. To strictly satisfy this boundary condition, another potential field source like a doublet (which has less strength) is added. Consequently, the image of the object also needs to be placed on the opposite side, and again the same problem will occur and the procedure will go on. It can be shown that the magnitude of these potential fields is decreasing (Carpenter, 1958), and although the series will be written up to infinity, usually up to three imaged objects will give results with very good accuracy. By these considerations, the added mass for cylinder as the frequency goes toward zero can be computed and calculated, and it can be written in the following form (Nath and Yamamoto, 2016):

$$F_x = -\rho \pi a^2 C_M \dot{U}$$

$$F_x = -\rho \pi a^2 C_M \dot{U}$$

$$C_M = 1 + 2 \sum_{j=2}^{\infty} (q_2^2 q_3^2 \ldots\ldots q_\infty^2) \qquad\qquad \text{Equation 4.52}$$

$$q_n = \frac{1}{\dfrac{2s}{a} - q_{n-1}} \ , q_1 = 1$$

where s and a are shown in Figure 4.8. Also, it can be noted that the boundary condition in the limit of zero frequency is different from the boundary condition in the limit of high frequency. Therefore, the added mass in the presence of a free surface should be a function of frequency, and it is not constant like the object in unbounded fluid.

So far, we discussed finding added mass and damping coefficient in the limit of zero and infinite frequency. Finding the added mass and damping coefficient in a wide range of wave frequencies usually requires advanced mathematical calculations. Usually for engineering application, numerical methods (Chapter 7) will be used. However, the analytical methods give very good insight about the physics of the problem of interest.

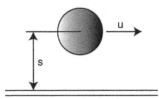

Figure 4.8 Forced surge oscillation of a circular cylinder in the limit of zero frequency is akin to oscillation near a plance boundary.

Here, we only briefly explain the finding of added mass and damping coefficients in surge direction for circular cylinders without details of the calculations: the model is solved by Rahman and Bhatta (1993) in a polar coordinate system. The separation of variables is used as the following for the potential field:

$$\phi(r,\theta,z) = R(r)\Theta(\theta)Z(z) \qquad \text{Equation 4.53}$$

Trying to satisfy all the boundary conditions on the free surface and body, the following potential function is obtained:

$$\phi(r,\theta,z) = \sum_{m=0}^{\infty} \left[\phi_m \cos(m\theta) \right] e^{-i\sigma t}$$

$$\phi_m = \alpha_{0m} H_m^1(kr) f_0(z) + \sum_{n-1}^{\infty} \alpha_{nm} H_m^1(k_n r) f_n(z) \qquad \text{Equation 4.54}$$

where:

$$f_n(z) = \frac{\cos k_n(z+h)}{D_n}$$

$$D_n = \sqrt{\frac{h}{2}\left(1 + \frac{\sin 2k_n h}{2k_n h}\right)} \qquad \text{Equation 4.55}$$

$$\alpha_{n1} = \frac{U}{k_n K_1'(k_n a)} \frac{\sin k_n h}{k_n D_n} \quad n = 1,2,\ldots$$

Based on the following solution for the potential function, the added mass and damping ratio can be found as follows:

$$C_m = i\rho a^3 \sigma \left(-\pi \frac{kd_0^2}{(ka)^2} \text{Im} \frac{H_1^1(ka)}{H_1^{1'}(ka)} \right)$$

$$C_d = \rho a^3 \left(-\pi \frac{kd_0^2}{(ka)^2} \text{Re} \frac{H_1^1(ka)}{H_1^{1'}(ka)} - \pi \sum_{n=1}^{\infty} \frac{k_n d_n^2}{(k_n a)^2} \frac{k_1(k_n a)}{k_1'(k_n a)} \right) \qquad \text{Equation 4.56}$$

where H_m^1 is the Hankel function of the first kind of order m, a is the radius of the cylinder, h is the water depth, σ is the wave frequency and K is the wave number. For more details about derivation of Equation 4.56, see Chakrabarti (1987) and Dean and Dalrymple (1991).

4.3.1.3 Considering Diffraction Effects on Calculating Wave Loads

So far, the wave force on a stationary body has been studied where the body is small compared to the wavelength. Also, the hydrodynamic forces on a body under forced oscillation have been discussed. When the body is large compared to the wavelength, the Froude–Krylov approximation is not very accurate anymore, and diffracted wave effects become important.

For studying diffraction effects for the case of a fixed structure, the potential flow theory similar to Equation 4.46 needs to be solved. The only difference is that the velocity at the body surface is equal to zero. Let's consider a surface piercing cylinder in limited water depth. We are interested to compute the forces on this cylinder by considering the diffraction effects. This problem is solved by MacCamy and Fuchs (1954) and is explained in this section (Chakrabarti, 1987; Dean and Dalrymple, 1991).

If the water depth is limited, then the normal velocity at the seabed also needs to be zero:

$$\frac{\partial \phi}{\partial z} = 0 \ at \ z = -d \qquad\qquad \text{Equation 4.57}$$

The potential field may be assumed to be a superposition of the incident wave plus the diffracted waves. Therefore, the potential field may be written as follows:

$$\phi = \phi_I + \phi_R \qquad\qquad \text{Equation 4.58}$$

For the incident wave, the potential field is known based on the Stokes theory. It can be written as follows:

$$\phi_I = \frac{gH}{2\omega} \frac{\cosh k(z+h)}{\cosh(kh)} \cos(kx - \omega t) \qquad\qquad \text{Equation 4.59}$$

This formula can be written in the complex form as follows:

$$\phi_I = \frac{gH}{2\omega} \frac{\cosh k(z+h)}{\cosh(kh)} \cos(kx - \omega t)$$

$$\phi_I = \mathrm{Re}\left\{ -\frac{gH}{2\omega} \frac{\cosh k(z+h)}{\cosh(kh)} e^{i(kx-\omega t)} \right\}$$

$$e^{i(kx-\omega t)} = e^{ikr\cos\theta} e^{-i\omega t} = \left[J_0(kr) + \sum_{m=1}^{\infty} 2i^m \cos(m\theta) J_m(kr) \right] e^{-i\omega t}$$

$$\phi_I = -\frac{gH}{2\omega} \frac{\cosh k(z+h)}{\cosh(kh)} \left[J_0(kr) + \sum_{m=1}^{\infty} 2i^m \cos(m\theta) J_m(kr) \right] e^{-i\omega t}$$

Equation 4.60

Radiation waves need to be symmetric and also need to damp in time. A Hankel-type equation that satisfies the Laplace equation can get the desired shape. Note that in contrast with the incident wave, where the amplitude of the wave is known, here the amplitude of the scattering wave is not known. The potential field is in the following form:

$$\phi_R = \frac{\cosh k(z+h)}{\cosh(kh)} \left[\sum_{m=1}^{\infty} A_m \cos(m\theta) H_m^{(1)}(kr) \right] e^{-i\omega t} \qquad\qquad \text{Equation 4.61}$$

where ϕ_R and ϕ_I satisfy the Laplace equation and boundary condition on the free surface. The remaining boundary condition is on the solid body, which can be written as follows:

$$\frac{\partial \phi_R}{\partial r} = -\frac{\partial \phi_I}{\partial r} \; at \; r = R$$

Equation 4.62

This formula states that the reflected potential field at the surface of a cylinder should create a velocity field in the normal direction equal to wave velocity and negative to make the total potential zero on the surface.

Satisfying this boundary condition results in knowing the amplitude of radiated waves; the final potential field can be written in the following form:

$$\phi = \phi_R + \phi_I = \frac{H\omega \cosh k(z+h)}{2k \cosh(kh)} \sum_{m=1}^{\infty} \delta_m i^{m+1} \left[J_m(kr) - \frac{J'_m(ka)}{H_m^{(1)'}(ka)} H_m^{(1)}(kr) \right] \cos(m\theta) e^{-i\omega t}$$

Equation 4.63

Using the Bernoulli equation, the force per unit length of the pile can be calculated as follows:

$$dF_I = \frac{2\rho gH}{k} \frac{\cosh(k(h+z))}{\cosh(kh)} G(D/L) \cos(\omega t - \alpha)$$

$$\tan \alpha = \frac{J'_1(ka)}{Y'_1(ka)}$$

Equation 4.64

$$G(D/L) = \frac{1}{\sqrt{J'_1(ka)^2 + Y'_1(ka)^2}}$$

which can be written in the form of a function of acceleration of the wave particle velocity, and the damping coefficient is equal to zero.

$$dF_x = \frac{4G(D/L)}{\pi^3 (D/L)^2} \rho \pi a^2 \dot{u}_\alpha$$

$$\dot{u}_\alpha = \frac{gHk}{2\omega} \frac{\cosh(k(z+h))}{\cosh(kh)} \cos(\omega t - \alpha)$$

Equation 4.65

Clearly, for very simple geometry like a cylinder, the close form of a diffraction problem is not very straightforward. Usually, numerical methods are used for calculating diffraction forces on offshore structures.

4.3.2 Morison Equation

Very frequently in offshore structures, it happens that the diameter of a cylinder is much smaller compared to the wavelength. In this case, the diffraction effects are negligible and the problem can be greatly simplified. One simple approximation is to only consider the Froude–Krylov force, which is a good initial estimation.

Morison (Morison *et al.*, 1950) gave a better approximation with the following argument. Assume that the wavelength is larger than the diameter of the offshore structure. If a wave with velocity U and frequency ω comes toward the cylinder, the problem of finding the diffracted wave by the cylinder can be approximated by assuming that the cylinder has forced oscillation with $-U$ velocity and the same frequency. In this case, we can think that the net velocity on the surface of the cylinder is close to zero. In this case, the force will be written as follows:

$$dF_x = \rho\pi a^2 \dot{u} + A_{11}\dot{u}$$

Equation 4.66

where the first term is the Froude–Krylov force, and the second term is the Morison approximation for diffraction effects. Equation 4.67 can be written as:

$$dF_x = C_M \dot{u}$$

Equation 4.67

Morison also included an approximation of viscous effects that, as described in the definition of the Reynolds number (Equation 4.28), should be proportional to the velocity square. Therefore, Morison concluded that the total force on a fixed cylinder can be calculated as follows:

$$dF_x = C_M \dot{u} + \frac{1}{2}\rho DC_d u|u|$$

Equation 4.68

The second term, which is the contribution of viscous force, is called the *drag force* on the cylinder. The drag term is written as $u|u|$ to take into account the direction of the drag force.

This formula, due to its simplicity and ease of use, is extremely popular in calculating wave loads on cylindrical-shaped offshore structures. However, in very large offshore structures, the approximation is not very accurate and may result in overprediction of wave forces on the structure.

4.4 Tides and Currents Kinematics

Currents are a steady movement of water in a specific direction. Coriolis effects of earth, and differences in temperature and wind, are among important sources of currents. Although there are different types of currents on earth with different periods, tidal currents and wind-stress currents (Wilson, 2003) are two types of flows that are important for analysis of offshore structures. Tidal currents are the horizontal movements of the water due to the water's rise and fall down (tides). Tidal current speeds vary from 1 to 2 m/s up to less than 10 m/s. The tidal currents usually vary with depth with a power law formula, and the following relation might be used.

$$u_{Tide-curr}(z) = \left(1 + \frac{z}{d}\right)^{1/7} u_{Tide-curr}(0)$$

Equation 4.69

where z is negative downward. Another important type of current is wind-induced current, which occurs due to steady blowing of water on the ocean surface. This type of current varies linearly with water depth and can be approximated with the following formula:

$$u_{wind-curr}(z) = \left(1 + \frac{z}{d}\right)u_{wind-curr}(0)$$

Equation 4.70

After calculating the total current on the structure and knowing the drag coefficients based on experimental or numerical data, the load on the structures can be calculated. Figure 4.9 shows the tidal current and wind-induced currents and the superposition of them.

The currents can lead to mean drift motion of the offshore structure such as a tension leg platform (TLP), which might result in mean tension on the mooring system. More importantly, the current usually leads to shedding of vorticity behind the structure. The vortices usually shed behind the structure with a frequency called the *shedding frequency*. A nondimensional number, called the Strouhal (*St*) number, is usually used for defining the shedding frequency, and it can be written as follows:

$$St = \frac{fL}{U}$$

Equation 4.71

where f is the shedding frequency, L is the characteristic length and U is the current velocity. The shedding frequency is mainly a function of the Reynolds number. For a circular cylinder, there are numerous experimental and numerical results in different flow regimes to define this dependency.

As the vortices shed behind the structure, it results in drag forces in the direction of the current and lift forces perpendicular to the current with the same frequency. If the structure is free to move like the tendons of a TLP, it will lead to oscillatory motion of the structure, which is called vortex-induced motion (VIM).

If the shedding frequency coincides with the offshore structure's natural frequencies, it might lead to resonant motion of the structure (called galloping) and should be carefully studied and avoided if possible.

Note that in contrast with wave load analysis, in which the potential functions are highly applicable regarding currents load analysis, this method is not very useful

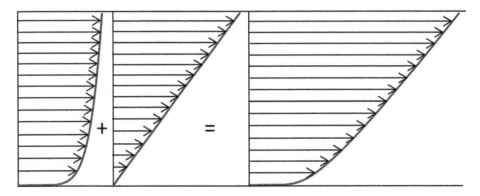

Figure 4.9 Profile of tidal currents (left), wind-induced currents (center) and the summation of these two (right).

anymore. The basic assumption in potential flow theory is irrotationality of flow, which is not valid here. According to potential flow theory, if the cylinder moves with a constant velocity in the water, the drag force applied on it is equal to zero! This is called the D'Alembert paradox and happens due to neglection of viscous effects. In the wave load analysis, since the viscous loads are usually negligible, this assumption is valid. However, here viscosity plays an important role and cannot be neglected. Therefore, estimations of shedding frequency and of drag and lift forces are usually based on computational fluid dynamics (CFD) methods or experimental tests. In Section 4.5, the current loads on offshore structures will be studied.

4.5 Current Loads on Offshore Structures

One of the simplest offshore structures, yet one of the most important ones, is circular cylindrical-shaped structures. The forces on a circular cylinder can be calculated, based on the Bernoulli equation. Assuming that a uniform flow from the infinity is moving toward the cylinder, the Bernoulli equation can be written as follows:

$$\frac{p_0}{\rho} + \frac{1}{2}V_\infty^{\ 2} = p + \frac{1}{2}V_\theta^{\ 2}$$

Equation 4.72

where the terms on the left-hand side are written for a point far before the cylinder, and those on the right-hand side are written for any point on the perimeter of the cylinder.

Based on elementary irrotational flow theory, superposition of a doublet and a uniform flow will result in potential field of a flow past a cylinder, and the tangential velocity can be obtained as follows:

$$V_\theta = 2U_\infty \sin\theta$$

Equation 4.73

Assuming that the pressure is calculated with respect to the far-field pressure, the pressure on the cylinder surface can be calculated as follows:

$$\frac{p}{\frac{1}{2}\rho u_\infty^2} = 1 - 4\sin^2\theta$$

Equation 4.74

The following pressure is by assuming that the flow is irrotational. However, in reality due to boundary layer effects, the viscous effects become important and the flow starts to separate from the surface. In the front edge of the cylinder, the agreement is quite good even for large Reynolds numbers; however, especially behind the cylinder, the difference is noticeable and nearly uniform pressure is measured. This is due to separation of the flow field.

The separation of flow usually occurs in the back portion of the cylinder and can be explained as follows. In the front of the cylinder (Figure 4.10) is the stagnation point, and both normal and tangential velocities are equal to zero. Based on the Bernoulli equation, the pressure is maximum there; as we move toward the top of the cylinder, the velocity outside of the boundary layer increases until it reaches its maximum value; therefore, the pressure is decreased, This is a favorable condition for the boundary layer, which is also

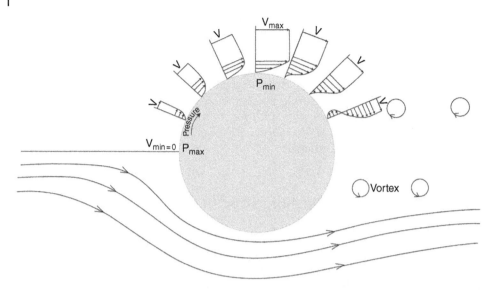

Figure 4.10 Boundary layer and separation of flow at the surface of a circular cylinder.

moving with slower velocity due to a no-slip boundary condition on the cylinder's surface. Therefore, the velocities both on the boundary layer and outside of the boundary layer are increasing. As the fluid passes the maximum velocity on top of the cylinder, the pressure tends to increase on the outer layer. The boundary layer will be also affected from this unfavorable pressure; the higher values of velocity in the boundary layer will only suffer by reduction in velocity. However, the velocities very close to the surface (inside the boundary layer) do not have enough momentum to keep moving, and the velocity will become negative to the direction of the flow. This will cause rotation of the flow near the surface, which is formally called separation of the flow. This flow separation will develop and affect the pressure field around the cylinder surface. The flow separation will result in a nearly uniform pressure field in the region of occurrence.

Because the irrotational assumption of fluid is violated in most of the practical applications for the flow field behind circular cylinders, experimental measurements or numerical techniques such as CFD methods will be used to measure the drag coefficient.

4.6 Wind Kinematics

Wind blows mainly due to the difference in amount of receiving energy from the sun. This difference results in difference in the temperature, hence the air pressure. The temperature and the air pressure become higher at the equator and lower at the polar regions. The wind starts to blow from the equator toward the poles to compensate the pressure difference. This creates the main mechanism of wind on earth. Another important mechanism is the rotation of earth, which leads to speeds of about 1670 km/h at the equator and zero at the poles, which highly affect the global wind patterns (Manwell *et al.*, 2010). Beside these global patterns, local patterns of earth such as uneven surface of earth and absorption of heat by oceans have considerable effects on the wind pattern.

4.6.1 Wind Data Analysis

The wind can be studied at time scales of inter-annual, annual, daily and short term (in the order of a fraction of a second) (Manwell *et al.*, 2010). The long-term prediction of highest wind speed is an important factor in design of offshore structures. The inter-annual and annual wind speed is more crucial for offshore wind energy harvesters to estimate the power output. To perform wind load analysis of offshore structures, usually wind speed is measured at the locations of interest and at different altitudes. The recorded values for each location are broken down into shorter time series, usually about 10 min. Usually, the mean value and standard deviation of wind speed during these 10 min are the characteristic values of each interval and are calculated as follows. Suppose N wind velocity (u) at a specific height (usually, 10 m above the surface) is recorded within 10 min. Then, the mean value and standard deviation are calculated as follows:

$$U_{10} = \frac{1}{N}\sum_{i=1}^{N} u_i \qquad\qquad \text{Equation 4.75}$$

$$\sigma_u = \sqrt{\frac{1}{N-1}\sum_{i=1}^{N}(u_i - U_{10})^2} \qquad\qquad \text{Equation 4.76}$$

To make the recorded data manageable in size within years of recording data, usually only the mean and standard deviation of every 10 min are stored. The pattern of wind speed during the 10 min typically follows some predictable models. One of the simplest models is the Rayleigh model in which the wind probability distribution only depends on the mean velocity; it is in the following form:

$$p(U) = \frac{\pi}{2}\exp\left(\frac{U}{U_{10}^2}\right)\exp\left(-\frac{\pi}{4}\left(\frac{U}{U_{10}}\right)^2\right) \qquad\qquad \text{Equation 4.77}$$

In Figure 4.11, the probability density distribution function is plotted for three different mean wind speeds.

Although Rayleigh distribution gives variation of wind speed from mean value, the variation of wind from mean wind speed should be evaluated within higher resolution. Therefore, a quantity called *turbulence intensity* is defined, which is standard deviation of the wind speed over the mean value for 10 min intervals:

$$I_u = \frac{\sigma_u}{U_{10}} \qquad\qquad \text{Equation 4.78}$$

The turbulence intensity usually varies between 0.1 and 0.4, and it is higher for lower wind speeds. The turbulence intensity usually follows a normal distribution, and the following formula can be represented for the probability density function of turbulence intensity.

$$p(u') = \frac{1}{\sigma_u\sqrt{2\pi}}\exp\left(-\frac{u'^2}{2\sigma_u^2}\right) \qquad\qquad \text{Equation 4.79}$$

The probability density function for turbulence intensity equal to 0.13 and mean wind speed equal to 15 m/s is plotted in Figure 4.12.

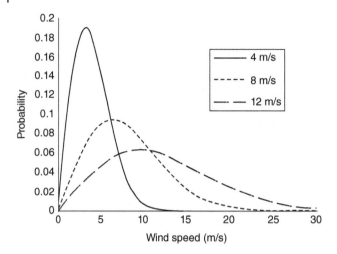

Figure 4.11 Distribution of mean wind speed in 10 min intervals based on Rayleigh model.

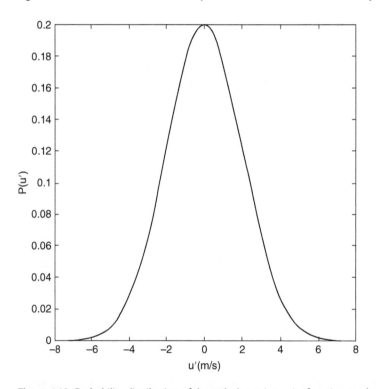

Figure 4.12 Probability distribution of the turbulence intensity function can be approximated by a normal distribution function.

Although the mean and variation of wind speed need to be taken into account, the frequency of wind is also important. There is a chance that one of the offshore structure natural frequencies coincides with wind frequency and leads to higher response of the structure. The same as wave numerical models for power spectral density, wind power

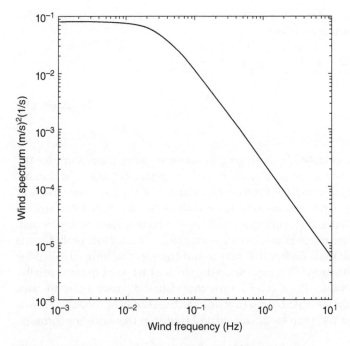

Figure 4.13 A power spectral density function of a wind condition.

spectral density also can be simulated by simple models. One of the famous and stand-ard ones is the von Karman model, which can be written in the following form:

$$S(f) = \frac{\sigma_u^2 4(L/U_{10})}{\left[1 + 70.8(fL/U_{10})^2\right]^{5/6}}$$

Equation 4.80

where f is the frequency of the wind and L is the turbulence length scale. The length scale can be calculated from the following:

$$L = \begin{cases} 2.45z & z < 30 \\ 73.5 & z \geq 30 \end{cases}$$

Equation 4.81

The wind power spectrum of a wind with 20 m/s mean wind speed and 0.15 turbulence intensity is plotted in Figure 4.13.

4.6.2 Extreme Wind Conditions

So far, the discussion has focused on estimating details of wind condition based on characteristic values of wind speed, which are mean wind speed and standard deviation. However, usually metrological data for specific location and height are available for a limited number of years. Typically, the worst wind conditions within 50 or 100 years are examined for design and survivability tests of the offshore structure. To overcome this difficulty, different models have been developed to estimate the extreme wind conditions based on limited data. One of the famous ones is the Gumbel probability distribution function, which is as follows:

$$p(U_e) = \frac{1}{\beta} \exp\left(\frac{-(U_e - \mu)}{\beta}\right) \exp\left(-\exp\left(\frac{-(U_e - \mu)}{\beta}\right)\right)$$

$$\beta = \left(\sigma_e \sqrt{6}\right) / \pi \qquad\qquad \text{Equation 4.82}$$

$$\mu = \overline{U_e} - 0.577\beta$$

where parameters $\overline{U_e}$ are obtained by averaging the extreme wave condition over the time where metrological data are available; and σ_e is the standard deviation for the corresponding U_e. Probability distribution function gives some insight about the chance of occurrence of a problem. However, in wind load analysis, we are very interested to know, for example, the chance that wind speed will reach a certain value or *higher* than that. If this chance of occurrence is more than once per 50 years, then probably the structure should survive in that condition. A very useful quantity can be used called the cumulative distribution function (F). It represents the chance that wind speed is smaller than or equal to a certain value. Then, $(1 - F)$ is the chance that it passes a certain value of wind speed. The cumulative distribution function for the Gumbel probability distribution function (Equation 4.82) can be calculated and results in the following formula:

$$F(U_e) = \int_0^{U_e} P(U_e)\,dU_e = \exp\left(-\exp\left(\frac{-(U_e - \mu)}{\beta}\right)\right) \qquad\qquad \text{Equation 4.83}$$

4.6.3 Wind Speed Variation with Height

Based on fluid mechanics theory, the wind that blows on the sea or land can be considered as a fluid passing above a surface, hence a boundary layer should be created. The velocity of the wind is equal to zero on the surface, and it increases continuously as it moves further away from the surface until it reaches a maximum speed; after that, the wind profile is nearly flat. Although the boundary layer is usually very thin and small in length, for the earth this boundary layer is not small anymore. Solving a simplified Navier–Stokes equation leads to a logarithmic profile of the wind with respect to height, which can be written as follows:

$$U(z) = \frac{U^*}{k_a} \ln\left(\frac{z}{z_0}\right) \qquad\qquad \text{Equation 4.84}$$

where U^* is the friction velocity and depends on the amount of drag induced on the velocity of the wind because of the presence of the surface; it can be written as follows:

$$U^* = \sqrt{\kappa} U_{10} \qquad\qquad \text{Equation 4.85}$$

where κ is the surface friction coefficient and can be calculated from the following formula:

$$\kappa = \frac{k_a^2}{\left(\ln\dfrac{H}{z_0}\right)^2} \qquad\qquad \text{Equation 4.86}$$

Table 4.2 Terrain roughness parameter z_0 and power law exponent α.

Terrain type	Roughness parameter z_0	Power-law exponent α
Plane ice	0.00001–0.0001	
Open sea without waves	0.0001	
Open sea with waves	0.0001–0.01	0.12
Coastal areas with onshore wind	0.001–0.01	
Snow surface	0.001–0.006	
Open country without significant buildings and vegetation	0.01	
Fallow field	0.02–0.03	
Long grass, rocky ground	0.05	
Cultivated land with scattered buildings	0.05	0.16
Pasture land	0.2	
Forests and suburbs	0.3	0.30
City centers	1–10	0.40

Source: Data from Det Norske Veritas (2000).

where H is the height of the location where the wind data are measured (10 m above the free surface is the common practice for measurements), z_0 is the terrain roughness parameter and can be found in Table 4.2, and k_a is the von Karman constant and equal to 0.4. Equation 4.84 is in fact trying to estimate the amount of drag applied to the wind velocity due to the presence of boundary; it needs to be related to the surface terrain roughness, given in Table 4.2, and one sample of measured wind in a specific height.

Another formula, which is simpler and popular for estimating wind speed as a function of height, is the power law formula. It can be written as follows:

$$U(z) = U(H)\left(\frac{z}{H}\right)^{\alpha}$$

Equation 4.87

where $U(H)$ is the velocity at H, where the wind data are available; and α depends on the surface terrain roughness and is given in Table 4.2.

Based on these two methods, Figure 4.14 shows the wind profile as a function of height for a wind blowing over open sea with a mean wind speed equal to 12 m/s measured at 10 m height of the water's free surface. It can be seen that the results predicted by the two approaches are in good agreement.

4.7 Wind Loads on Offshore Structures

For offshore structures whose main purpose is not harvesting wind energy, and for offshore structures whose natural frequency is far from wind frequencies, static wind load analysis usually gives a good estimate.

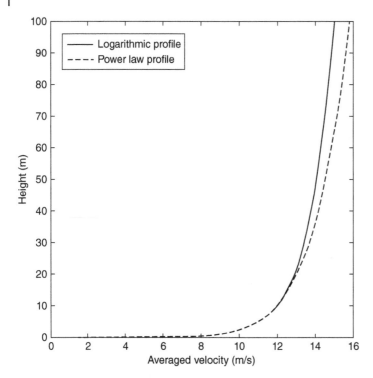

Figure 4.14 Wind speed as a function of height based on logarithmic and power law approaches.

Static wind load analysis can be calculated very similar to water current load analysis, since wind and water are both fluids and have nearly steady velocities. The wind load on an offshore structure can be calculated by:

$$q = 1/2\rho u^2 C_s \qquad\qquad \text{Equation 4.88}$$

where u is the wind's mean speed at the interested height; ρ is the density of air; C_s is the shape function; C_s is the function of the geometry of the body for which the induced wind load needs to be calculated; and C_s, for some geometries that can be seen frequently in offshore engineering, is given in Table 4.3.

In modeling offshore structures, structures can be decomposed to simpler structures in which the shape functions are known and the results may be superimposed. In superposition, in fact, we are neglecting the coupling effects between different subsections of the structure.

After calculating wind load by Equation 4.88, the force in different directions can be calculated by:

$$F = qA \qquad\qquad \text{Equation 4.89}$$

where A is a vector with the magnitude of the projected area in the normal direction of the surface. The normal direction follows the direction where the force needs to be calculated.

Table 4.3 Shape function for different geometries.

Geometry	Shape function (C_s)
Spherical shapes	0.4
Cylindrical shapes (all sizes) 0.5	0.5
Large flat surfaces (hulls, deckhouses, smooth deck areas) 1.0	1.0
Drilling derrick 1.25	1.25
Wires 1.2	1.2
Exposed beams and girders under deck	1.3
Small parts	1.4
Isolated shapes (cranes, beams etc.)	1.5
Clustered deckhouses and similar structures	1.1

Source: Data from Det Norske Veritas (2000).

For studying wind loads on offshore structures, if two structures are tandemly placed and the distance is relatively short compared to the characteristic length, the second structures will be shielded behind the first one; therefore, the wind loads on these structures are considerably less. This is called the shielding effect, and it may reduce the loads by a factor of 2.

In all the wind load analyses of offshore structures, shape functions and superposition and shielding effects are approximated. In the case that actual wind tunnel tests are available, they are preferred for estimation.

In all the above-calculated conditions, the goal is not to harvest the wind energy. Therefore, relatively simplified models are used. In the next section, we turn our attention to offshore wind turbines, where wind loads on wind turbine blades require more detailed studies.

4.8 Aerodynamic Analysis of Offshore Wind Turbines

In this section, we are going to study the aerodynamic load analysis of wind turbines at three levels of complexity. First, we will talk about 1D momentum theory. Then, we will include the effect of blades' rotation; and, at the end, we will study a very practical approach for wind load analysis of wind turbines called blade element momentum (BEM) theory.

4.8.1 1D Momentum Theory

This is one of the simplest models, yet it is insightful to analyze wind loads on both offshore and onshore wind turbines. By assuming that there are no rotation effects from the wind turbine, and that the wind speed is uniform along the wind turbine, conservation of momentum for the control volume shown in Figure 4.15 can be written in the following form (Manwell *et al.*, 2010):

$$T = U_1\left(\rho A_1 U_1\right) - U_4\left(\rho A_4 U_4\right)$$

Equation 4.90

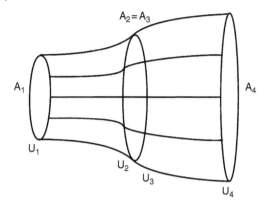

Figure 4.15 Control volume for studying thrust force on a wind turbine; linear momentum theory approach. *Source:* This figure is made available under the Creative Commons CC0 1.0 Universal Public Domain Dedication.

Since the control volume for writing mass conservation is plotted along the stream line, the flow rate is conserved at the very upstream and downstream of Figure 4.15.

$$\rho A_1 U_1 = \rho A_4 U_4 \qquad \text{Equation 4.91}$$

In addition, writing the Bernoulli equation, considering both sides of the wind turbine and assuming that the velocities at both sides of the turbine are the same ($U2 = U3$) and the pressures from far upstream and far downstream the turbine are similar, the thrust on the wind turbine can be rewritten in the following form:

$$T = A_2 \left(p_2 - p_3 \right) \qquad \text{Equation 4.92}$$

Comparing Equation 4.90 and Equation 4.92 provides a relation between the velocity at the wind turbine location with very upstream and downstream velocity.

$$U_2 = U_3 = \frac{1}{2} \left(U_1 + U_4 \right) \qquad \text{Equation 4.93}$$

Equation 4.93 is simply saying that it is assumed that the velocity at the wind turbine location is average of upstream and downstream velocity.

Using Equation 4.93, it can be concluded that if the velocity at the wind turbine is reduced by a percent of very upstream velocity, the velocity at the very downstream will be reduced by $2a$ percent. In other words:

$$U_2 = \left(1 - a \right) U_1$$
$$U_4 = \left(1 - 2a \right) U_1 \qquad \text{Equation 4.94}$$

where a is an important parameter in wind load analysis of wind turbines. It shows the percentage of velocity reduction due to the presence of wind turbine, and it is called the *induction factor*.

$$a = \left(U_1 - U_2 \right) / U_1 \qquad \text{Equation 4.95}$$

Writing the thrust force in terms of induction factors, the thrust force on the wind turbine can be calculated in the following form:

$$T = \frac{1}{2}\rho A_2 (U_1 - U_4)(U_1 + U_4) = \frac{1}{2}\rho A4aU_1^2 (1-a)$$

Equation 4.96

The power output of the wind turbine is equal to:

$$P = \frac{1}{2}\rho A_2 (U_1 - U_4)(U_1 + U_4)U_2 = \frac{1}{2}\rho A4aU_1^3 (1-a)^2$$

Equation 4.97

Knowing that the whole energy of the wind (with velocity U_1) coming toward the wind turbine is equal to:

$$P = \frac{1}{2}\rho A U_1^3,$$

Equation 4.98

the efficiency of the wind turbine can be written in the form of:

$$c_P = 4a(1-a)^2$$

Equation 4.99

where the maximum value of c_P for Equation 4.99 is 0.5926, which is called the Betz limit of a wind turbine. Although this relation is obtained with very simplified approaches, the Betz limit is very robust and universal, and it is true for nearly any type of wind turbine.

4.8.2   Effects of Wind Turbine Rotation on Wind Thrust Force

In calculations discussed in Section 4.8.1, the effects of wind turbine rotation are neglected. Blades of horizontal wind turbines rotate due to the incoming waves. Therefore, the air particles apply a rotational speed Ω to the blades. Hence, the blades should apply the same force (not velocity) with equal magnitudes and negative direction to the air particles, which can be denoted as ω. We still assume that the velocity at the normal velocity along the wind turbine is not varying and that the amount of angular velocity of wind is small compared to the rotational speed of the wind turbine (Manwell *et al.*, 2010).

Using the control volume shown in Figure 4.16 and writing the conservation of angular momentum (very similar to writing linear momentum conservation in Section 4.8.1) result in:

$$P_2 - P_3 = \rho(\Omega + 1/2\omega)\omega r^2$$

Equation 4.100

and the thrust force on the wind turbine can be calculated by:

$$dT = \rho\left(\Omega + \frac{1}{2}\omega\right)\omega r^2 (2\pi r dr)$$

Equation 4.101

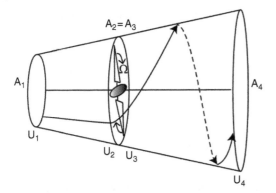

Figure 4.16 Control volume for calculating thrust force and torque for a horizontal axis wind turbine by considering the rotational wake effects.

In Section 4.8.1, we introduced the induction factor a. Similar to that, we define a' as the ratio of the wind particle angular velocity to the rotor angular wind speed.

$$a' = \frac{\omega}{2\Omega} \qquad \text{Equation 4.102}$$

Since only rotation is added to the model, therefore the linear momentum should remain valid and the thrust force based on that theory is still valid for every cross section of wind turbine blade. Equating two relations, the relation between the linear and angular induction factor can be obtained ($U_1 = U$):

$$\frac{a(1-a)}{a'(1+a')} = \frac{\Omega^2 r^2}{U^2} = \lambda_r^2 = \left(\lambda \frac{r}{R}\right)^2, \qquad \text{Equation 4.103}$$

where λ is the tip speed ratio and is defined as the ratio of tip speed rotation with respect to the wind velocity:

$$\lambda = \frac{\Omega R}{U} \qquad \text{Equation 4.104}$$

The power produced by the wind turbine can be calculated by knowing that the torque applied to the wind turbine blades should be equal and negative to the torque applied to the wind particles. Based on that, the power generated by the wind turbine can be calculated.

$$dp = \frac{1}{2}\rho A U^3 \left(\frac{8}{\lambda^2} a'(1-a)\lambda_r^3 d\lambda_r \right) \qquad \text{Equation 4.105}$$

Integrating this equation over the radius r:

$$P = \frac{1}{2}\rho A U^3 \left(\frac{8}{\lambda^2} \int_0^\lambda a'(1-a)\lambda_r^3 d\lambda_r \right)$$

$$\qquad \qquad \text{Equation 4.106}$$

$$C_p = \frac{8}{\lambda^2} \int_0^\lambda a'(1-a)\lambda_r^3 d\lambda_r$$

Using the relation between a and a' based on the tip speed ratio, independent variables for calculating the power of wind turbines will reduce to only two. For each tip speed

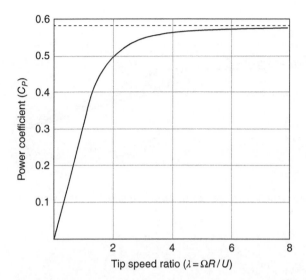

Figure 4.17 Comparison of power efficiency by considering or neglecting the rotational effects of a wind turbine. Dashed line shows the Betz limit in which rotational effects are neglected.

ratio, one can differentiate power with respect to *a* and find the maximum power that can be generated by the wind turbine.

Figure 4.17 shows the maximum power of a wind turbine as a function of the tip speed ratio. It can be seen that as the tip speed ratio increases, the maximum power becomes closer to the wind turbine's Betz limit.

4.8.3 Blade Element Momentum Theory

In Sections 4.8.1 and 4.8.2, the induction factor was not known. We tried in this chapter to find the induction factor that leads to maximum power generation. Very often, we have a wind turbine with specific blades and shapes, and we are interested in computing the thrust force and torque moment on the wind turbine. To do that, we add the blade element theory to the momentum theory to find the relevant induction factor, and hence be able to calculate the thrust and torque on a wind turbine (Manwell *et al.*, 2010).

First, we need to review the basic concepts regarding airfoil analysis. Figure 4.18 shows a wind turbine blade and an airfoil-shaped cross section of it.

According to Figure 4.18, the drag force on the airfoil is the force in the direction of the airflow, and lift is in the normal direction to the airflow. The angle of attack is the angle between the airflow and the chord line of the airfoil. Usually for airfoil, the lift and drag coefficients are given based on angle of attack and Reynolds number.

The lift and drag coefficients are defined for the airfoil as follows:

$$C_l = \frac{L/l}{\frac{1}{2}\rho U^2 c}$$

$$C_d = \frac{D/l}{\frac{1}{2}\rho U^2 c}$$

Equation 4.107

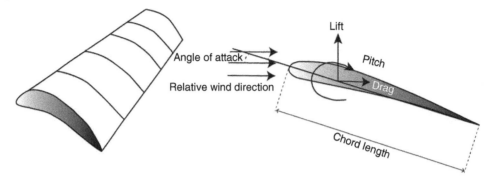

Figure 4.18 Wind turbine blade and cross section. Main characteristics of a cross section of a wind turbine blade are shown.

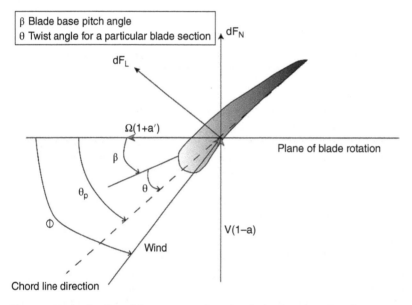

Figure 4.19 Angle of attack for a cross section of a wind turbine based on the twist angle and wind relative velocity.

where l is the length of the cross section of the airfoil, and c is the chord length of the airfoil. For calculating the lift and drag coefficients of an airfoil, we need to know the angle of attack, which depends on the ratio between the rotation speed (depends on the blade rotational speed and wind particle rotational speed) and incoming wind speed (Figure 4.19), and also the angle in which the blade is twisted (Figure 4.20). For wind turbines' airfoil cross sections close to the blade root because the arm of the rotation angle is not very large, the rotational speed is comparable with the incoming wind. Therefore, a considerable twisting angle is required for the airfoil to get the maximum torque. While close to the tip of the blade, the rotation speed is much larger than incoming wind speed, so the blades are more vertical to get the maximum torque. In the wind turbine analysis for each cross section, we need to know the twisting angle of the blade and the angle of the wind direction, which in total give the pitching angle of each blade cross section.

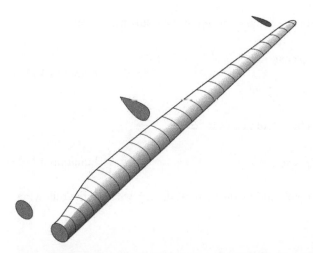

Figure 4.20 Different shapes of a wind turbine's cross sections at different radii of the wind turbine.
Source: Figure is licensed under the Creative Commons Attribution-Share Alike 3.0 Unported license.

The pitching angle of the blade can be written with respect to the tip pitching angle plus an additional twist for maximizing the torque in that cross section.

$$\theta_p = \theta_{p0} + \theta_T \qquad \text{Equation 4.108}$$

The angle of incoming wind can be described as:

$$\tan\varphi = \frac{U_{Thrust}}{U_{Rot}} \qquad \text{Equation 4.109}$$

Based on analysis of wind speed in the presence of wake formation (Section 4.8.2), the rotational speed of particles at the wind turbine can be written as:

$$U_R = \Omega r + \frac{\omega}{2}r = \Omega r\left(1+a'\right) \qquad \text{Equation 4.110}$$

where the first component is rotational speed due to blade rotation, and the second component is due to the wake effects. The thrust force as discussed in this chapter is equal to $U(1-a)$. Therefore, Equation 4.109 can be written in the following form:

$$\tan\varphi = \frac{U\left(1-a\right)}{\Omega r\left(1+a'\right)} = \frac{1-a}{\left(1+a'\right)\lambda_r} \qquad \text{Equation 4.111}$$

Assuming that at each cross section dr, the twist angle and type of airfoil are constant, the lift and drag forces can be written as follows:

$$dF_L = C_l \frac{1}{2}\rho U_{rel}^2 c\,dr$$
$$dF_D = C_d \frac{1}{2}\rho U_{rel}^2 c\,dr \qquad \text{Equation 4.112}$$
$$dF_N = dF_L\cos\varphi + dF_D\sin\varphi$$
$$dF_T = dF_L\sin\varphi - dF_D\cos\varphi$$

Therefore, the thrust force and torque can be written in the following form:

$$dF_N = B\frac{1}{2}\rho U_{rel}^2\left(c_l\cos\varphi + c_d\cos\varphi\right)cdr$$

$$dF_T = B\frac{1}{2}\rho U_{rel}^2\left(c_l\sin\varphi - c_d\cos\varphi\right)cdr$$

Equation 4.113

The torque per cross section can be calculated as follows:

$$dQ = B\frac{1}{2}\rho U_{rel}^2\left(c_l\sin\varphi - c_d\cos\varphi\right)crdr$$

Equation 4.114

Hence, using both momentum theory and blade element theory, we have the following four equations:

$$dF_N = B\frac{1}{2}\rho U^2\frac{(1-a)^2}{\sin^2\varphi}\left(c_l\cos\varphi + c_d\cos\varphi\right)cdr$$

$$dQ = B\frac{1}{2}\rho U^2\frac{(1-a)^2}{\sin^2\varphi}\left(c_l\sin\varphi - c_d\cos\varphi\right)crdr$$

$$dT = dF_N = \rho U^2 4a(1-a)\pi rdr$$

$$dQ = 4a'(1-a)\rho U\pi r^3\Omega dr$$

Equation 4.115

We also know that the φ depends on linear and angular induction factors given in Equation 4.111. Therefore, only two independent unknowns, a and a', are present in Equation 4.115. By equating the thrust and torque from momentum theory and blade element theory (Equation 4.115), obtaining unknown variables is possible.

For calculating a and a', iterative techniques need to be used, since lift and drag coefficients depend on the φ. In practice, for most of the airfoils drag coefficients are small, and (it is assumed) zero for the iterative method. Assuming that the drag coefficient is equal to zero greatly simplifies the iteration procedure. So, we can start with guessing an a and a', and based on that the pitch angle (and, hence, lift coefficient) can be calculated. This guessing will be repeated until thrust and torque based on Equation 4.115 converge. Since a and a' and φ depend on the distance from the blade root, this iteration needs to be performed at different cross sections of wind turbine blades. After finding the thrust and torque for each cross section of a wind turbine, by integration, the total thrust and torque of the wind turbine blades can be calculated.

References

Carpenter, L.H. 1958. "On the Motion of Two Cylinders in an Ideal Fluid." *Journal of Research of the National Bureau of Standards* 61 (2): 83–87.

Chakrabarti, S.K. 1987. *Hydrodynamics of Offshore Structures*. Southampton, UK: WIT Press.

Dean, R.G. and R.A. Dalrymple. 1991. *Water Wave Mechanics for Engineers and Scientists*. Singapore: World Scientific.

Det Norske Veritas (DNV). 2000. *Environmental Conditions and Environmental Loads.* DNV-RP-C205. Oslo: DNV.

Faltinsen, O. 1993. *Sea Loads on Ships and Offshore Structures.* Cambridge: Cambridge University Press.

Faltinsen, O.M., J.N. Newman and T. Vinje. 1995. "Nonlinear Wave Loads on a Slender Vertical Cylinder." *Journal of Fluid Mechanics* 289 (April): 179–198. doi:10.1017/S0022112095001297

Fox, R.W. and A.T. Mcdonald. 1994. *Introduction to Fluid Mechanics.* New York: John Wiley & Sons.

Hasselmann, K., T.P. Barnett, E. Bouws, H. Carlson, D.E. Cartwright, K. Enke, J.A. Ewing *et al.* 1973. "Measurements of Wind-Wave Growth and Swell Decay during the {Joint North Sea Wave Project}." *Deut. Hydrograph. Zeit.* 8 (12): 1–95.

Kochin, N.E., I.A. Kibel, N.V. Roze and W.R. Dean. 1966. "Theoretical Hydrodynamics." *Journal of Applied Mechanics* 33: 237. doi:10.1115/1.3625011

MacCamy, R.C. and R.A. Fuchs. 1954. "Wave Forces on Piles: A Diffraction Theory." Washington, DC: US Army Corps of Engineers.

Manwell, J.F., J.G. McGowan and A.L. Rogers. 2010. *Wind Energy Explained: Theory, Design and Application.* Hoboken, NJ: John Wiley & Sons.

Morison, J.R., J.W. Johnson and S.A. Schaaf. 1950. "The Force Exerted by Surface Waves on Piles." *Journal of Petroleum Technology* 2 (5): 149–154. doi:10.2118/950149-G

Musial, W. and B. Ram. 2010. "Large-Scale Offshore Wind Power in the United States. Assessment of Opportunities and Barriers." NREL/TP--500-40745. Golden, CO: National Renewable Energy Lab. (NREL). http://www.osti.gov/scitech/biblio/1219151

Nath, J.H., and T. Yamamoto. 2016. "Forces from Fluid Flow around Objects." In *Coastal Engineering 1974,* 1808–27. American Society of Civil Engineers. http://ascelibrary.org/doi/abs/10.1061/9780872621138.109

Newman, J.N. 1977. *Marine Hydrodynamics.* Cambridge, MA: MIT Press.

Rahman, M., and D.D. Bhatta. 1993. "Evaluation of Added Mass and Damping Coefficient of an Oscillating Circular Cylinder." *Applied Mathematical Modelling* 17 (2): 70–79. doi:10.1016/0307-904X(93)90095-X

Sclavounos, P.D. 2012. "Nonlinear Impulse of Ocean Waves on Floating Bodies." *Journal of Fluid Mechanics* 697 (April): 316–335. doi:10.1017/jfm.2012.68

Wilson, J.F. 2003. *Dynamics of Offshore Structures.* Hoboken, NJ: John Wiley & Sons.

5

Fundamentals of Structural Analysis

5.1 Background

Design and analysis of offshore structures are highly influenced by environmental loads actions and their effects. The structural analysis considering external fluid loads is the most common load-response analysis in offshore engineering. Depending on the structure and nature of the loads, different approaches may be applied for structural analysis. This covers a wide range of possible mathematical expressions for physical phenomena and nature, including load and structural representation.

Different analytical/numerical mathematical models with various fidelity, accuracy and resource/computational time requirements have been developed for centuries. The load calculation and analysis can be static, quasi-static or dynamic. Also, the analysis with respect to interaction of fluid and structure can be coupled or uncoupled. The analysis may be linearized or nonlinear, and it may be solved with time-domain, frequency-domain or hybrid methods.

In particular, due to the rapid growth of offshore renewable energy structures such as offshore wind and ocean energy devices (e.g. wave energy converters and tidal current turbines), the science, technology and engineering in this field are seeing phenomenal development. To assess the functionality and structural integrity of these systems, one must predict the dynamics, vibration and structural responses. A reliable and robust design should be based on accurate calculation of loads and responses.

Offshore energy structures are complicated, regarding the dependency of loads and load effects. In these cases, the response itself may also be important for the loads (i.e. hydroelastic effects and coupled effects between floater and mooring system). The wave- and wind-induced loads are highly connected to instantaneous wave elevation, relative motions and responses. Hence, the instantaneous positions should be considered for updating the hydrodynamic and aerodynamic forces.

Depending on the structure and its characteristics, the accelerations, velocities and motions at the instantaneous position should be applied. In some cases, the geometrical updating adds some nonlinear loading that can excite the natural frequencies of the structure. The relative velocity should be applied to the hydro loads, and the updated wave acceleration at the instantaneous position is required for analysing some concepts. This highlights the importance of coupled time-domain aero-hydro-servo-elastic dynamic and vibration analyses for offshore energy structures.

Offshore Mechanics: Structural and Fluid Dynamics for Recent Applications, First Edition.
Madjid Karimirad, Constantine Michailides and Ali Nematbakhsh.
© 2018 John Wiley & Sons Ltd. Published 2018 by John Wiley & Sons Ltd.

Hence, dynamic response analysis is the basis for design of offshore structures; and, normally, limit states analyses are based on combinations of individual dynamic analyses, such as to consider a FLS (fatigue limit state) which is based on accumulated damages. This shows the importance of performing correct dynamic and vibration analyses for offshore energy structures, including the wave power, wind energy and hybrid energy devices.

These issues are not limited to offshore energy structures. Most offshore and marine structures require proper structural analyses considering accurate load calculation. As an example, consider an oil platform subjected to wave loads. One approach to analyse the structural responses is to apply computational fluid dynamic (CFD) simulations to find the loads over time; the structure is modelled rigidly in such hydrodynamic analysis. Afterward, the pressure/loads are sequentially applied using a finite element model (FEM), and responses are calculated in a quasi-static manner. If the dynamics are found to be important, the pressure/loads varying by time can be applied in a dynamic FEM analysis. The most advanced approach is to account for the elastic deformations and rigid displacements when calculating the loads and load effects. This is called hydroelastic (wave-induced), aero-elastic (wind-induced) or aero-hydroelastic (wave-wind-induced).

In this chapter, the basics of structural analyses are introduced, while in subsequent chapters, the advanced structural analysis for recent applications in offshore engineering is discussed. The basic materials provided in this chapter are essential to understand the following analytical and numerical formulations expressed in the rest of this book when it comes to structural analysis. The most relevant and required preliminaries are discussed in this chapter. Although most readers may have a good structural engineering background, still the present chapter ensures the integrity of the book as a self-contained material for offshore mechanics.

5.1.1 Structural Components

Structural analysis describes the relations between external forces, internal forces and deformation of structural materials. It is necessary to make clear the various terms that are usually used to describe forces and deformation and their relations. Normally, structural mechanics refers to solid mechanics because a solid can sustain loads parallel to the surface. Fluid cannot resist such loads; still, some fluid-like behaviour (e.g. creep) is also part of structural mechanics (Wolf *et al.*, 2003). In offshore technology, although lots of analyses are based on elastic formulation, plasticity and elastic–plastic mechanics are widely applied (e.g. gradual plastification strain hardening).

Different types of structures and structural components, simple and complex, with varieties of functions and purposes are used in offshore engineering. But, all of them should resist the loads with an acceptable range of deflections/deformations in a correct relative position without collapsing. Offshore structures are basically made of simple structural parts like beams, trusses, frames, plates, shells, panels and so on. In this chapter, the main structural components are introduced, while analytical structural analysis is discussed and failure modes related to them are explained. Some of the main structures and structural elements are as follows:

- Bars (tie rods): Rods and bars are slender structural members subjected to tensile loads.
- Beams: Structural members used generally to carry transverse loads; may be designed from various materials (concrete, metal etc.) with different cross sections.

- Columns/pillars: These members are usually designed to resist axial compressive loads, and they are subjected to buckling.
- Trusses: These are composed of slender rods commonly designed in triangular fashion. Plane trusses are manufactured of members that are in the same plane.
- Frames: These structures are usually composed of beams and columns connected rigidly or by hinge connections.
- Plane structures: These structures (e.g. plates and walls) have two major dimensions, and the other dimension is relatively small (thickness).
- Surface structures: These structures (like shells) have three-dimensional (3D) shape; similar to plane structures, one dimension (thickness) is small compared to the others.

Loads on structures appear as forces and moments. These loads can be internal, such as gravity and centrifugal forces; and/or the structure may be subjected to different external loads. For example, hydrodynamic loads are the result of fluid pressure on the structure. The pressure integration over the surface results in forces and moments. Forces are vector quantities: they are defined by their magnitude and direction. Normally, we categorize them considering their relationship to a reference plane: compressive and tensile forces act normal into the plane and out of the plane, respectively. However, shear forces act parallel to the reference plane. A pair of forces acting in opposite directions produces moments. If the external and body loads are in balance, the body is in static equilibrium. Otherwise, accelerations will present and inertia forces appear. Based on D'Alembert's principle, the result of the inertial forces is such that the equation of equilibrium is satisfied when they are added to the original system (Wolf *et al.*, 2003). If the loads, condition/status of the structural system and structural properties (i.e. boundary conditions, supports, stiffness, damping or mass) do not vary by time or their variation in time is negligible, the static structural analysis can be performed. Quasi-static analysis in which several static analyses are performed sequentially is another approach to perform structural analysis. The final displacement/deflections and responses of the previous step are considered as initial conditions for the new analysis in the quasi-static analysis. Otherwise, a dynamic analysis considering any possible time dependency of loads and system/structure properties and status is needed. If the loads do not vary by time or can be assumed to be static, they are so-called *dead load*. There are different types of static loads, such as concentrated force or moment, distributed load, displacement load and so on. To perform structural analysis, one must idealize the structure with respect to loads, supports, boundary conditions and connections of its structural members. Here, the connections and supports are discussed.

- Fixed (rigid) connections: These connections carry moments, shear and axial forces between different structural elements. The nodal rotation and displacements are the same for all members in such cases (see Figure 5.1).
- Hinged (pin) connections: These connections can carry shear and axial forces, but moments between different structural elements cannot be transferred. Jointed members have different rotations but the same displacements in such case (refer to Figure 5.1).

Every 3D deformable element has six degrees of freedom (three displacements and three rotations) of each end node. These degrees of freedom are controlled using supports, and hence, the elements cannot move on the limited direction, or they move

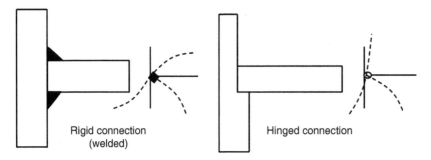

Figure 5.1 Rigid and hinged connection of beams; a hinged connection does not carry the moments between different elements, while the rigid connections (e.g. welded beams) carry moments between elements. Dashed lines present the deformed beams under external loads.

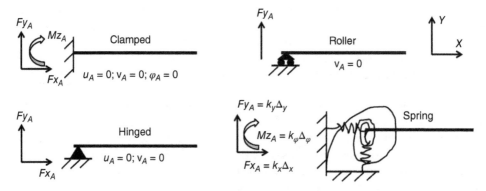

Figure 5.2 Different types of boundary conditions, supports and their corresponding reaction forces and displacements. Fx_A, Fy_A, Mz_A are support reaction forces (in the x and y directions) and moment at point A. u_A, v_A, φ_A are displacements (in x and y directions) and slope at point A. k_x, k_y, k_φ denote linear horizontal, vertical and rotational stiffness. Δ_x, Δ_y, Δ_φ denote horizontal and vertical deflections and rotation (e.g. $\Delta_\varphi = \varphi_2 - \varphi_1$).

with controlled values (the limitations are boundary conditions). There are different boundary conditions, such as free, hinged, clamped, sliding and so on. In the following, some of the main supports and boundary conditions are discussed (see Figure 5.2).

- Fixed (clamped) support: This support carries moment, shear and axial forces and does not allow any displacements of the support point.
- Hinged (pin) support: This support carries shear and axial forces but not moment. The hinged support allows rotation of the support point, but the two displacements are zero.
- Roller support (sliding): The sliding support allows rotation and one displacement.
- Spring supports: These supports account for real stiffness, but they are not fully rigid.

In Figure 5.3, different approaches for expressing equilibrium conditions for a beam under loads are shown. Cantilever beams can be a simplified model for several offshore structures. For example, monopile wind turbines or gravity-based oil platforms may be considered as beams with distributed mass, inertia and sectional elastic properties (such as axial, bending, shear and torsion stiffness). In the former case, the monopile is

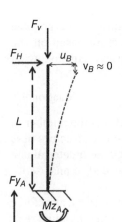

Case 1. Equilibrium condition for deformed shape:

$$\sum M = 0 \longrightarrow Mz_A = F_v u_B + F_H (L - v_B)$$

Case 2. Equilibrium condition for deformed shape, small deformation:

$$v_B \approx 0 \longrightarrow Mz_A = F_v u_B + F_H L$$

Case 3. Equilibrium condition for initial shape, small displacement and small deformation:

$$v_B \text{ \& } v_B \approx 0 \longrightarrow Mz_A = F_H L$$

Figure 5.3 Different approaches for expressing equilibrium conditions for a beam under loads.

Figure 5.4 Superposition principal illustration.

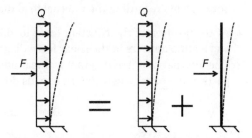

driven to soil, and the soil stiffness provides resistance to handle environmental loads as well as dead loads. In the latter case, the gravity of the bottom part (usually, the concrete part) provides stability against external loads (e.g. loads from drilling at topside, wave loads etc.) as well as dead loads (e.g. topside mass and inertia). In both cases, the structure can be assumed to be a beam, to simplify the problem to a large extent. For sure, the stiffness of the soil in both cases affects the deflections and structural responses; hence, other boundary conditions and supports may be applied to account for the soil stiffness effects (see Figure 5.2).

For linear elastic material and when the deformations/displacements are small, it is possible to use the principal of superposition. By using the superposition principal, the total displacement (or internal forces) for a structure subjected to several external loadings can be obtained by summation of the displacements (or internal forces) due to each of the external loads (see Figure 5.4).

Each body placed in one plane has three degrees of freedom, two translational degrees of freedom (in the X and Y directions) and one rotational degree of freedom; in offshore technology, they are called *surge*, *heave* and *pitch*. These motions can be controlled or limited by supports. So, three special supports can be arranged to stabilize the body (without any movement possibility). If the body is able to carry loads, it is called structure. The load actions appear as support reactions which can be determined. If the body has the possibility of movement due to insufficient supports, it is called mechanism.

The structure is called indeterminate if it has extra supports; see Figure 5.5. It is possible to introduce some formula to determine the status of the structure based on the number of connected bodies (elements), joints and supports. For trusses, the following formula may be used:

$$\iota = 2K - \Delta - A \qquad\qquad\qquad \text{Equation 5.1}$$

where K is the number of hinges, Δ is the number of elements and A is the number of support links. If $\iota > 0$, the system is a mechanism; if $\iota = 0$, a determinate structure (it is possible to analyse it with only equilibrium conditions); and, if $\iota < 0$, an indeterminate structure (equilibrium conditions and additional equations are needed to analyse it). Refer to Figure 5.5.

In a 2D plane for 2D elements (i.e. a beam), three load components – shear, axial force and bending moment – appear as internal loads. Figure 5.6 shows different joints realizing bending moments (hinge or pin joint), releasing normal axial force and releasing shear force. Positive sign conventions also are shown in Figure 5.6: tension axial forces on the section, shears that produces clockwise moments and bending moments that produce compression in the top.

Some points regarding the internal load diagrams are:

- At the position of the constant force load, the internal moment diagram has a peak, and a jump appears in the shear force diagram.
- The internal moment diagram is linear and the shear force diagram is a constant for parts of the beam where external loads do not exist.

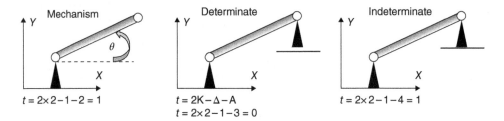

$t = 2\times2-1-2 = 1$

$t = 2K-\Delta-A$
$t = 2\times2-1-3 = 0$

$t = 2\times2-1-4 = 1$

Figure 5.5 Mechanism, determinate and indeterminate structures.

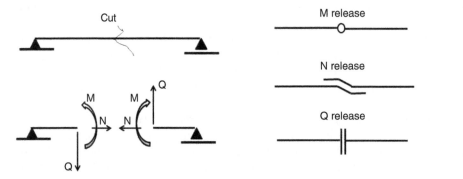

Figure 5.6 Different types of joints; moment (hinge), axial force and shear force releases; N, Q and M are the normal axial force, shear force and bending moment, respectively; positive sign conversion is shown.

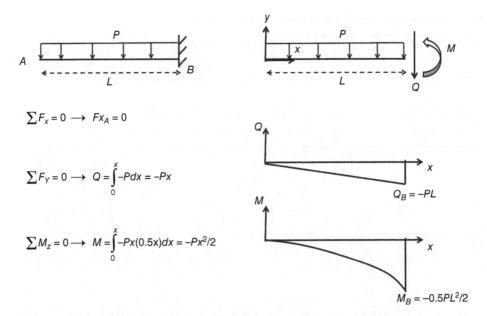

$$\sum F_x = 0 \longrightarrow Fx_A = 0$$

$$\sum F_y = 0 \longrightarrow Q = \int_0^x -Pdx = -Px$$

$$\sum M_z = 0 \longrightarrow M = \int_0^x -Px(0.5x)dx = -Px^2/2$$

Figure 5.7 Reaction forces and internal loads diagram for a beam under uniform pressure.

- For a distributed load (e.g. see Figure 5.7), the internal moment diagram is parabolic and the shear force diagram is linear.
- If a local moment is applied, the moment diagram has a jump; however, the shear diagram is not affected.

Example 5.1 For a cantilever beam subjected to uniform pressure, calculate the reaction forces as well as internal loads.

It is very useful to consider a cantilever beam under uniform load or pressure. This can be simply a balcony under its weight. The weight can be assumed to be normally distributed along the beam. The free end does not have any reaction force; so, the shear and moments are zero at $x = 0$. At the fixed point ($x = L$), the reaction in the horizontal direction is zero as there is not external load in this direction. However, the vertical reaction force is simply the total weight of the beam or the integrated pressure over the length. Also, the moment can be found by integrating the force (pressure integration results in forces). The shear and moment diagrams are shown in Figure 5.7. The shear diagram is linear, and the moment diagram is parabolic. Note the sign convention shown in Figure 5.6.

It is clear that there are relations between external loads and internal forces. Here, some of the important relations will be discussed. In general, the relations are either integrative or derivative.

The slope of a shear diagram is the intensity of the distributed load. For example, in Figure 5.7, $dQ/dx = -P \equiv q$. Note that the positive load distribution is in the Y-direction. Here, in this example, the load (weight) is negative $q = -P$. The slope of a moment diagram is equal to the shear at the point; in the current example: $Q = dM/dx = -Px$. In Table 5.1, these relations are listed. In Figure 5.8, the shear and bending moment

Table 5.1 Relations between distributed load, shear and moment.

Distributed load intensity	q	$\dfrac{dQ}{dx}$	Slope of shear diagram
Shear function	Q	$\dfrac{dM}{dx}$	Slope of moment diagram
Change in moment	ΔM	$\int Q(x)dx$	Area under shear function
Change in shear	ΔQ	$\int q(x)dx$	Area under distributed load

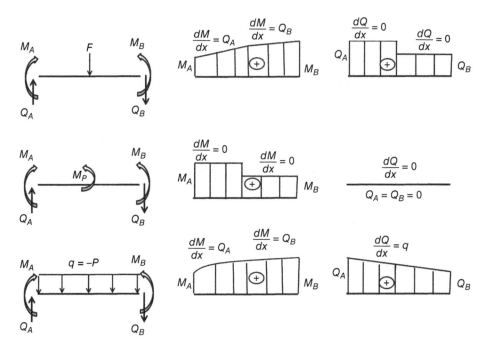

Figure 5.8 Internal force diagram for a part of a beam subjected to point force, bending moment and distributed load (positive value is in the Y-direction).

diagram for a part of a beam subjected to local force, moment and distributed load, and their relation as explained before (see Table 5.1), are shown.

For linear analysis, it is possible to use the superposition method for composing the internal force diagrams (web.aeromech.usyd.edu.au, 2015). So, by knowing the moment and shear diagrams of the simple cases, it is practical to apply the superposition method and calculate the diagrams for the complicated loading cases.

5.1.2 Stress and Strain

Deformation or strain is the change of element length and shape when displaced to a curve in the final deformed shape. For example, the strain for a rod under tension is defined by $\varepsilon = \Delta L/L$, in which ΔL is the change of element length and L is the initial length. If the displacements are small compared to the element dimensions and if the

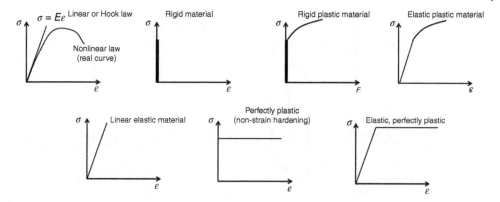

Figure 5.9 Stress–strain curve representation.

deformations are insignificant, it is possible to write the equilibrium conditions for the initial shape of the structure and neglect the small displacement of the structure; see Figure 5.9.

The structural integrity can be assessed by relating the loads and load effects to the capacity of structures. To define the strength and direction of the internal forces on a given plane, the term *stress* is used. The stress components are compared to their corresponding material strengths. Stress is one of the parameters defined to help designers evaluate the capacity and strength of the structures. For a bar subjected to tensile load, stress is simply defined as force divided by the cross-sectional area. So, its dimension is Newton per square meter (same as pressure). In real life, the relation between stress and strain for elastic materials (like steel) is nonlinear. For small deflections, the relation of stress and strain is linear, so Hook's law is valid. Hook's law states that the stress is proportional to the strain within the elastic limits.

For most materials, it is difficult to describe the entire stress–strain relation with a simple expression. So, depending on the problem and the behaviours which are important in that problem, the material behaviour is represented by an idealized stress–strain relation; see Figure 5.9. In Figure 5.9, the relation between stress and strain is shown. For linear cases, the relation between stress and strain is defined as $\sigma = E\varepsilon$, where σ is stress, ε is strain and E denotes the modulus of elasticity (Young modulus), which for steel is around $210E+9$ Pa (N/m^2). This value for concrete varies from $30E+9$ Pa to $40E+9$ Pa, depending on the characteristics of the concrete.

Mathematically, a stress vector can be defined as (Crandall *et al.*, 1972):

$$\overset{\vec{n}}{\vec{T}} = \lim_{dA \to 0} \frac{d\vec{F}}{dA} \qquad \text{Equation 5.2}$$

where $\vec{n}$ is the normal of the surface on which the force vector $(\overline{dF})$ is applied, and dA is the area. Stress is a point value vector, meaning that the stress vector is calculated at a defined point (normal of the surface $(\vec{n})$ passes this point). Usually, the stress vector is resolved into two components: normal (σ) and shear stresses (τ).

Stress is a second-order tensor quantity, as the selection of the cutting plane (orientation of the plane) results in stresses differing in both direction and magnitude. The magnitudes and directions of stresses on all possible planes through a point establish

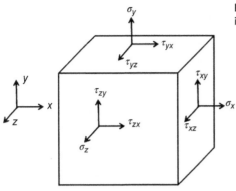

Figure 5.10 Infinitesimal cube about point of interest and stress state.

the state of stress for that point. To evaluate the strength of a member, the maximum stresses alone are not sufficient, and orientation of these stresses is also important. So, the overall stress state should be found to derive the maximum stress values. The state of stress can be computed by calculating the stresses acting on conveniently oriented planes passing the selected point. Then, standardized analytical or graphical methods can be used to determine the stresses acting on any other planes. So, the stress state is usually determined at faces of a cube of infinitesimal size surrounding the point of interest (see Figure 5.10).

Normal stresses $(\sigma_x, \sigma_y, \sigma_z)$ have a single subscript corresponding to their plane normal. For example, σ_x denotes the normal stress in the x-direction. Shear stresses $(\tau_{xy}, \tau_{xz}, \tau_{yz}, \tau_{yx}, \tau_{zx}, \tau_{zy})$ have double-subscript notation. The first designates the plane of the face, and the second denotes the stress direction. The plane of the face is represented by the normal axis. For example, τ_{xy} is the shear stress in the y-direction (second subscript), and it is acting the yz-plane (x is the normal, the first subscript). Normal stresses are positive for tensile stress; all stresses shown in Figure 5.10 are positive. A face is positive when its outward-normal vector points in the positive-coordinate axis direction. Equation 5.3 represents the stress state as a tensor. The tensor S is a symmetric tensor: symmetrical about its principal diagonal. Note that the shear stresses have pairs, for example $\tau_{xy}, = \tau_{yx}$ (Wolf *et al.*, 2003).

$$S = \begin{pmatrix} \sigma_x & \tau_{xy} & \tau_{xz} \\ \tau_{yx} & \sigma_y & \tau_{yz} \\ \tau_{zx} & \tau_{zy} & \sigma_z \end{pmatrix}$$

Equation 5.3

$$\tau_{xy}, = \tau_{yx}; \tau_{xz} = \tau_{zx}; \tau_{yz} = \tau_{zy}$$

For a symmetric tensor, an orthogonal set of axes 1, 2, 3 (so-called principal axes) exists in which the off-diagonal tensor elements are all zero; see Equation 5.4. The diagonal tensor elements are called *principal stresses* $(\sigma_1, \sigma_2, \sigma_3)$, the planes of zero shear stress are called *principal planes* and the directions of outer normal of the principal planes are called *principal directions*. If the faces of an element are principal planes, then, it is called a *principal element*. Generally, the principal axes are titled to fulfil $\sigma_3 \leq \sigma_2 \leq \sigma_1$.

$$S_p = \begin{pmatrix} \sigma_1 & 0 & 0 \\ 0 & \sigma_2 & 0 \\ 0 & 0 & \sigma_3 \end{pmatrix}$$

Equation 5.4

Stress is not a directly measurable quantity. Hence, for engineers, strain is very important as it can be directly measured. As discussed before, the fundamental element of deflection and deformation is "strain." Based on the serviceability limit state (SLS), the structural integrity of a structure or element may be insufficient due to excessive deformations, even if the corresponding stresses are acceptable compared to the allowable limits based on yielding and fracture. Strain at a point is the intensity and direction of the deformation with respect to a specific plane passing that point. In this respect, stress and strain are similar; the state of strain is a tensor-like stress state. So, strains are resolved into normal and shear components, ε and γ. In Figure 5.11, the plane deformations resulting from normal and shear strains are shown. On the left, the body is deformed where the x-dimension is extended and the y-dimension is contracted. The normal and shear strains are defined by:

$$\varepsilon_x = \lim_{x \to 0} \frac{dx}{x} \quad and \quad \varepsilon_y = \lim_{y \to 0} \frac{dy}{y}$$

$$\gamma_{yx} = \lim_{y \to 0} \frac{dx}{y} \equiv \tan\theta \approx \theta$$

Equation 5.5

Subscript notations of strain and stress are similar: ε_x and ε_y are normal strains, and γ_{yx} is shear strain resulting from taking adjacent planes normal to the y-axis and displacing them in the x-direction. For clockwise rotation, shear strain γ_{yx} is positive ($\gamma_{yx} > 0$); see Figure 5.11. Sign conventions of strain and stress are consistent, and positive stress produces positive strain. Equation 5.6 represents the strain state as a tensor.

$$T = \begin{pmatrix} \varepsilon_x & 0.5\gamma_{xy} & 0.5\gamma_{xz} \\ 0.5\gamma_{yx} & \varepsilon_y & 0.5\gamma_{yz} \\ 0.5\gamma_{zx} & 0.5\gamma_{zy} & \varepsilon_z \end{pmatrix}$$

Equation 5.6

Let us consider a 2D case or plane stress. The stress tensor for such a case is expressed as follows:

$$S = \begin{pmatrix} \sigma_x & \tau_{xy} \\ \tau_{yx} & \sigma_y \end{pmatrix}$$

$$\tau_{xy} = \tau_{yx}$$

Equation 5.7

Figure 5.11 Plane deformations resulting in normal and shear strains.

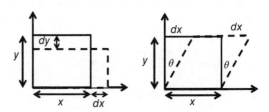

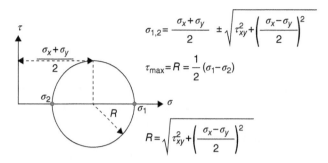

Figure 5.12 Mohr stress circle.

Shear stress is a continuous function of the cutting plane angle, and an intermediate plane with zero shear stress should exist. The principal stress tensor is:

$$S_P = \begin{pmatrix} \sigma_1 & 0 \\ 0 & \sigma_2 \end{pmatrix} \ and \ \sigma_3 = 0; \ \sigma_2 \leq \sigma_1 \qquad \text{Equation 5.8}$$

As mentioned, if the state of stress is known at this point, standardized analytical or graphical methods can be used to determine the stresses acting on any other planes. For example, the Mohr circle (see Figure 5.12) can be used to find the stress components acting on a differently oriented plane passing through that point. The radius and centre of the Mohr circle are:

$$cc = \frac{\sigma_x + \sigma_y}{2}$$
$$R = \sqrt{\tau_{xy}^2 + \left(\frac{\sigma_x - \sigma_y}{2}\right)^2} \qquad \text{Equation 5.9}$$

The principal stresses are defined by:

$$\sigma_{1,2} = \frac{\sigma_x + \sigma_y}{2} \pm \sqrt{\tau_{xy}^2 + \left(\frac{\sigma_x - \sigma_y}{2}\right)^2} \qquad \text{Equation 5.10}$$

The maximum shear stress is the radius of the circle; $\tau_{max} = R = \frac{1}{2}(\sigma_1 - \sigma_2)$. If the principal stresses are known, the stresses acting on an oriented plane can be found as follows:

$$\sigma_\alpha = \frac{\sigma_1 + \sigma_2}{2} \pm \frac{\sigma_1 - \sigma_2}{2}\cos 2\alpha$$
$$\tau_\alpha = \frac{\sigma_2 - \sigma_1}{2}\sin 2\alpha \qquad \text{Equation 5.11}$$

As the strain state and stress state have similarities, we can write:

$$\varepsilon_{1,2} = \frac{\varepsilon_x + \varepsilon_y}{2} \pm \sqrt{\left(0.5\gamma_{xy}\right)^2 + \left(\frac{\varepsilon_x - \varepsilon_y}{2}\right)^2}$$

$$\gamma_{max} = \pm 2\sqrt{\left(0.5\gamma_{xy}\right)^2 + \left(\frac{\varepsilon_x - \varepsilon_y}{2}\right)^2}$$

Equation 5.12

General stress–strain relation was briefly discussed in this chapter. Here, we recall the mathematical relations for elastic stress–strain, as such relationships are very important for engineering, design and stress analysis. For example, if the strain state is known at a point, the stress state can be determined; this is useful for testing components and material strength. Conversely, the state of strain can be determined by knowing the stress state. The scope of this section just covers solids loaded in the elastic range and considers isotropic materials (same elastic properties in all directions).

The main focus of this chapter is steel structures; note that concrete is not isotropic, while steel can be assumed to be isotropic. For a normal stress in the x-direction, the normal strain is presented by $\varepsilon_x = \sigma_x / E$, and the lateral strains are $\varepsilon_y = \varepsilon_z = -\vartheta\varepsilon_x = -\vartheta\sigma_x / E$. ϑ is the Poisson ratio (Roylance, 2000b). Similar results are obtained from strains due to stress in y- and z-directions. However, the shear stress produces only its corresponding shear strain (in which G is the shear modulus); $\gamma_{xy} = \tau_{xy}/G; \gamma_{zy} = \tau_{zy}/G; \gamma_{xz} = \tau_{xz}EG$. Hence, for a linear-elastic isotropic material with all components of stress present, the generalized Hooke's law is as follows (Wolf *et al.*, 2003):

$$\varepsilon_x = \frac{1}{E}\left[\sigma_x - \vartheta\left(\sigma_y + \sigma_z\right)\right]$$

$$\varepsilon_y = \frac{1}{E}\left[\sigma_y - \vartheta\left(\sigma_x + \sigma_z\right)\right]$$

$$\varepsilon_z = \frac{1}{E}\left[\sigma_z - \vartheta\left(\sigma_x + \sigma_y\right)\right]$$

$$\gamma_{xy} = \tau_{xy}/G; \gamma_{zy} = \tau_{zy}/G; \gamma_{xz} = \tau_{xz}/G$$

Equation 5.13

Also, for an isotropic material, it is possible to relate shear modulus, modulus of elasticity and Poisson ratio as: $G = E/[2(1+\vartheta)]$. For steel and stainless steel, the Poisson ratio (ϑ) is around 0.3.

5.2 Structural Analysis of Beams

5.2.1 Introduction

In offshore engineering, beams are the most common type of structural component. Similar to bars, one of the dimensions of beams is significantly bigger than the others. The larger dimension is the so-called beam axis (longitudinal dimension). The main function of beams is to support transverse loading and carry it to the designed supports and foundations (Beer *et al.*, 2012).

Frame structures like jackets and tripods are made from several slender structural members jointed together. The braces and legs are practically modelled as beams in engineering problems. Even the columns and pontoon of a semisubmersible can be assumed as beams for general design. If the local design is needed, a shell model can be applied. For conceptual design of several offshore structures, when one dimension dominates, beam theory is practical and quite useful to provide basic structural mechanics information like shear and bending moments. For example, consider a floating bridge, a submerged tunnel or a gravity-based structure oil platform; it is obvious that one dimension is much larger compared to the other two.

A general beam is a bar-like structural element that can resist a combination of loading actions, including bending moments, transverse shear forces, axial tension or compression forces as well as torsion moment. If the beam is subjected to compression forces, like pillars, the buckling analysis should be performed. A spatial beam (3D) supports transverse loads acting on random directions along the cross section, while a plane beam (2D) resists primarily transverse loading on a longitudinal plane.

A beam subjected to transverse loads resists primarily against the loads by bending actions; the bending moment results in compressive longitudinal stresses on one side and tensile stresses on the other side of the beam (Damkilde, 2000). These regions are separated by a "neutral surface" which has zero stress. The action of tensile and compressive stresses results in an internal bending moment which is the principal mechanism that transports loads to the supports; see Figure 5.13 (University of Colorado at Boulder, 2014).

Figure 5.13 shows a beam subjected to transverse distributed loading. As shown in the figure, the intersection of normal planes to the longitudinal dimension with the beam is called the cross section. A beam is prismatic if its cross section is constant. Please note that a beam may have a variable cross section; the tower of a wind turbine is an example of a beam with a variable cross section (the diameter and thickness of the tower are reduced with height). The plane passing the beam axis is the so-called longitudinal plane. A beam is called straight if its longitudinal axis is straight.

As an example, Figure 5.14 shows a schematic layout of a ship-shaped structure in waves; sagging and hogging load cases are presented. The structure can be roughly presented as a beam subjected to weight and buoyancy forces distributed along the beam.

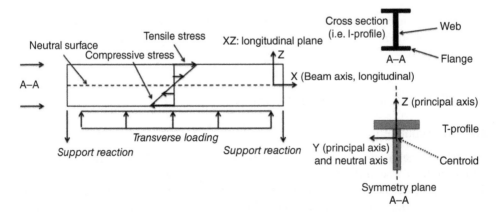

Figure 5.13 Schematic layout of a beam subjected to transverse loading.

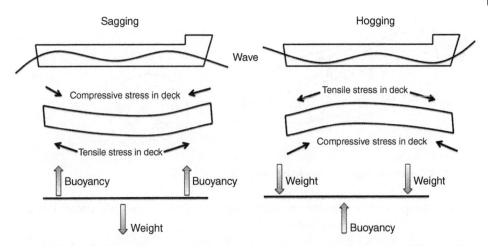

Figure 5.14 Schematic layout of a ship-shaped structure in waves; sagging and hogging load cases; the structure can be roughly presented as a beam subjected to weight and buoyancy forces distributed along the beam.

5.2.2 Beams under Torsion

In this section, the twisting, or torsion of beams is briefly discussed. A practical reference to design of steel beams subjected to torsion is Hughes *et al.* (2011). Generally, torsion in beams appears due to the action of shear loads with application points different from the shear centre of the beam section. The solution of torsion problems is complex for arbitrary shapes for which exact solutions do not exist. In such cases, empirical formulas may be used which are developed in terms of correction factors based on the geometry of a particular shape of the beam cross section (Megson, 1996).

The stress distributions and twist angles due to the applied torsion rely on the St. Venant warping (Haukaas, 2012) or Prandtl stress methods (Iowa State University of Science and Technology, 2011) based on the theory of elasticity (e.g. refer to Timoshenko and Goodier, 1951). A structural member subjected to torsion may warp in addition to twisting. The applied torque is resisted entirely by torsional shear stresses if the member is allowed to warp freely; hence, just the St. Venant torsional shear stress appears (uniform or pure torsion). However, the applied torque is resisted by St. Venant torsional shear stress and warping torsion if the member is not allowed to warp freely. This is called non-uniform torsion consisting of St. Venant torsion and warping torsion. In the following, first pure torsion is discussed for both closed and open sections. Afterward are notes regarding warping and non-uniform torsion.

Torsion of a shaft or bar (solid cylinder) is the simplest case, while hollow sections, thin-walled open-section beams and thin-walled closed-section beams are more challenging. Figure 5.15 shows a circular-section beam under torsion. Here, it is assumed that cross sections remain plane during twisting.

The shear strain at the surface is $\gamma_R = AA'/L = R\theta/L$ and at any radius is given by $\gamma_r = r\theta/L$. The shear stress and shear strain are related as follows (Crandall *et al.*, 1972):

$$\gamma_r = \frac{\tau}{G} = \frac{r\theta}{L} \rightarrow \frac{\tau}{r} = G\frac{\theta}{L} \qquad \text{Equation 5.14}$$

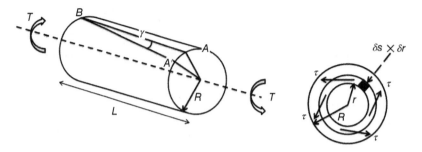

Figure 5.15 Torsion of a solid circular-section beam.

The total torque can be obtained by summing the torques from each element in the beam's cross section; refer to Figure 5.15.

$$T = \int_0^{R} \int_0^{2\pi r} \tau r \delta r \delta s = \int_0^{R} 2\pi r^2 \tau \delta r \qquad \text{Equation 5.15}$$

$$\tau = G\frac{r\theta}{L} \quad \rightarrow \quad T = \int_0^{R} 2\pi r^3 G\frac{\theta}{L} dr \quad \rightarrow$$

$$T = \frac{\pi r^4}{2} G\frac{\theta}{L} \quad \text{or} \quad T = JG\frac{\theta}{L} \qquad \text{Equation 5.16}$$

$$\frac{T}{J} = \frac{\tau}{r} = G\frac{\theta}{L}$$

In Equation 5.16, J is the polar second moment of area of the beam cross section. For a hollow circular beam, similar relations can be obtained (Bucciarelli, 2002b). However, the polar second moment of area for a hollow cylinder is $\frac{\pi}{2}(R_0^4 - R_i^4)$.

$$T = \int_{R_i}^{R_o} 2\pi r^3 G\frac{\theta}{L} dr = \frac{\pi}{2}\left(R_0^4 - R_i^4\right)G\frac{\theta}{L} \qquad \text{Equation 5.17}$$

Example 5.2   For the beam shown in Figure 5.15, calculate the strain energy due to torsion.

The energy due to a gradually applied torque is equal to the area under the torque angle of rotation. From Equation 5.16 ($T = JG\theta/L$), it is clear that the relation between torque and rotation angle is linear. Hence, the strain energy is $E = T\theta/2$. By using relations between torque and stress in Equation 5.16, $\left(\frac{T}{J} = \frac{\tau}{r} = G\frac{\theta}{L}\right)$, and substituting torque and angle in terms of the maximum shear stress:

$$E = \frac{1}{2}\frac{\tau_{max}J}{R}\frac{\tau_{max}L}{RG} = \frac{1}{4}\frac{\tau_{max}^2}{G}\pi R^2 L$$

$$\qquad \text{Equation 5.18}$$

$$E = \frac{\tau_{max}^2}{4G}\pi R^2 L = \frac{\tau_{max}^2}{4G}\nabla$$

where ∇ is the volume of the beam. It is more convenient to write the energy in terms of the applied torque, hence:

$$E = \frac{T\theta}{2} = \frac{T^2 L}{2GJ}$$

Equation 5.19

Equation 5.16 is applicable if the stress–strain relation is linear. If the stress exceeds a certain value (so-called yield stress), then plasticity in the outer region of the circular beam is induced, which will be extended inward by increasing the loading. In Figure 5.9, the stress–strain relation for elastic–perfectly plastic material is shown. For such material, it is assumed that after yield, the shear strain increases at a constant value of shear stress. This means that the shear stress in the plastic region is constant and equal to yield shear stress (τ_y). By some mathematics, it is shown that for a cylindrical beam which is partly yielded:

$$T = \frac{2\pi R^3}{3}\tau_y\left(1 - \frac{r_e^3}{4R^3}\right)$$

Equation 5.20

where r_e denotes the radius of the elastic core of the beam (the outer part is yielded and has τ_y). If the entire section is yielded, then fully plastic torque is $T_P = 2\pi R^3 \tau_y / 3$. From Equation 5.16, if the maximum stress at the outer surface reaches the yield, then $T_y = 0.5\pi R^3 \tau_y$; and it is clear that $T_P / T_y = 4/3$. This means that, after yielding, only a one-third increase in torque is required to bring the circular solid beam (or a bar) to its ultimate load-carrying capacity. Practically, in the fully plastic state, twisting continues with no increase in torque (Megson, 1996).

A thin-walled beam has been widely applied in offshore structures. Their structural behaviour under torsion loads is discussed here. Figure 5.16 shows a thin-walled closed-section beam subjected to torsion; an arbitrary shape of the cross section is presented. A pure torque loading is applied. Due to structural or loading discontinuities or supports, a system of direct stresses in the walls of the beam can be produced even though the loading consists of torsion only. These effects (known as axial constraint effects) are out of the scope of this discussion. Here, it is assumed that the induced stress in the beam walls only consists of shear stresses.

We will present the shear stress system on an element of the beam wall in terms of shear flow. The thickness is constant along the beam, but it may vary in the cross

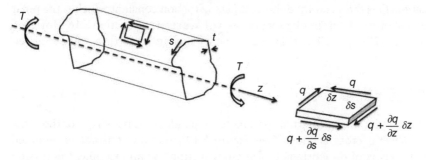

Figure 5.16 Schematic of a thin-walled closed-section beam subjected to torsion.

section. The variation of thickness over the sides of the element is neglected. Considering equilibrium in the z and s directions:

$$\left. \begin{array}{l} \left(q+\dfrac{\partial q}{\partial s}\delta s\right)\delta z-q\delta z=0 \rightarrow \dfrac{\partial q}{\partial s}=0 \\[2mm] \left(q+\dfrac{\partial q}{\partial z}\delta z\right)\delta s-q\delta s=0 \rightarrow \dfrac{\partial q}{\partial z}=0 \end{array} \right\} \rightarrow q \equiv const. \qquad \text{Equation 5.21}$$

So, a pure torque to a thin-walled closed-section beam results in a constant shear flow in the beam wall. However, the shear stress may vary around the cross section of the beam, as the wall thickness can be variable. The relationship between the torque and constant shear flow is derived as follows (Noels, 2013–2014):

$$\begin{aligned} T &= \oint \rho q\,ds = q\oint \rho\,ds \\ \oint \rho\,ds &= 2A \rightarrow T = 2Aq \\ &\rightarrow \tau = \frac{q}{t} = \frac{T}{2At} \end{aligned} \qquad \text{Equation 5.22}$$

where ρ is the distance between shear flow and applied external torque, and A is the area enclosed by the mid-line of the beam wall. The theory of torsion of thin-walled closed-section beams (refer to Equation 5.22) is recognized as the Bredt–Batho theory (Jennings, 2004). The rate of twist of a closed-section beam under torsion if the shear flow is constant can be presented as:

$$\frac{d\theta}{dz} = \frac{q}{2A}\oint\frac{ds}{Gt} = \frac{T}{4A^2}\oint\frac{ds}{Gt} \qquad \text{Equation 5.23}$$

If the walls of beam sections do not form a closed-loop system, such beams are generally called solid-section beams; two examples are shown in Figure 5.13. We have seen that the applied external torque and the rate of twist of the beam are related by the following equation:

$$T = GJ\frac{d\theta}{dz} \qquad \text{Equation 5.24}$$

In this equation, G is the shear modulus and J is the torsion constant which is the polar second moment of area for a circular section, as we have seen in Equation 5.16. However, from Equation 5.23, it is clear that the torsion constant of a thin-walled closed section is:

$$J = 4A^2 / \oint (ds/t) \qquad \text{Equation 5.25}$$

The torsion constant is an important parameter dependent on the shape of the cross section. For several cross sections (see Figure 5.13), torsion constant is obtained empirically in terms of dimensions of the cross section. As an example, the torsion constant of an I-section (see Figure 5.17) is defined by (Megson, 1996):

Figure 5.17 I-profile terms for calculation of torsion constant; see Equation 5.26.

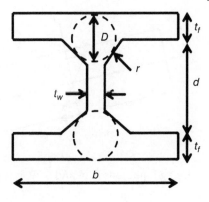

$$J = 2J_1 + J_2 + 2\alpha D^4$$

$$J_1 = \frac{bt_f^3}{3}\left[1 - 0.63\frac{t_f}{b}\left(1 - \frac{t_f^4}{12b^4}\right)\right]$$

$$J_2 = \frac{1}{3}dt_w^3$$

Equation 5.26

$$\alpha = \frac{t_1}{t_2}\left(0.15 + 0.1\frac{r}{t_f}\right)$$

$$\text{for } t_f < t_w \rightarrow t_1 = t_f \ \& \ t_2 = t_w$$

$$\text{for } t_f > t_w \rightarrow t_1 = t_w \ \& \ t_2 = t_f$$

The torsion constant for the complete section is the sum of the torsion constants of the components plus a contribution from the material at the web and flange junction. The contribution from the latter can be assumed negligible for thin-walled sections with relatively small D.

$$J \approx 2\frac{1}{3}bt_f^3 + \frac{1}{3}dt_w^3$$

Equation 5.27

For thin-walled sections, the torsion constant is presented as follows (Department of Aerospace Engineering Sciences, 2014):

$$J = \frac{1}{3}\Sigma lt^3$$

Equation 5.28

where l is the length of the component, and t is its thickness. The distribution of shear stress in a thin-walled open-section beam is a function of the twist rate:

$$\tau = 2Gn\frac{d\theta}{dz}$$

Equation 5.29

where n is the distance measured normally from the mid-line of the section wall. So, across the thickness, the distribution is linear, and it is zero at the mid-line; see Figure 5.18.

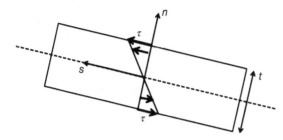

Figure 5.18 Distribution of shear stress in part of a thin-walled open section subjected to torsion (truncated component just for illustration and clarification of terms used in equations).

Using Equations 5.24 and 5.29, it is possible to derive:

$$\tau = 2n\frac{T}{J}$$
$$\tau_{max} = \pm\frac{Tt}{J}$$

Equation 5.30

where τ_{max} is the maximum shear stress which occurs at the outer surface of the wall (Beer *et al.*, 2012). Both closed- and open-section beams subjected to torsional loads twist and develop internal shear stresses. The development of shear stress is different in open and closed sections. In an open-section beam, it can only develop within the thickness of the walls; see Figure 5.19. Hence, shear stresses in an open section are limited by the wall's thickness. So, the torsional stiffness of thin-walled open sections is relatively low compared to that of closed sections (Hughes *et al.*, 2011).

In pure torsion, it is assumed that the shapes of beam sections (open and closed) remain undistorted during torsion. However, they may not remain plane, meaning that some of the cross sections warp. Warping of the cross section contradicts the assumption that "plane cross sections remain plane." This leads to compatibility problems at joints between angled beam elements. As for closed cells, the torsional deformations do not fit together with the flexural and axial deformations. For sufficiently thick-walled beam elements, the torsional deformations will usually be significantly smaller than the other deformations. This means that, for most practical purposes, the incompatibility concern has no significant effect (Andreassen, 2012).

In the classical beam theory, free torsion is assumed. This means that warping (distortion out of the plane) is allowed without restrictions. However, if the member is not allowed to warp freely, the applied torque is resisted by a combination of St. Venant torsional shear stress and warping torsion. Hence, warping of cross sections under torsion appears which should be included in non-uniform torsion. The calculation of warping effect is important to investigate the structural integrity. For example, if a beam is cantilevered, the beam is not freely warping. Therefore, tensile and compressive stresses presented are essential, particularly for concrete beams, as even low tensile stresses make severe cracking which should be investigated in design (Megson, 1996).

If the member carries the applied external torque by axial stresses plus shear stresses, warping torsion occurs. This happens during twisting of the beam when the cross section does not freely warp, for example the beam is prevented from displacing axially (has some constraints). Note that not all cross sections warp. Also, those that warp do not carry torque by axial stresses unless they are axially constrained at some locations.

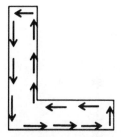 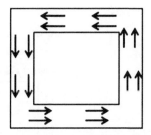

Figure 5.19 Distribution of shear stress in open and closed sections.

Cross sections that do not warp include (a) axisymmetric cross sections (e.g. solid and hollow circular-section bars and square box sections of constant thickness), and (b) thin-walled cross sections with straight parts that intersect at one point of the cross section (such as X-shaped, T-shaped and L-shaped cross sections). All torque is carried by shear stresses (St. Venant torsion), regardless of the boundary conditions for these cross sections (Haukaas, 2012). To study the warping torsion in more details, the governing principal of bending should be introduced. In Section 5.2.3, the bending of beams is presented.

5.2.3 Bending of Beams

In offshore technology, beams are primarily designed to take bending moments. For example, consider a gravity-based structure (GBS) used as an oil platform or as a substructure for a wind turbine; the wave loads, wind loads and wave/wind-induced loads result in distributed transverse loads along the structure. The structure can be assumed as a beam or a few beams (depending on variations of structural stiffness and dimensions of the cross sections) subjected to external loads. Figure 5.20 shows the schematic layout of a gravity-based bottom-fixed wind turbine and a possible simplified modelling using a few beams with different cross sections and section stiffness.

In contrast to torsion, in which the shear stresses are governing, the main stresses induced due to bending are normal stresses, tension or compression stresses. The stress state in a beam subjected to bending is complex, as there are shear stresses generated as well as normal stresses. However, the shear stresses are generally of smaller order compared to the bending stresses. In the case of non-uniform bending, shear force produces warping (out-of-plane distortion), which means that the plane section no longer remains plane after bending. But the normal stress calculated from the flexure formula (pure bending) is not significantly altered by the presence of shear force and warping. Still, the shear stresses should be included to check the structural integrity of the design. The theory of pure bending can be applied even for non-uniform bending. The flexure formula is applied if the stress distribution is not disrupted by irregularities in the shape of the beam, or by discontinuities in the loading; if not, stress concentration occurs.

In the previous section, a relationship between the applied torque and the rotation (twist) has been derived. A similar relationship between the moment and the radius of curvature of the beam in bending is obtained to define the normal stress distribution over the beam's cross section. Furthermore, the transverse displacement and slope of the beam's longitudinal axis are determined. The deflected shape of the beam will

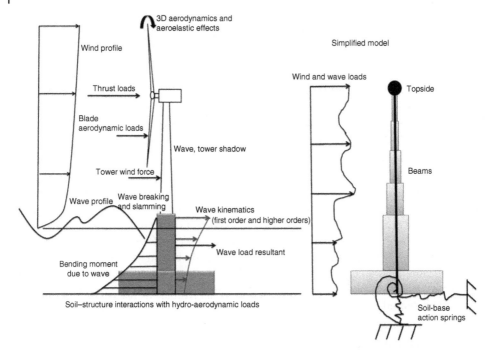

Figure 5.20 Schematic layout of a gravity-based bottom-fixed wind turbine and a simplified modelling using a few beams with diffident cross sections and section stiffness.

usually vary along the axis, which depends upon how the loads are applied over the span. First, the deformations/displacements of a beam in pure bending are presented. When the bending moment is constant over a portion, pure bending takes place and the shear force over that portion is zero (referring to Table 5.1, $dM/dx = Q$).

Displacement fields are obtained using symmetry arguments, based on a compatible strain state and its corresponding stress state. This is done by relating the displacement field to the bending moment, requiring that the stress distribution over a cross section and the bending moment are equivalent. Hence, a moment–curvature relationship is obtained which is a stiffness relationship. The moment–curvature relationship is a differential equation for the transverse displacement. Two main assumptions for dealing with bending problems are: (a) plane cross sections remain normal to the longitudinal axis of the beam, and (b) plane cross sections should remain plane (Bucciarelli, 2002c). Figure 5.21 shows a portion of a beam before and after deflection due to pure bending load. The relation between curvature and strain state is derived and presented as well.

The strain varies linearly for the pure bending case presented in Figure 5.21: $\varepsilon_x(y) = -y/\rho$; ρ is the radius of curvature of the neutral axis (surface). The elements at the top of the beam are in compression, and the elements below the neutral axis are in extension (see Figure 5.13) when the beam is subjected to a positive bending moment (refer to Figure 5.6). The strain–displacement relation for pure bending is given by: $u(x,y) = -\kappa xy$; $v(x,y) = 0.5\kappa x^2$, where κ is curvature (Kelly, 2015).

Referring to Equation 5.13, for pure bending the shear tensions in the x and y directions are zero ($\tau_{xy} = \tau_{xz} = 0$). Based on the plane stress assumption, we assume that σ_z and τ_{yz}

On the neutral surface:

$\rho \Delta \phi = \Delta s = \Delta x$

$\rho = \Delta s / \Delta \phi$

At a surface above the neutral surface:

$\varepsilon_x(y) = \lim_{\Delta s \to 0} (a'c' - ac)/ac$

$\varepsilon_x(y) = \lim_{\Delta s \to 0} ((\rho - y)\Delta \phi - \Delta s)/\Delta s$

$\varepsilon_x(y) = \lim_{\Delta s \to 0} -y\Delta \phi / \Delta s$

$\varepsilon_x(y) = -y/\rho$

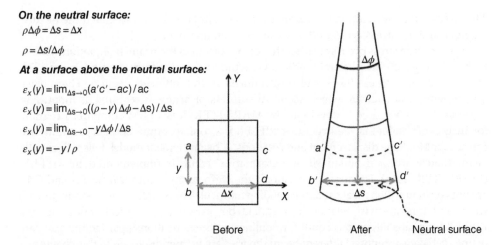

Before   After   Neutral surface

Figure 5.21 Pure bending of beam; radius of the curvature is shown for the deformed case.

can be neglected. Also, σ_y is assumed to be negligible. This is difficult to be justified for a beam carrying a distributed load in the y direction (if the load is applied at the top, stress for the load-bearing surface is proportional to the load, and it is zero on the surface below). Finally, by several assumptions, we can write the stress distributed over the cross section of the beam $\sigma_x = E\varepsilon_x = -Ey/\rho$ (Hibbeler, 2013). An example of distribution of stress is shown in Figure 5.13.

The resultant force in the axial direction is zero for the pure bending case. Equation 5.31 shows that the neutral axis passes through the centroid of the beam's cross section.

$$F_x = \int_{area} \sigma_x dA = \int_{area} -Ey/\rho dA = -\frac{E}{\rho} \int_{area} y dA$$

$$F_x = 0 \rightarrow \int_{area} y dA = 0$$

Equation 5.31

In the following, the relation between the bending moment and curvature is derived.

$$M = -\int_{area} y\sigma_x dA \rightarrow M = \frac{E}{\rho} \int_{area} y^2 dA \rightarrow M = \frac{EI}{\rho} \rightarrow M = EI\kappa$$

$$\sigma_x = -My/I$$

$$S = \frac{I}{y_{\text{max or min}}} \rightarrow |\sigma| = |M/S|$$

Equation 5.32

where I is the area moment of inertia or the second moment of area, and S is the section modulus. Using the moment–curvature relationship, the radius of curvature of an initially straight and uniform beam with a symmetric cross section subjected to a bending moment can be found (Beer *et al.*, 2012). The moment–curvature relationship is similar to stiffness relationships for beams under axial or torsion loads, that is:

$$M = \frac{EI}{\rho}, \quad T = GJ\frac{d\phi}{dz}, \quad F = AE\frac{du}{dx}$$

Equation 5.33

The stress–moment relation is written as: $\sigma_x = -M(x)y/I$. This is so-called the flexure formula, and the stress is called flexural stress (bending stress).

The maximum stress occurs at either the top or bottom of the beam. In Equation 5.33, EI is the bending stiffness (flexural rigidity), GJ is the torsional stiffness and EA is the axial stiffness. Shear stiffness (GA) is also introduced in the following section. There exist several numerical codes in which sectional stiffness of beams is required as input. For example, RIFLEX is a code developed by MARINTEK; it is a FEM computer program for analysis of slender marine structures. RIFLEX is a nonlinear time-domain program that can account for large displacement and rotations. A beam element model, Euler–Bernoulli beam theory with an added shear deformation term, is implemented in RIFLEX (MARINTEK, 2014). For some applications, the stiffness properties $(EI, GJ, EA$ and $GA)$ are not constant which means that the bending moment–curvature, tension–strain, force–strain and torsion–twist relationships should be provided. Numerical codes such as RIFLEX have capabilities to account for nonlinear stiffness relationships, and the user can define such data as inputs. Different beam theories are further discussed in this chapter.

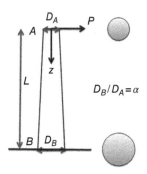

Figure 5.22 A tapered cantilever beam with a circular section subjected to point load.

Example 5.3 For the tapered cantilever beam (shown in Figure 5.22) with a circular section subjected to point load, determine the maximum stress. Also, find the ratio between the maximum stress and the stress at the support, in general. What is the stress ratio for the area ratio of three?

$$D_z = D_A + (D_B - D_A)z/L$$

$$S_z = \frac{I_z}{D_z/2} = \frac{\pi}{32}\left[D_A + (D_B - D_A)z/L\right]^3$$

$$\sigma_z = M_z / S_z = 32Pz / \left(\pi\left[D_A + (D_B - D_A)z/L\right]^3\right)$$

$$\sigma_B = 32PL / \left(\pi D_B^3\right)$$

Equation 5.34

$$\frac{d\sigma_z}{dz} = 0 \rightarrow \sigma_z \text{ is max} \rightarrow z = \frac{D_A}{D_B - D_A}\frac{L}{2}$$

$$z = \frac{D_A}{\alpha D_A - D_A}\frac{L}{2} = \frac{L}{2(\alpha - 1)}$$

$$\sigma_{max} = \frac{32PL}{2\pi(\alpha - 1)\left[1.5D_A\right]^3} = \frac{4.741}{(\alpha - 1)}\frac{PL}{\pi D_A^3}$$

Equation 5.35

Now, by having the maximum stress in the beam (Equation 5.35) and the stress at the support, the ratio of the stresses can be found as follows:

$$\frac{\sigma_{max}}{\sigma_B} = \frac{\alpha^3}{6.749(\alpha - 1)}$$

Equation 5.36

$$\text{if } \alpha = 3 \rightarrow \frac{\sigma_{max}}{\sigma_B} = 2$$

Figure 5.23 Shear stress and shear force in a cantilevered beam subjected to a point load.

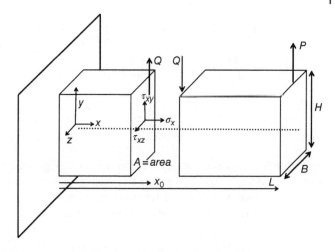

Up to now, the pure bending is studied in which we assumed that the shear force has a small effect on the normal stress distribution and deflection of the beam. For most engineering design when the length of the beam is much greater than the cross section's dimensions, these assumptions are acceptable. The shear stresses in beams due to transversal loads are discussed herein.

Let us consider a cantilevered beam subjected to a point load at its free end; see Figure 5.23. Based on static equilibrium, there is a shear force at each section equal to the applied point load. The shear force is distributed over the plane cross section in the form of a shear stress (τ_{xy}). Due to a resultant bending moment, there is a normal stress (σ_x) as we have seen before (in the pure bending case). The shear stress (τ_{xy}) is only a function of y (in the current case, the shear stress in the z direction is zero, $\tau_{xz} = 0$). Also, the shear stress is zero at the top and bottom of the beam $\tau_{xy}|_{y=\pm H/2} = 0$. The shear stress grows continuously to some value at some point in the interior. The distribution is relatively smooth without jumping up and down as it varies with y. In the rectangular beam shown in Figure 5.23, the maximum value should not be much different from its mean value: $\tau_{xy}|_{mean} \simeq \dfrac{Q}{A} = \dfrac{P}{BH}$. As the maximum stress due to bending is $\sigma_x|_{max} = \dfrac{6PL}{BH^2}$, hence the ratio of τ_{xy}/σ_x is in the order of H/L. In general, the shear stress due to transverse loads is small relative to the normal stress due to bending. However, this does not mean that we can neglect shear stresses. Particularly, excessive shear can be a reason of failure for composite beams.

To present a more detailed picture of the shear stress distribution, we consider an element with length of Δx and height of Δy at (x, y) location (running from y up to the top of the beam) with breadth of B. The equilibrium of forces in the x-direction results in the following expressions (Bucciarelli, 2002c):

$$\tau_{yx}(y)B\Delta x = \int_A \sigma_x(x+\Delta x, y)dA - \int_A \sigma_x(x, y)dA$$

$$\tau_{yx}(y) = \int_A \frac{\sigma_x(x+\Delta x, y) - \sigma_x(x, y)}{B\Delta x}dA \qquad \text{Equation 5.37}$$

$$\tau_{yx}(y) = \int_A \frac{1}{B}\frac{\partial}{\partial x}\sigma_x(x, y)dA$$

$$\sigma_x(x,y) = -M(x)y/I \rightarrow \tau_{yx}(y) = \int_A \frac{1}{B}\frac{\partial}{\partial x}\left[-M(x)y/I\right]dA$$

$$Q = -\frac{dM(x)}{dx} \rightarrow \tau_{yx}(y) = \int_A \frac{yQ}{BI}dA$$

Equation 5.38

$$\tau_{yx}(y) = \tau_{xy}(y) = \frac{Q}{BI}\int_A y dA$$

Equation 5.39

Example 5.4   For the cantilever beam (shown in Figure 5.23) with a rectangular section subjected to point load, determine the shear stress distribution. Find out the maximum shear stress in the section and where it happens.

Using Equation 5.39 and integrating over the area, the shear stress is presented as follows:

$$\tau_{xy}(y) = \frac{Q}{BI}\int_A y dA = \frac{P}{BI}\int_y^{H/2} yBdy = \frac{P}{I}\int_y^{H/2} ydy$$

$$\tau_{xy}(y) = \frac{P}{2I}\left(\frac{H^2}{4} - y^2\right)$$

Equation 5.40

The maximum shear stress happens at the mid-section when $y = 0$.

$$Max(\tau_{xy}(y)) = \tau_{xy}(0) = \frac{P}{2I}\frac{H^2}{4} = \frac{3}{2}\frac{P}{BH}$$

Equation 5.41

This means the shear stress is parabolic with its maximum in the mid-section, and it reduces to zero at edges of the cross section. The maximum of the shear stress for a rectangular section is 50% more than the average value (Gruttmann and Wagner, 2001).

From Equations 5.13 and 5.39, it is possible to define the shear stiffness (GA) as follows:

$$\tau_{xy}(y) = \frac{Q}{BI}\int_A y dA \quad and \quad \gamma_{xy} = \tau_{xy}/G$$

$$\gamma_{xy}G = \frac{Q}{BI}\int_A y dA \approx \frac{Q}{A} \Rightarrow Q = GA\gamma_{xy}$$

Equation 5.42

5.2.4   Beam Deflections

Stress analyses are performed to ensure that the structural elements do not fail due to stress levels exceeding the allowable values. Since most structural elements and components are deformable (note that *rigidity* is a relative term and some elements may be assumed rigid depending on the application), deflections and deformations should be considered.

Deflections of beam under torsion are studied earlier in this chapter. Also, from basics we know that a uniform beam under an axial load (P) has a deflection of PL/EA, in which L is the length of the beam and EA is the axial stiffness. E is the young modulus

Figure 5.24 Curvature of the deformed neutral axis and transverse displacement.

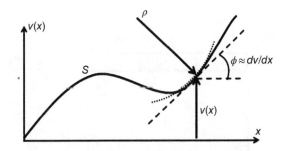

of elasticity which is 210 GPa for steel, and A is the area of the beam/bar cross section. Deflections of beams under bending moments and transversal loads are studied herein.

Considering the pure bending, the moment–curvature relationship is presented as the following differential equation (refer to Figure 5.21 and Equation 5.32):

$$\frac{M}{EI} = \frac{d\phi}{ds}$$ Equation 5.43

Using classical calculus (Adams and Essex, 2010), the curvature of the deformed neutral axis in terms of the transverse displacement is expressed as (see Figure 5.24):

$$\frac{d\phi}{ds} = \frac{d^2v}{dx^2} \Bigg/ \left(1 + \left(\frac{dv}{dx}\right)^2\right)^{1.5}$$ Equation 5.44

The expression in Equation 5.44 is a nonlinear, second-order, ordinary differential equation (ODE). Leonhard Euler (eighteenth century) attacked and resolved it for some sophisticated end-loading conditions. Under some assumptions (i.e. small displacement and small rotations), it is more straightforward to solve than Equation 5.44. We assume that $(dv/dx)^2$ is small which means that the rotation of the beam cross-section (dv/dx) is small. So, the moment–curvature relation can be expressed as follows (Budynas and Nisbett, 2014):

$$\frac{M}{EI} = \frac{d^2v}{dx^2}$$ Equation 5.45

In Figure 5.25, load intensity, reaction forces, shear force, bending moment, slope and deflection of a beam and their relation are shown for a simple case. If we integrate four times the load intensity function (applying the boundary conditions to evaluate the integration constants), the deflection of the beams can be found. More practically, the moment governing equation is integrated to get the slope and deflection.

Superposition has been introduced in this chapter. If the deflections are small and the resultant deformation from one type of load does not alter the geometry, superposition can be applied. The load and load effects should have a linear relation for each loading type. Different methods for beam deflections have been developed such as (a) singularity functions, (b) strain energy and (c) Castigliano's theorem, among others. We have used strain energy method in Example 5.2. In Table 5.2, the strain energy for different loading types is listed. Castigliano's theorem is explained herein.

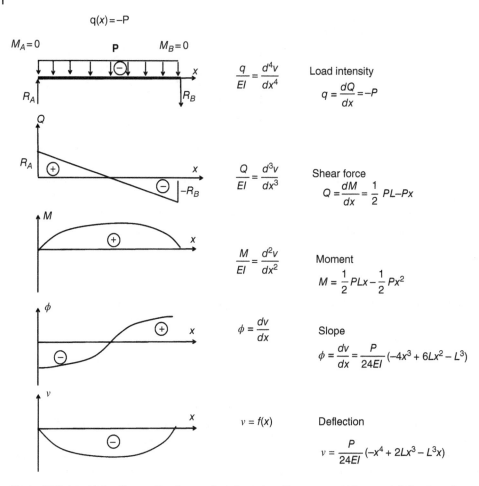

Figure 5.25 Load intensity, reaction forces, shear force, bending moment, slope and deflection of a beam and their relation; positive moment and shear force convention are shown in Figure 5.6.

Table 5.2 Relations between distributed load, shear and moment and strain energy.

Loading	Parameters	Equation	Strain energy
Axial (tension/compression)	F, E, A	$F = AE\dfrac{du}{dx}$	$U = \displaystyle\int_0^L \dfrac{F^2}{2EA}\,dx$
Torsion (twisting)	T, G, J	$T = GJ\dfrac{d\phi}{dz}$	$U = \displaystyle\int_0^L \dfrac{T^2}{2GJ}\,dx$
Shear force (transverse)	Q, G, A	$Q = GA\gamma_{xy}$	$U = \displaystyle\int_0^L c\,\dfrac{Q^2}{2GA}\,dx$
Bending moment	M, E, I	$M = \dfrac{EI}{\rho}$	$U = \displaystyle\int_0^L \dfrac{M^2}{2EI}\,dx$

Note: c is the cross-sectional correction factor: 1.2 for rectangular, 1.11 for circular and 2 for thin-walled sections.

One of the energy methods for solving the deflection problems is Castigliano's theorem which was developed based on strain energy and could be used for a wide range of problems. The theory can be applied to both translational and rotational deflections. Based on Castigliano's theorem (for a structural element elastically deformed by loads), the deflection at any point A in any direction $\vec{r}$ is expressed as the partial derivative of strain energy (U) with respect to a load at A in the direction $\vec{r}$:

$$\delta_j = \frac{\partial U}{dF_j}$$
$$\theta_j = \frac{\partial U}{dM_j}$$

Equation 5.46

where U is the strain energy; δ_j is the translational displacement of the point of application of the force F_j in the direction of F_j; and, θ_j is the rotationa displacement (in radians) of the moment M_j in the direction of M_j (Budynas and Nisbett, 2014). If there is no load at a desired point, then a dummy force will be applied to use the method (which will be set to zero).

Example 5.5 For a wind turbine, calculate the horizontal displacement at the hub due to the weight of the rotor.

Note: Neglect the transverse shear effects; also, assume that the nacelle weight is passing the tower axis and hence its effects on the desired deflection are negligible.

The wind turbine is simplified, as is shown in Figure 5.26. First of all, we define the strain energy based on the loads in each part. Bending in the shaft is mgy, and bending in the tower is $mgh + Qx$, where Q is a dummy force at the hub (the dummy force is applied towards right, negative y direction). The axial load in the tower is mg, and the axial load in the shaft is $-Q$.

The strain energy for the loads mentioned above is:

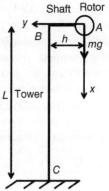

Figure 5.26 Simplified structural beam-model of a bottom-fixed turbine.

$$U = \int_0^h \frac{(mgy)^2}{2EI_{shaft}} dy + \int_0^L \frac{(mgh+Qx)^2}{2EI_{tower}} dx + \frac{Q^2 h}{2EA_{shaft}} + \frac{(mg)^2 L}{2EA_{tower}}$$

Equation 5.47

The horizontal deflection at A (the hub) is:

$$\delta_A = \left[\frac{\partial U}{\partial Q}\right]_{Q=0} = \int_0^L \frac{x(mgh+Qx)}{EI_{tower}} dx + \frac{Qh}{EA_{shaft}} = \frac{mghL^2}{2EI_{tower}}$$

Equation 5.48

5.2.5 Buckling of Beams

Slender and thin-walled members under compression are likely prone to buckling. Hence, beams subjected to compressive loads should be checked for buckling. In structural engineering, a straight and long beam (relative to its cross section) subjected to compressive loading is called a *column*. The design and analysis of compression members should include a buckling check. Buckling is a sudden, large lateral deflection, and the member fails before the compressive stress exceeds the yield (allowable) value. This is an example of a failure mode in which the structural member does not experience any fracture or plastic flow, so it is often called *elastic buckling*. Buckling is an "elastic instability" that may cause terrible failure of structures and structural components.

Although columns are more exposed to elastic buckling failure, in practice all structures may buckle for special boundary and loading conditions. In structural analysis, it is necessary to find out what kind of load and boundary conditions may result in possible elastic instabilities, and hence determine the load levels and conditions resulting in buckling. So, it is important to review the equilibrium regimes: (a) stable equilibrium, (b) neutral equilibrium and (c) unstable equilibrium (Akin, 2009). By increasing the compressive loads, the equilibrium state of a column may change from stable to neutral and then to unstable. The load that causes a member to become unstable is called the *critical buckling load*, or Euler buckling load. When the condition changes to unstable equilibrium, a small lateral movement can result in buckling and dramatic failure.

First, the buckling of a long beam under a central compressive load is briefly studied here to explain the buckling behaviour. The critical buckling load of such a beam (e.g. column, pillar or pile) is predicted using the Euler formula. We will show that the elastic buckling load (Euler buckling load) depends on: (a) boundary conditions (fixed-fixed, pinned-pinned, pinned-fixed and free-fixed), (b) material properties (modulus of elasticity) and (c) geometry (i.e. length and cross-section) of the beam. Figure 5.27 shows a pinned-pinned beam under compressive load. We assume that the beam slightly bent under the action of the load.

Using Equation 5.45 and applying the relation between the moment and the load at equilibrium conditions (see Figure 5.27), we can write (Lagace, 2009):

$$\frac{M}{EI} = \frac{d^2 y}{dx^2} \rightarrow = \frac{d^2 y}{dx^2} + \frac{Py}{EI} = 0 \qquad \text{Equation 5.49}$$

Equation 5.49 is a homogeneous, second-order differential equation (SODE) with the following general solution:

$$y = a \sin\left[\sqrt{\frac{P}{EI}} x\right] + b \cos\left[\sqrt{\frac{P}{EI}} x\right] \qquad \text{Equation 5.50}$$

We use the boundary conditions to find out constants. The transverse displacement at both ends is zero; hence, the cosine term should be removed $b = 0$. And $a \sin\left[\sqrt{\frac{P}{EI}} L\right] = 0$.

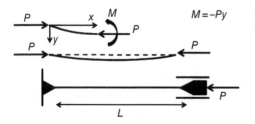

$$M = -Py$$

Figure 5.27 A pinned-pinned beam subjected to compressive load.

If constant a is zero, this means the beam does not buckle (which is not what we are looking for); hence:

$$\sin\left[\sqrt{\frac{P}{EI}}L\right] = 0 \rightarrow \sqrt{\frac{P}{EI}}L = n\pi \quad for \; n = 1,2,3,\ldots.$$

$$\Rightarrow P = \frac{n^2\pi^2 EI}{L^2}$$

Equation 5.51

The first mode of buckling occurs for $n = 1$, corresponding to the smallest load or critical load, for the current boundary conditions (pinned-pinned): $P_{cr} = \pi^2 EI/L^2$, which is the Euler formula. Using Equation 5.50 and the Euler formula for a critical load (Beer *et al.*, 2012):

$$\left. \begin{array}{l} y = a\sin\left[\sqrt{\dfrac{P}{EI}}x\right] \\ P_{cr} = \pi^2 EI \,/\, L^2 \end{array} \right\} \rightarrow y = a\sin\left[\dfrac{\pi}{L}x\right]$$

Equation 5.52

The beam buckles about the axis with the smallest area moment of inertia (weakest axis), as this gives the smallest critical buckling load $P_{cr} = \pi^2 EI_{Smallest}/L^2$. If the area moment of inertia is written in the format of $I = k^2 A$, in which A is the cross-sectional area and k is the radius of gyration, then it is possible to define the critical buckling stress (the stress that corresponds to unstable equilibrium):

$$\sigma_{cr} = \frac{P_{cr}}{A} = \frac{\pi^2 EI}{AL^2} = \frac{\pi^2 EAk^2}{AL^2} = \frac{\pi^2 Ek^2}{L^2} = \frac{\pi^2 E}{(L/k)^2}$$

Equation 5.53

where L/k is the slenderness ratio which is usually used as the reference length rather than the actual length of the beam. There is another important parameter called *effective length* which reflects the boundary conditions. The same procedure that applied for a pinned-pinned beam can be applied to solve the buckling differential equation with different boundary conditions (support end points). The Euler formula for different support conditions can be presented as:

$$P_{cr} = \frac{\pi^2 EI}{L_e^2}$$

Equation 5.54

where L_e is the effective length; see Figure 5.28. We can write the Euler formula and critical buckling stress as follows: λ_{EC} is the end-condition constant reflecting the effect of the support conditions in the formula (colorado.edu, 2015b):

$$P_{cr} = \frac{\lambda_{EC}\pi^2 EI}{L^2}$$

$$\sigma_{cr} = \frac{P_{cr}}{A} = \frac{\lambda_{EC}\pi^2 E}{(L/k)^2}$$

Equation 5.55

Practically, even for welded connections, fixing an end is not possible. Hence, more realistic values should be applied which can be found in the literature. Presenting such

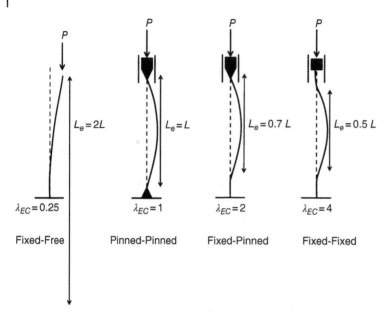

Figure 5.28 End-condition effect on effective length and end-condition constant for buckling of a beam with different support conditions.

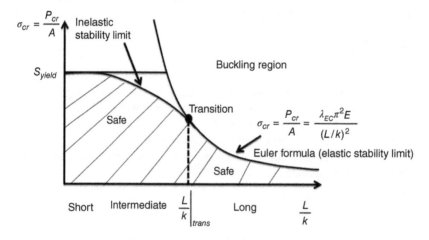

Figure 5.29 Critical stress plotted versus the slenderness ratio, highlighting the safe and buckling (unsafe) regions based on elastic (Euler formula) and inelastic stability limits.

tables defining the realistic values for effective length is out of the scope of the present discussion (Domokos *et al.*, 1997). In Figure 5.29, the safe and unsafe regions based on elastic (Euler) and inelastic stability limits are shown. The critical buckling stress (σ_{cr}) versus the slenderness ratio (L/k) is shown. For long structural members, the Euler formula can predict the buckling; however, for short members (small slenderness ratio), the inelastic stability limit states governs.

For a long slender member subject to a compressive force, Euler buckling governs. The theoretical Euler formula leads to infinite forces in very short columns which clearly exceeds the yield stress. For slender elements, the stability loss occurs at stresses far below the yield stress. In reality, the Euler column-buckling formula is applicable for special cases. In general, semi-analytical and empirical equations are required to accurately estimate the buckling behaviour; for example, refer to NS-EN 1993-1-1:2005 (European Committee for Standardization, 2005).

In practice, to predict the material linear failure, finite element analysis (FEA) may be used in which a linear algebraic system for the unknown displacements can be solved. Then, the strains and stresses obtained are compared to allowable design strain and stress. The load resulting in buckling is influenced by structural member stiffness rather than the material strength. Buckling is usually independent of material strength; buckling is loss of stability of a component, and the stability loss typically happens within the elastic range of the material.

As buckling failure is mainly governed by structural stiffness loss, it is not possible to analyse buckling by linear finite element modelling. Thus, a finite-element eigenvalue–eigenvector formulation (Lund, 2014), accounting for the buckling load factor (BLF) for each mode and geometric stiffness due to the stresses caused by the loading, should be solved to present the associated buckling displacement shape for each mode.

Abaqus/Standard (SIMULIA, 2015) is capable of estimating elastic buckling by eigenvalue analysis. The approach is normally useful for stiff structures when the pre-buckling response is linear. The buckling load estimation is found as a multiplier of the pattern of perturbation loads that are added to a set of base state loads. The base state of the structure results from a response history accounting for the nonlinear effects (the perturbation loads are added to this initial state). To have reasonable eigenvalue estimates, the response to the perturbation loads should be elastic up to the estimated buckling load values (SIMULIA, 2015).

5.3   Mathematical Models for Structural Dynamics of Beams

As beams are actually 3D bodies, all analytical mathematical models include some approximations to present the physics, and reasonable assumptions are considered. The main theories are the (a) Bernoulli–Euler, (b) Timoshenko, (c) Rayleigh and (d) shear beam theories. These theories are widely applied in offshore, mechanical and civil engineering. Early researchers recognized that the bending effect is the most important factor in a transversely vibrating beam. The Bernoulli–Euler beam theory and the Timoshenko beam theory are the most common beam theories for straight and prismatic beams, in particular the Bernoulli–Euler beam theory as it is the simplest and provides approximations for various engineering problems. The differences between the Bernoulli–Euler model and the other models decrease with increasing slenderness ratio (ratio of the beam length to radius of gyration of the beam cross section). More advanced theories, for example second-order beam theory, are introduced (Stephen and Levinson, 1979). Still, the Bernoulli–Euler and Timoshenko beam theories are widely applied in engineering problems. The key assumptions needed for beam theories presented in this chapter are as follows:

- Axial direction is significantly larger compared to the others.
- Symmetric cross-sectional area: So, the neutral and centroidal axes coincide.

- Normality of plane sections: Planes originally perpendicular to the neutral axis remain perpendicular after deformation.
- Neglecting the Poisson effect
- Small angle of rotation
- Linear elastic material.

In this section, each of these models is briefly explained. Figure 5.30 shows the natural frequency percentage deviation compared to the experimental values for the first and second modes; non-slender beams were studied in which the shear and rotary effects were important (Traill-Nash and Collar, 1953).

1) Bernoulli–Euler beam model: The Bernoulli–Euler model appeared for the first time in the eighteenth century. Bernoulli found that the curvature of an elastic beam is proportional to the bending moment and formulated the differential equation of motion of a vibrating beam. Later, Euler investigated the shape of elastic beams under various loading conditions and accomplished research about the elastic curves. Bernoulli–Euler beam theory includes the strain energy caused by bending and the kinetic energy due to lateral displacement (Han *et al.*, 1999).
   The Bernoulli–Euler beam theory is also called Euler–Bernoulli beam theory, classical beam theory, Euler beam theory, Bernoulli beam theory and engineering beam theory (University of Colorado at Boulder, 2014). This theory is simple and practical for many structural engineering problems; however, the Bernoulli–Euler model tends to slightly overestimate the natural frequencies, in particular for higher modes. The results for non-slender beams are not very accurate either.
2) Rayleigh beam model: The Rayleigh beam theory includes the effect of rotation of the cross-section and hence provides some improvement on the Bernoulli–Euler theory. So, it moderately corrects the overestimation of natural frequencies in the Bernoulli–Euler model. However, the natural frequencies are still overestimated (Han *et al.*, 1999).
3) Shear beam model: The shear model adds shear distortion to the Bernoulli–Euler model to considerably improve the estimate of the natural frequencies.

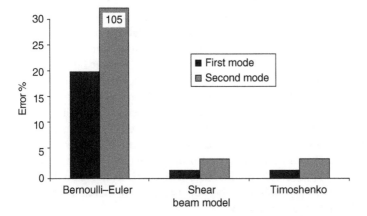

Figure 5.30 The natural frequency percentage deviation compared to the experimental values for first and second modes; non-slender beams were studied. The Bernoulli–Euler model has almost 105% error for the second-mode natural frequency value compared to experiments. *Source:* Han *et al.* (1999).

4) Timoshenko beam model: The Timoshenko model adds the effect of shear as well as the effect of rotation to the Bernoulli–Euler beam model. To account for transverse shear effects, the Timoshenko beam model incorporates a first-order kinematic correction. For non-slender beams and for high-frequency responses where shear or rotary effects are not negligible, the Timoshenko model should be applied. The Timoshenko beam model has been widely applied in dynamics and vibration analysis. The method of eigenfunction expansion can be applied to solve the response of a Timoshenko beam due to initial conditions and the external forces (Reismann and Pawlik, 1974). As the shear is not constant over the cross section, a parameter (the so-called shape factor) is introduced in the Timoshenko beam theory. The shape factor is also called the *area reduction factor* or *shear coefficient*, which is an important parameter in this theory. The shape factor is a function of the cross-sectional shape, the Poisson ratio and the frequency of vibration (the dependency on frequency of vibration is usually disregarded).

5.3.1 Bernoulli–Euler Beam Theory

The detailed derivations for the Bernoulli–Euler model are presented in textbooks such as Thomson (1993) and Rao (1995). Here, Hamilton's variational principle (Han *et al.*, 1999) is used to derive the equation of motion. Referring to Table 5.2 and Equation 5.45, the potential energy (strain energy due to bending, $U^{\Diamond}_{Bending}$) of a uniform beam subjected to bending load is as follows (Yu, 2012):

$$U^{\Diamond}_{Bending} = \frac{1}{2}\int_{0}^{L^{\Diamond}} E^{\Diamond} I^{\Diamond} \left[\frac{\partial^2 v^{\Diamond}\left(x^{\Diamond},t^{\Diamond}\right)}{\partial x^{\Diamond 2}} \right]^2 dx^{\Diamond} \qquad \text{Equation 5.56}$$

where $E^{\Diamond}$ is the elasticity modulus, $I^{\Diamond}$ is the area moment of inertia (of the cross section about the neutral axis), $v^{\Diamond}(x^{\Diamond},t^{\Diamond})$ is the transverse deflection at the axial-location $x^{\Diamond}$ and time $t^{\Diamond}$, and $L^{\Diamond}$ is the beam total length. Superscript $^{\Diamond}$ symbols indicate that dimensional quantities of parameters are set in the equation. We present the dimensionless potential energy as follows using the dimensionless quantities:

$$U_{Bending} = \frac{1}{2}\int_{0}^{1} \left[\frac{\partial^2 v(x,t)}{\partial x^2} \right]^2 dx$$

$$U_{Bending} = \frac{U^{\Diamond}_{Bending}}{E^{\Diamond} I^{\Diamond}/L^{\Diamond}}; \; L = L^{\Diamond}/L^{\Diamond}; \; v = v^{\Diamond}/L^{\Diamond}; \; x = x^{\Diamond}/L^{\Diamond} \qquad \text{Equation 5.57}$$

The kinetic energy is presented as:

$$K^{\Diamond}_{bending} = \frac{1}{2}\int_{0}^{L^{\Diamond}} \underbrace{\rho^{\Diamond} A^{\Diamond}}_{mass} \underbrace{\left[\frac{\partial v^{\Diamond}\left(x^{\Diamond},t^{\Diamond}\right)}{\partial t^{\Diamond}} \right]^2}_{Velocity^2} dx^{\Diamond} \qquad \text{Equation 5.58}$$

where $\rho^\lozenge$ is the density of the beam, and $A^\lozenge$ is the cross-sectional area. The dimensionless kinetic energy is presented as:

$$K_{bending} = \frac{1}{2}\int_0^1 \rho A \left[\frac{\partial v(x,t)}{\partial t}\right]^2 dx$$

$$K_{bending} = \frac{K_{bending}^\lozenge}{E^\lozenge I^\lozenge / L^\lozenge}; \rho = \frac{\rho_{bending}^\lozenge L^{\lozenge 6} \omega_1^{\lozenge 2}}{E^\lozenge I^\lozenge}; A = A^\lozenge / L^{\lozenge 2}; t = t^\lozenge \omega_1^\lozenge$$

Equation 5.59

The dimensionless Lagrangian is written in the form of (Preumont, 2013):

$$L = K_{bending} - U_{Bending} = \frac{1}{2}\int_0^1 \rho A \left[\frac{\partial v(x,t)}{\partial t}\right]^2 dx - \frac{1}{2}\int_0^1 \left[\frac{\partial^2 v(x,t)}{\partial x^2}\right]^2 dx$$

$$L = \frac{1}{2}\int_0^1 \left\langle \rho A \left[\frac{\partial v(x,t)}{\partial t}\right]^2 - \left[\frac{\partial^2 v(x,t)}{\partial x^2}\right]^2 \right\rangle dx$$

Equation 5.60

The virtual work $(\delta W_{NC}^\lozenge)$ of a non-conservative transverse force per unit length $(F^\lozenge(x^\lozenge,t^\lozenge))$ is defined by (Bathe, 2009):

$$\delta W_{NC}^\lozenge = \int_0^{L^\lozenge} F^\lozenge(x^\lozenge,t^\lozenge)\delta v^\lozenge(x^\lozenge,t^\lozenge) dx^\lozenge$$

Equation 5.61

Note: A force is conservative if the work that force does is independent of the path; for example, gravity is a conservative force (Keeports, 2006). The non-dimensional work is presented by the following expression:

$$\delta W_{NC} = \int_0^1 F(x,t)\delta v(x,t) dx$$

$$\delta W_{NC} = \frac{\delta W_{NC}^\lozenge L^\lozenge}{E^\lozenge I^\lozenge}; F(x,t) = \frac{F^\lozenge(x,t)E^\lozenge I^\lozenge}{L^{\lozenge 3}}$$

Equation 5.62

The extended Hamilton's principle is represented as (Preumont, 2013):

$$\left.\begin{array}{l}\int_{t_1}^{t_2}(\delta K_{bending} + \delta W)dt = 0 \\ \delta W = \delta W_C + \delta W_{NC} \\ = -\delta U_{bending} + \delta W_{NC}\end{array}\right\} \Rightarrow \int_{t_1}^{t_2}(\delta K_{bending} - \delta U_{bending} + \delta W_{NC})dt = 0 \Rightarrow$$

$$\int_{t_1}^{t_2}(\delta L + \delta W_{NC})dt = 0$$

Equation 5.63

Inserting Equations 5.60 and 5.62 in Equation 5.63, we can set up the following integral expression based on an extended Hamilton's principle for transverse deflection of a uniform beam:

$$\int_{t_1}^{t_2}\left(\delta\left[\frac{1}{2}\left\langle\int_0^1\rho A\left[\frac{\partial v(x,t)}{\partial t}\right]^2-\left[\frac{\partial^2 v(x,t)}{\partial x^2}\right]^2\right\rangle dx\right]+\int_0^1 F(x,t)\delta v(x,t)dx\right)dt=0$$

$$\int_{t_1}^{t_2}\int_0^1\left\{\rho A\left[\frac{\partial v(x,t)}{\partial t}\right]\delta\left[\frac{\partial v(x,t)}{\partial t}\right]-\frac{\partial^2 v(x,t)}{\partial x^2}\delta\left[\frac{\partial^2 v(x,t)}{\partial x^2}\right]+F(x,t)\delta v(x,t)\right\}dxdt=0$$

Equation 5.64

By some mathematics considering the weak formulation of the above equation, we can find out the differential equation of motion with its corresponding boundary conditions as (Tiwari, 2010):

$$\rho A\frac{\partial^2 v(x,t)}{\partial t^2}+\frac{\partial^4 v(x,t)}{\partial x^4}=F(x,t)$$

$$\frac{\partial^2 v}{\partial x^2}\delta\left(\frac{\partial v}{\partial x}\right)\Big|_0^1=0$$

$$\frac{\partial^3 v}{\partial x^3}\delta v\Big|_0^1=0$$

Equation 5.65

where v is the dimensionless displacement, $\partial v/\partial x$ is the dimensionless slope, $\partial^2 v/\partial x^2$ is the dimensionless moment and $\partial^3 v/\partial x^3$ is the dimensionless shear; see Figure 5.25. δv is the variation of displacement; $\delta v=0$ means that the displacement is known. Moreover, except for base-excited or end-forcing problems, $\delta v=0$ or $\delta(\partial v/\partial x)=0$ means that the displacement or the slope is zero (Han *et al.*, 1999). Four combinations of boundary conditions are possible to satisfy Equation 5.65:

$$\partial^2 v/\partial x^2=0,\ v=0 \qquad\qquad \textit{hinged}$$
$$\partial v/\partial x=0,\ v=0 \qquad\qquad \textit{clamped}$$
$$\partial v/\partial x=0,\ \partial^3 v/\partial x^3=0 \qquad \textit{sliding}$$
$$\partial^2 v/\partial x^2=0,\ \partial^3 v/\partial x^3=0 \quad \textit{free}$$

Equation 5.66

We apply separation of variables and eigenfunction expansion methods to solve the initial-boundary value problem. The separation of variables method relies on the idea of representing the solution of the general boundary-value problem as a linear combination of separated solutions $v(x,t)=W(x)T(t)$; for example, refer to Peirce (2014). By setting external force to zero, a homogeneous problem is considered to obtain the natural frequencies and eigenfunctions. So, the equation of motion is separated into two ODEs:

$$\frac{d^2 T(t)}{dt^2}+\omega^2 T(t)=0$$

$$\frac{d^4 W(x)}{dx^4}-a^4 W(x)=0$$

Equation 5.67

$$a^4=\rho A\omega^2$$

Equation 5.68

Equation 5.68 is called the dispersion relationship, where ω is the angular frequency and $a^{\diamond}$ is the wave number defined as $a^{\diamond} = 2\pi/wavelength^{\diamond}$. Therefore, the dimensionless wave number is given by $a = 2\pi/wavelength$. From Equation 5.67, it is clear that $W(x)$ and $T(t)$ can be presented as:

$$W(x) = A\sin ax + B\cos ax + C\sinh ax + D\cosh ax$$
$$T(t) = E\sin \omega t + F\cos \omega t$$

Equation 5.69

The constant coefficients can be found using the boundary conditions. The boundary conditions can be stated in terms of $W(x)$ (the spatial function):

$$\frac{\partial^2 W}{\partial x^2}\delta\left(\frac{\partial W}{\partial x}\right)\Big|_0^1 = 0$$
$$\frac{\partial^3 W}{\partial x^3}\delta W\Big|_0^1 = 0$$

Equation 5.70

By applying the boundary conditions to the spatial solution, the wave number and frequency of vibration can be found.

Example 5.6 For a beam hinged at both ends, derive the frequency equation.
From Equation 5.66: $\partial^2 W/\partial x^2 = 0, W = 0$ *hinged*
Using Equation 5.69, apply the boundary conditions as follows:

$$W(x) = A\sin ax + B\cos ax + C\sinh ax + D\cosh ax$$
$$\frac{\partial W(x)}{\partial x} = a(A\cos ax - B\sin ax + C\cosh ax + D\sinh ax)$$

Equation 5.71

$$\frac{\partial^2 W(x)}{\partial x^2} = a^2(-A\sin ax - B\cos ax + C\sinh ax + D\cosh ax)$$

$$\left.\begin{array}{l} W(0) = B + D = 0; \\ \partial^2 W(x=0)/\partial x^2 = a^2(-B+D) = 0 \end{array}\right\} \Rightarrow B = D = 0$$

$$W(x) = A\sin ax + C\sinh ax$$

Equation 5.72

$$W(1) = A\sin a + C\sinh a = 0 \Rightarrow A\sin a + C\sinh a = 0 \quad (I)$$

$$\partial^2 W(x=1)/\partial x^2 = a^2(-A\sin a + C\sinh a) = 0 \quad (II)$$

$$I, II \rightarrow \begin{bmatrix} \sin a & \sinh a \\ -\sin a & \sinh a \end{bmatrix}\begin{bmatrix} A \\ C \end{bmatrix} = 0$$

Equation 5.73

$$\Rightarrow \sin a \sinh a = 0$$

where $a_1 = \pi$ corresponds to the first eigenmode and it is half-wave in this hinged-hinged beam ($a = 2\pi/wavelength$; the dimensionless wave length in this conditions is 2, twice the beam length).

Example 5.7 For a ship-shaped offshore structure, derive the hydroelastic dynamic equation of motion; see Figure 5.31.

$$\sum F_y = \mathfrak{M}\frac{d^2 y}{dt^2} \rightarrow Q - \left(Q + \frac{\partial Q}{\partial x}dx\right) + qdx = mdx\frac{d^2 y}{dt^2}$$

Equation 5.74

$$\Rightarrow -\frac{\partial Q}{\partial x} + q = m\frac{d^2 v}{dt^2} \quad (i)$$

where $\mathfrak{M}$ is the mass of the element; $\mathfrak{M} = mdx$ in which m is the mass per length. q is the external force, including the wave loads (hydrodynamics), weight of the structure, deadweight and buoyancy force (hydrostatic). This may include restoring forces from the mooring system and station keeping as well.

$$\sum M = 0 \rightarrow \left(M + \frac{\partial M}{\partial x}dx\right) - M - Qdx - qdx\frac{dx}{2} = 0$$

Equation 5.75

$$\Rightarrow \frac{\partial M}{\partial x} = Q + q\frac{dx}{2} \rightarrow \frac{\partial M}{\partial x} \approx Q \quad (ii)$$

$$(Eq.5-45)\ M = EI\frac{\partial v^2}{\partial x^2} \bigg\} \rightarrow \frac{\partial Q}{\partial x} = \frac{\partial^2 M}{\partial x^2} = \frac{\partial^2}{\partial x^2}\left[EI\frac{\partial v^2}{\partial x^2}\right] \quad (iii)$$

$$(ii) \qquad \frac{\partial M}{\partial x} \approx Q$$

Equation 5.76

$$(i),(iii) \Rightarrow m\frac{d^2 v}{dt^2} + \frac{\partial^2}{\partial x^2}\left[EI\frac{\partial v^2}{\partial x^2}\right] = q$$

Equation 5.77

Equation 5.77 is the hydroelastic dynamic formulation. If the structural damping is considered, the equation of motion is modified as follows (see Equation 5.79):

$$M = EI\frac{\partial}{\partial x}\left(\frac{\partial v}{\partial x} + \xi\frac{\partial^2 v}{\partial x \partial t}\right)$$

Equation 5.78

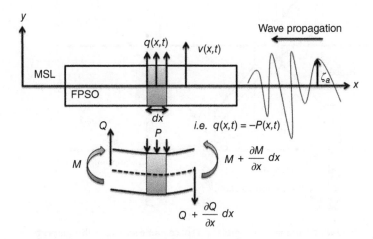

Figure 5.31 The schematic layout of a ship-shaped structure (i.e. a floating production storage and offloading [FPSO] vessel) subjected to wave loads.

where ξ is the structural damping coefficient.

$$m\frac{d^2v}{dt^2} + \frac{\partial^2}{\partial x^2}\left[EI\frac{\partial v^2}{\partial x^2}\right] + \frac{\partial^2}{\partial x^2}\left[EI\xi\frac{\partial^3 v}{\partial x^2 \partial t}\right] = q$$

Equation 5.79

5.4 Frame Structures and Matrix Analysis

Frame structures are widely applied in offshore technology (a structure built up of beams is called a *frame*). A space frame support-structure of a bottom-fixed wind turbine, for example the tripod shown in Figure 5.32, is an example of this type of structure. Normally, if the structure members are intended to carry the applied loads via bending, then the structure is a frame. Truss structures are designed to support the externally applied loads via

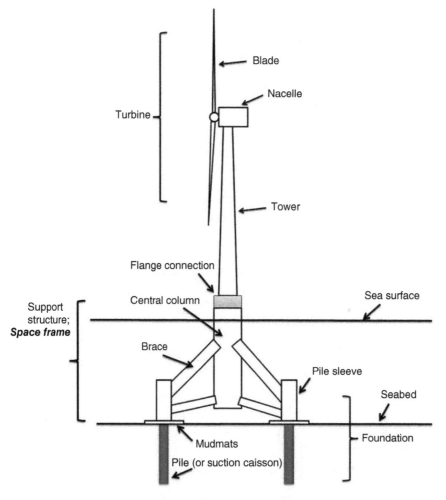

Figure 5.32 Tripod bottom-fixed wind turbine; an example of a space frame structure; the support structure is made of a few beams connected at joints.

tension and compression. Hence, the members are pinned at joints, and they cannot transmit the bending moment. But, in a frame, members are rigidly fixed to one another at the joints, and bending moments can be transmitted from one structural element to another.

The frames are normally modelled as a collection of discrete beams. The number of elements is arbitrary and depends on several parameters, such as the externally applied load and its spatial variability, the homogeneity of the material of the structure (frame and beams) and the acceptable accuracy of the results.

The geometric boundary conditions at the ends of a beam element include both slope and displacement. At each end of the beam (for a 3D frame), six degrees of freedom exist (three translational and three rotational). To satisfy the equilibrium of displacement, linear simultaneous equations should be solved. The whole structure's stiffness matrix is constructed by assembling the stiffness matrix of individual beams. Hence, the first step is to construct the stiffness matrix of a beam.

Figure 5.33 shows an element in a plane with three degrees of freedom (two translations and one rotation). At each of its end nodes, axial force, transverse force and moment are applied. These loads result in axial deformation (u), transverse deformation (v) and slope (ϕ), as indicated in the figure.

In the following, the entries in the stiffness matrix for this beam element are obtained.

$$
\begin{bmatrix}
N_1 \\
Q_1 \\
M_1 \\
N_2 \\
Q_2 \\
M_2
\end{bmatrix}
=
\begin{bmatrix}
k_{11} & . & . & . & . & k_{16} \\
. & . & . & . & . & . \\
. & . & . & . & . & . \\
. & . & . & . & . & . \\
. & . & . & . & . & . \\
k_{61} & . & . & . & . & k_{66}
\end{bmatrix}
\begin{bmatrix}
u_1 \\
v_1 \\
\phi_1 \\
u_2 \\
v_2 \\
\phi_2
\end{bmatrix}
\qquad \text{Equation 5.80}
$$

Assuming $u_1 = 1$ and the other deformations to be zero, we have the following expression:

$$
\begin{bmatrix}
N_1 \\
Q_1 \\
M_1 \\
N_2 \\
Q_2 \\
M_2
\end{bmatrix}
=
\begin{bmatrix}
k_{11} \\
k_{21} \\
k_{31} \\
k_{41} \\
k_{51} \\
k_{61}
\end{bmatrix}
\qquad \text{Equation 5.81}
$$

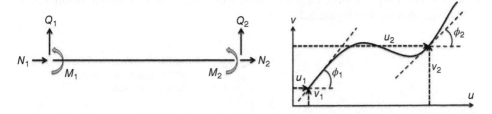

Figure 5.33 External loads and deflections of a beam; for 3D frames, each beam has six degrees of freedom at each end. Here, a beam in a plane is assumed which has three degrees of freedom at each end (two translations and one rotation).

When $u_1 = 1$ and the other deflections are zero, the beam is just subjected to compression load $N_1 = EA/L$. To satisfy the equilibrium, $N_2 = -N_1 = -EA/L$. Hence, $k_{11} = -k_{41} = EA/L$. If the same procedure is applied for $u_2 = 1$, and setting the other deformations to be zero, then the fourth column of the stiffness matrix can be investigated: $N_2 = EA/L = -N_1$, which results in $k_{44} = -k_{14} = EA/L$. The other elements of the matrix in columns 1 and 4 are zero, so:

$$
\begin{bmatrix} N_1 \\ Q_1 \\ M_1 \\ N_2 \\ Q_2 \\ M_2 \end{bmatrix} = \begin{bmatrix} EA/L & . & . & -EA/L & . & . \\ 0 & . & . & 0 & . & . \\ 0 & . & . & 0 & . & . \\ -EA/L & . & . & EA/L & . & . \\ 0 & . & . & 0 & . & . \\ 0 & . & . & 0 & . & . \end{bmatrix} \begin{bmatrix} u_1 \\ v_1 \\ \phi_1 \\ u_2 \\ v_2 \\ \phi_2 \end{bmatrix}
$$

Equation 5.82

Now, if $v_1 = 1$ while the other deflections are assumed to be zero (see Figure 5.34), the combination of moment and transverse loading appears. For small rotations and displacements, the axial force does not affect the bending. Also, bending of beam does not make any axial distortion (Bucciarelli, 2002a). However, for large deflections, the axial load affects bending (e.g. in case of buckling).

From Figure 5.34, it is clear that the beam is cantilevered at the right-hand side (the vertical displacement and rotation are zero). To determine the transverse force and bending moment at the left-hand side that give $v_1 = 1$, we use superposition of two cases: (a) cantilever beam subjected to bending moment, and (b) cantilever subjected to transverse force. Then, superimpose the two cases to ensure $v_1 = 1$ and $\phi_1 = 1$.

For a cantilevered beam subjected to transverse load, the deflections and deformations are:

$$
v_1 = \frac{L^3}{3EI} Q_1 ; \quad \phi_1 = -\frac{L^2}{2EI} Q_1
$$

Equation 5.83

For a cantilevered beam subjected to bending moment, the deflections and deformations are:

$$
v_1 = -\frac{L^2}{2EI} M_1 ; \quad \phi_1 = -\frac{L}{EI} M_1
$$

Equation 5.84

Now, we need to superimpose both transverse force and bending moment to construct the loading presented in Figure 5.34.

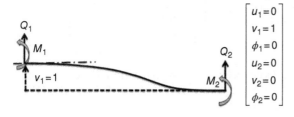

$$
\begin{bmatrix} u_1 = 0 \\ v_1 = 1 \\ \phi_1 = 0 \\ u_2 = 0 \\ v_2 = 0 \\ \phi_2 = 0 \end{bmatrix}
$$

Figure 5.34 Transverse and bending loads to present $v_1 = 1$, while other deflections are zero.

$$v_1 = \frac{L^3}{3EI}Q_1 - \frac{L^2}{2EI}M_1$$

$$\phi_1 = -\frac{L^2}{2EI}Q_1 - \frac{L}{EI}M_1$$

Equation 5.85

$$\left.\begin{array}{l} \phi_1 = -\dfrac{L^2}{2EI}Q_1 - \dfrac{L}{EI}M_1 \equiv 0 \\[2ex] v_1 = \dfrac{L^3}{3EI}Q_1 - \dfrac{L^2}{2EI}M_1 \equiv 1 \end{array}\right\} \Rightarrow \left\{\begin{array}{l} Q_1 = \dfrac{12EI}{L^3} \\[2ex] M_1 = \dfrac{6EI}{L^2} \end{array}\right.$$

Equation 5.86

To satisfy the equilibrium:

$$Q_2 = -\frac{12EI}{L^3}$$

$$M_2 = \frac{6EI}{L^2}$$

Equation 5.87

Using Equations 5.86 and 5.87, the elements of the second column are defined. For the fifth column, the same approach can be applied and elements are found. So, the stiffness matrix of the beam becomes:

$$\begin{bmatrix} N_1 \\ Q_1 \\ M_1 \\ N_2 \\ Q_2 \\ M_2 \end{bmatrix} = \begin{bmatrix} EA/L & 0 & . & -EA/L & 0 & . \\ 0 & 12EI/L^3 & . & 0 & -12EI/L^3 & . \\ 0 & 6EI/L^2 & . & 0 & -6EI/L^2 & . \\ -EA/L & 0 & . & EA/L & 0 & . \\ 0 & -12EI/L^3 & . & 0 & 12EI/L^3 & . \\ 0 & 6EI/L^2 & . & 0 & -6EI/L^2 & . \end{bmatrix} \begin{bmatrix} u_1 \\ v_1 \\ \phi_1 \\ u_2 \\ v_2 \\ \phi_2 \end{bmatrix}$$

Equation 5.88

The last task to complete the stiffness matrix is defining the third and sixth columns. We consider the third column corresponding to $\phi_1 = 1$, and all other displacements are zero. Same as what we did to obtain Equation 5.85, we can express:

$$\left.\begin{array}{l} \phi_1 = -\dfrac{L^2}{2EI}Q_1 + \dfrac{L}{EI}M_1 \equiv 1 \\[2ex] v_1 = \dfrac{L^3}{3EI}Q_1 - \dfrac{L^2}{2EI}M_1 \equiv 0 \end{array}\right\} \Rightarrow \left\{\begin{array}{l} Q_1 = \dfrac{6EI}{L^2} \\[2ex] M_1 = \dfrac{4EI}{L} \end{array}\right.$$

Equation 5.89

To satisfy the equilibrium:

$$Q_2 = -\frac{6EI}{L^2}$$

$$M_2 = \frac{2EI}{L}$$

Equation 5.90

Finally, the stiffness matrix of the beam is presented as follows (Weaver and Gere, 1980):

$$K_{6\times6} = \begin{bmatrix} EA/L & 0 & 0 & -EA/L & 0 & 0 \\ 0 & 12EI/L^3 & 6EI/L^2 & 0 & -12EI/L^3 & 6EI/L^2 \\ 0 & 6EI/L^2 & 4EI/L & 0 & -6EI/L^2 & 2EI/L \\ -EA/L & 0 & 0 & EA/L & 0 & 0 \\ 0 & -12EI/L^3 & -6EI/L^2 & 0 & 12EI/L^3 & -6EI/L^2 \\ 0 & 6EI/L^2 & 2EI/L & 0 & -6EI/L^2 & 4EI/L \end{bmatrix}$$

Equation 5.91

The coordinate transformation matrix $[T]$ is used to express the relation between local $[r]$ and global $[R]$ deflections (see Figure 5.35); $[r] = [T][R]$.

It is easy to show that the coordinate transformation matrix for the beam shown in Figure 5.35 $[T]$ is defined by:

$$[T] = \begin{bmatrix} \cos\theta & \sin\theta & 0 & 0 & 0 & 0 \\ -\sin\theta & \cos\theta & 0 & 0 & 0 & 0 \\ 0 & 0 & 1 & 0 & 0 & 0 \\ 0 & 0 & 0 & \cos\theta & \sin\theta & 0 \\ 0 & 0 & 0 & -\sin\theta & \cos\theta & 0 \\ 0 & 0 & 0 & 0 & 0 & 1 \end{bmatrix}$$

Equation 5.92

The stiffness matrix of a beam in global coordinates is expressed as: $[K] = [T]^T[k][T]$. If the mathematics is performed, the final matrix is presented as (only the upper triangle is shown, as the stiffness matrix is symmetric):

$$[K] = \begin{bmatrix} \frac{EA}{L}c^2 + \frac{12EI}{L^3}s^2 & \frac{EA}{L}sc - \frac{12EI}{L^3}sc & -\frac{6EI}{L^2}s & -\frac{EA}{L}c^2 - \frac{12EI}{L^3}s^2 & -\frac{EA}{L}sc + \frac{12EI}{L^3}sc & -\frac{6EI}{L^2}s \\ & \frac{EA}{L}s^2 + \frac{12EI}{L^3}c^2 & \frac{6EI}{L^2}c & -\frac{EA}{L}sc + \frac{12EI}{L^3}cs & -\frac{EA}{L}s^2 - \frac{12EI}{L^3}c^2 & \frac{6EI}{L^2}c \\ & & \frac{4EI}{L} & \frac{6EI}{L^2}s & -\frac{6EI}{L^2}c & \frac{2EI}{L} \\ & & & \frac{EA}{L}c^2 + \frac{12EI}{L^3}s^2 & \frac{EA}{L}sc - \frac{12EI}{L^3}sc & \frac{6EI}{L^2}s \\ & & & & \frac{EA}{L}s^2 + \frac{12EI}{L^3}c^2 & -\frac{6EI}{L^2}c \\ & & & & & \frac{4EI}{L} \end{bmatrix}$$

Equation 5.93

Figure 5.35 Global and local deformations and deflections of end 1 of a beam.

Example 5.8 Consider a submerged tunnel, and derive a stiffness matrix of the cross section shown in Figure 5.36.

The submerged tunnel shown in Figure 5.36 is supported by buoyancy structures that are floating at the mean water level surface. The columns connect the tunnels to these buoyancy structures. We may consider the floating structures relatively stiff compared to the columns. In Figure 5.37, a proposed simplified beam model is shown. Using matrix analysis, the stiffness matrix is calculated as follows. The stiffness matrix of the horizontal and vertical beams (left and right) is found using Equation 5.93. Equation 5.94 represents the stiffness matrix of the horizontal beam. Equations 5.95 and 5.96 present the stiffness matrix of the left and right vertical beams, respectively. Equation 5.97 is the total stiffness matrix obtained by using Equations 5.94, 5.95 and 5.96.

$$
[K]_{hor} =
\begin{bmatrix}
\dfrac{EA}{a} & 0 & 0 & -\dfrac{EA}{a} & 0 & 0 \\
 & \dfrac{12EI}{a^3} & \dfrac{6EI}{a^2} & 0 & -\dfrac{12EI}{a^3} & \dfrac{6EI}{a^2} \\
 & & \dfrac{4EI}{a} & 0 & -\dfrac{6EI}{a^2} & \dfrac{2EI}{a} \\
 & & & \dfrac{EA}{a} & 0 & 0 \\
 & & & & \dfrac{12EI}{a^3} & -\dfrac{6EI}{a^2} \\
 & & & & & \dfrac{4EI}{a}
\end{bmatrix}
\qquad \text{Equation 5.94}
$$

Figure 5.36 Schematic layout of a submerged tunnel.

Figure 5.37 Simplified frame representing the submerged tunnel section shown in Figure 5.36.

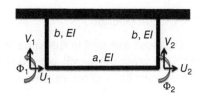

$$[K]_{ver\_left} = \begin{bmatrix} \dfrac{12EI}{b^3} & 0 & -\dfrac{6EI}{b^2} & -\dfrac{12EI}{b^3} & 0 & -\dfrac{6EI}{b^2} \\ & \dfrac{EA}{b} & 0 & 0 & -\dfrac{EA}{b} & 0 \\ & & \dfrac{4EI}{b} & \dfrac{6EI}{b^2} & 0 & \dfrac{2EI}{b} \\ & & & \dfrac{12EI}{b^3} & 0 & \dfrac{6EI}{b^2} \\ & & & & \dfrac{EA}{b} & 0 \\ & & & & & \dfrac{4EI}{b} \end{bmatrix}$$

Equation 5.95

$$[K]_{ver\_right} = \begin{bmatrix} \dfrac{12EI}{b^3} & 0 & \dfrac{6EI}{b^2} & -\dfrac{12EI}{b^3} & 0 & \dfrac{6EI}{b^2} \\ & \dfrac{EA}{b} & 0 & 0 & -\dfrac{EA}{b} & 0 \\ & & \dfrac{4EI}{b} & -\dfrac{6EI}{b^2} & 0 & \dfrac{2EI}{b} \\ & & & \dfrac{12EI}{b^3} & 0 & -\dfrac{6EI}{b^2} \\ & & & & \dfrac{EA}{b} & 0 \\ & & & & & \dfrac{4EI}{b} \end{bmatrix}$$

Equation 5.96

$$[K] = \begin{bmatrix} \dfrac{EA}{a}+\dfrac{12EI}{b^3} & 0 & -\dfrac{6EI}{b^2} & -\dfrac{EA}{a} & 0 & 0 \\ & \dfrac{EA}{b}+\dfrac{12EI}{a^3} & \dfrac{6EI}{a^2} & 0 & -\dfrac{12EI}{a^3} & \dfrac{6EI}{a^2} \\ & & \dfrac{4EI}{b}+\dfrac{4EI}{a} & 0 & -\dfrac{6EI}{a^2} & \dfrac{2EI}{a} \\ & & & \dfrac{12EI}{b^3}+\dfrac{EA}{a} & 0 & -\dfrac{6EI}{b^2} \\ & & & & \dfrac{EA}{b}+\dfrac{12EI}{a^3} & -\dfrac{6EI}{a^2} \\ & & & & & \dfrac{4EI}{b}+\dfrac{4EI}{a} \end{bmatrix}$$

Equation 5.97

5.5 Plate Theories

5.5.1 Introduction

Plates are widely used in ships and in other ocean, offshore and coastal structures. They are basic structural members to construct complicated modules like: pontoons, stiffened panels, topside of offshore platforms, bulkheads, drilling decks platforms and similar structures and structural parts. Plates appear in different forms; for example, they can be uniform and homogeneous metallic structures, or composite (e.g. lamination, sandwich structures or fibre reinforcement), thin, relatively thick or very thick. This means different approaches to study the behaviour of plates under static and dynamic loads, stress and deformations, dynamic structural responses and failure modes (such as buckling), depending on the type of plate and its characteristics, are needed.

In Section 5.1.2, the relation of stress and strain was introduced. All the analysis and theories discussed in this chapter are based on the mechanics of materials approach. Moreover, advanced structural analysis can be presented based on the theory of elasticity. The main difference between conventional methods (engineering approaches based on mechanics of materials) and the theory of elasticity is that in conventional methods, a proper hypothesis for geometry of deflections/deformations is included. However, the theory of elasticity does not require such assumptions. For example, in the previous sections discussing analysis of beams based on Bernoulli–Euler theory, the cross section should remain plane during deformation.

Concepts of equilibrium, continuum mechanics and material constitutive relationships (Mase, 1970) introduce the theory of elasticity to analyse the structures for finding the displacements, stress and strain within the elements for specified loading and boundary conditions. The mathematical format of the theory is expressed by field equations: (a) equations of equilibrium, (b) strain–displacement relations and (c) stress–strain law. For the stress state shown in Figure 5.10, assuming a linearly elastic isotropic body with small deformations in the absence of body forces, the field equations are as follows.

Equilibrium equations $\nabla[S] = 0$, in which the stress matrix S is defined in Equation 5.3, are presented as follows (Roylance, 2000a):

$$\frac{\partial \sigma_x}{\partial x} + \frac{\partial \tau_{xy}}{\partial y} + \frac{\partial \tau_{xz}}{\partial z} = 0$$

$$\frac{\partial \tau_{xy}}{\partial x} + \frac{\partial \sigma_y}{\partial y} + \frac{\partial \tau_{yz}}{\partial z} = 0 \qquad\qquad \text{Equation 5.98}$$

$$\frac{\partial \tau_{xz}}{\partial x} + \frac{\partial \tau_{yz}}{\partial y} + \frac{\partial \sigma_z}{\partial z} = 0$$

Strain–displacement relations:

$$\varepsilon_x = \frac{\partial u}{\partial x}; \; \gamma_{yz} = \frac{\partial v}{\partial z} + \frac{\partial w}{\partial y}$$

$$\varepsilon_y = \frac{\partial v}{\partial y}; \; \gamma_{xz} = \frac{\partial u}{\partial z} + \frac{\partial w}{\partial x} \qquad\qquad \text{Equation 5.99}$$

$$\varepsilon_z = \frac{\partial w}{\partial z}; \; \gamma_{xy} = \frac{\partial u}{\partial y} + \frac{\partial v}{\partial x}$$

Stress–strain law is given in Equation 5.13; a more generalized form of it can be presented as (Timoshenko and Goodier, 1951):

$$\sigma_i = 2G\varepsilon_i + \lambda e; \quad \tau_{ij} = G\gamma_{ij} \quad for \quad i,j = x,y,z$$

$$\lambda = \frac{vE}{(1+v)(1-2v)}; \quad G = \frac{E}{2(1+v)}; \quad e = \varepsilon_x + \varepsilon_y + \varepsilon_z \qquad \text{Equation 5.100}$$

The field equations presented in Equations 5.98 through 5.100 can be used via the *displacement approach* to find the displacements and consequently stress and strain using the stain–displacement and stress–strain relations. Another method is the *stress approach*, in which the stresses are found first, and afterwards strains and displacements are calculated. In the latter case, compatibility conditions are needed to ensure the continuum nature of the structure after deformation.

5.5.2 Plane Stress

Plate is a 3D structural member with special geometric features: (a) it is flat, and the mid-surface of the plate is a plane; and (b) it is thin, and one of the plate dimensions (thickness) is considerably smaller than the other two. Thickness is the distance between the plate faces, and the mid-plane is halfway between the top and bottom faces. The transverse direction is normal to the mid-plane, and "in-plane" directions are parallel to the mid-plane. In structural mechanics, a relatively thick flat sheet of material is also called a *slab*, but not usually for plane stress conditions (Felippa, 2015).

If the external loads are assumed to act on the plate mid-surface (see Figure 5.38), the plates are in the *plane stress state*, also called the membrane state or lamina state. If the distribution of stresses and strains over the thickness is assumed to be uniform, analysis using the 3D theory of elasticity is not necessary, and the problem can be reduced to 2D. If the plate behaviour is linearly elastic (for range of applied loads), the problem is reduced to 1D or 2D analysis.

Here, we consider a plate loaded in its mid-plane; see Figure 5.38. The plate is in membrane state if: (a) all loads act in the mid-plane direction, and are symmetric with respect to the mid-plane; (b) all support conditions are symmetric about the mid-plane; (c) in-plane displacements, strains and stresses are uniform across the thickness; (d) normal[1] and shear stress components in the transverse direction are negligible (Chakrabarty, 2010); and (e) the plate should be transversely homogeneous, having the same material through the thickness.

For the plane stress state, the thickness of plate should be small (i.e. smaller than 10% of the shortest in-plane dimension). If the plate thickness varies, such variations should be gradual. Also, the plate fabrication should be symmetric with respect to mid-plane (Felippa, 2015).

In the plane stress state, the plate is projected onto its mid-plane and represented as a 2D boundary value problem (BVP). The plane stress problem of a plate is defined by: (a) domain geometry, or boundary; (b) thickness: it can be a function of in-plane

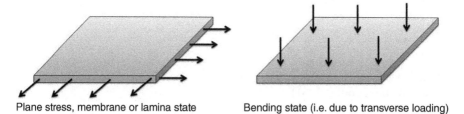

Plane stress, membrane or lamina state   Bending state (i.e. due to transverse loading)

Figure 5.38 Plane stress, membrane state versus bending state.

---

1 If the transverse stresses are considered, this results in a *generalized plane stress state*.

dimensions, but in most cases it is constant, and sudden changes of thickness are not accepted for the plane stress state, as discussed in this chapter; (c) material data, defined by the constitutive equations for plate material, are linearly elastic but not necessarily isotropic; (d) specified interior forces, such as body forces or volume forces (e.g. the plate weight), and face forces (e.g. the friction force); (e) specified surface forces which act on the boundary of the plate; and (f) displacement boundary conditions specifying how the plate is supported (Felippa, 2015).

For the plane stress state, the unknown fields (displacements, strains and stresses) are functions of in-plane dimensions (x and y), and dependency on the transverse direction (z) disappears due to the assumed wall fabrication homogeneity.
Displacements are defined by:

$$U(x,y) = \begin{bmatrix} u(x,y) \\ v(x,y) \end{bmatrix}$$

Equation 5.101

Strains are given by:

$$\varepsilon(x,y) = \begin{bmatrix} \varepsilon_x \\ \varepsilon_y \\ \gamma_{xy} \end{bmatrix}$$

Equation 5.102

The shear strain components γ_{xz} and γ_{zy} vanish. Also, the transverse normal strain ε_z is generally not zero (same as displacement in the z-direction) due to the Poisson ratio effect. However, as the stress in the transverse direction is zero, the strain in the transverse direction will not appear in governing equations (Chakrabarty, 2010).

Stresses are given by:

$$\sigma(x,y) = \begin{bmatrix} \sigma_x \\ \sigma_y \\ \tau_{xy} \end{bmatrix}$$

Equation 5.103

The governing equations for the plane stress problem can be presented by relating displacements, strains and stresses and applying kinematic, constitutive and internal equilibrium equations (assuming initial strain effects are ignored):

$$\begin{bmatrix} \varepsilon_x \\ \varepsilon_y \\ \gamma_{xy} \end{bmatrix} = \begin{bmatrix} \partial/\partial x & 0 \\ 0 & \partial/\partial y \\ \partial/\partial y & \partial/\partial x \end{bmatrix} \begin{bmatrix} u \\ v \end{bmatrix}$$

$$\begin{bmatrix} \sigma_x \\ \sigma_y \\ \tau_{xy} \end{bmatrix} = \begin{bmatrix} E_{11} & E_{12} & E_{13} \\ E_{12} & E_{22} & E_{23} \\ E_{13} & E_{23} & E_{33} \end{bmatrix} \begin{bmatrix} \varepsilon_x \\ \varepsilon_y \\ \gamma_{xy} \end{bmatrix}$$

Equation 5.104

$$\begin{bmatrix} \partial/\partial x & 0 & \partial/\partial y \\ 0 & \partial/\partial y & \partial/\partial x \end{bmatrix} \begin{bmatrix} \sigma_x \\ \sigma_y \\ \tau_{xy} \end{bmatrix} + \begin{bmatrix} B_x \\ B_y \end{bmatrix} = \begin{bmatrix} 0 \\ 0 \end{bmatrix}$$

where B_x and B_y are body forces. Note that $\gamma_{xy} = 2\varepsilon_{xy}$. Alternatively, the governing equations can be written as:

$$\varepsilon = SU; \quad \sigma = K\varepsilon; \quad S^T\sigma + B = 0 \qquad \text{Equation 5.105}$$

For isotropic material with a modulus of elasticity of E and a Poisson ratio of ϑ, it is possible to show that $E_{11} = E_{22} = E/(1-\vartheta^2)$, $E_{33} = 0.5E/(1+\vartheta) = G$, $E_{12} = \vartheta E_{11}$ and $E_{13} = E_{23} = 0$. Hence, we may write:

$$\begin{bmatrix} \sigma_x \\ \sigma_y \\ \tau_{xy} \end{bmatrix} = E/(1-\vartheta^2)\begin{bmatrix} 1 & \vartheta & 0 \\ \vartheta & 1 & 0 \\ 0 & 0 & 0.5(1-\vartheta) \end{bmatrix}\begin{bmatrix} \varepsilon_x \\ \varepsilon_y \\ \gamma_{xy} \end{bmatrix} \qquad \text{Equation 5.106}$$

The partial differential equations for plane stress problems are:

$$\frac{\partial^2 u}{\partial x^2} + \frac{\partial^2 u}{\partial y^2} = \frac{1+\vartheta}{2}\left(\frac{\partial^2 u}{\partial y^2} - \frac{\partial^2 v}{\partial x \partial y} \right)$$
$$\frac{\partial^2 v}{\partial x^2} + \frac{\partial^2 v}{\partial y^2} = \frac{1+\vartheta}{2}\left(\frac{\partial^2 v}{\partial y^2} - \frac{\partial^2 u}{\partial x \partial y} \right) \qquad \text{Equation 5.107}$$

Example 5.9 For a plane stress problem with isotropic material, derive the relation between the strain and stress matrix (similar to what is presented in Equation 5.106).

Refer to Section 5.1.2 and Equation 5.13, considering the fact that for plane stress $\sigma_z = \tau_{zy} = \tau_{zx} = 0$, we can write:

$$\varepsilon_x = \frac{1}{E}\left[\sigma_x - \vartheta\sigma_y \right]$$
$$\varepsilon_y = \frac{1}{E}\left[\sigma_y - \vartheta\sigma_x \right]$$
$$\varepsilon_z = \frac{-\vartheta\left(\sigma_x + \sigma_y \right)}{E} \qquad \text{Equation 5.108}$$
$$\gamma_{xy} = \tau_{xy}/G; \quad G = E/\left[2(1+\vartheta) \right]$$

$$\begin{bmatrix} \varepsilon_x \\ \varepsilon_y \\ \gamma_{xy} \end{bmatrix} = \frac{1}{E}\begin{bmatrix} 1 & -\vartheta & 0 \\ -\vartheta & 1 & 0 \\ 0 & 0 & 2(1+\vartheta) \end{bmatrix}\begin{bmatrix} \sigma_x \\ \sigma_y \\ \tau_{xy} \end{bmatrix} \qquad \text{Equation 5.109}$$

5.5.3 Mathematical Models for Bending of Plates

In offshore and marine structures, plates are usually subjected to loads that cause bending (see Figure 5.38). Examples are decks of platforms and ships, bridges and so on. In these structures, plates carry out lateral loads by a combination of moment and shear forces. When the plates are subjected to transverse loads (the loads normal to mid-surface), the plate deflects out of its plane (the mid-surface). Hence, the distribution of stress and strains is not uniform across the thickness. This is called the *bending state* (which is shown in Figure 5.38) versus the membrane state (Hartsuijker and Welleman, 2008). Usually, the plates are stiffened by beams to increase the structural integrity, for example with respect to bending and/or buckling.

Due to transverse applied loading, a plate may be bent in the following ways: (a) plate bending (in-extensional bending), if the plate does not experience considerable stretching/contractions; and (b) extensional bending (combined bending-stretching, coupled membrane-bending or shell-like behaviour), if the mid-surface has major stretching/contractions. The applied loading may comprise both in-plane and bending components. In FEMs, this is handled using flat shell models (a superposition of membrane and bending elements) (see e.g. colorado.edu, 2015a).

The 2D continuum mechanics models (see Section 5.5.2, "Plane Stress") can effectively present the membrane state of the plates; see Figure 5.38. Coupling of membrane and bending actions appears in many structures and structural parts. Hence, bending of plates has been widely studied, and several models have been presented by scientists during the past decades. Some of these mathematical/analytical models are listed here (see e.g. the lecture notes for "Advanced Finite Element Methods" by Professor Carlos Felippa, Department of Aerospace Engineering Sciences, University of Colorado at Boulder, USA; Felippa, 2015):

1) Membrane shell theory (Le Dret and Raoult, 1996): Applicable for very thin plates dominated by membrane actions (e.g. sails).
2) von Karman theory (Neukamm and Velcic, 2010): Practical for post-buckling analysis. Suitable for very thin bent plates with strong interaction of membrane and bending effects affecting finite lateral deflections.
3) Kirchhoff–Love theory (Reddy, 2006): Applicable and practical for several engineering problems for thin plates. It is assumed that the membrane and bending actions are uncoupled, shear energy is negligible and deflections are small. The Kirchhoff–Love theory is based on the Bernoulli–Euler beam theory "extension." Love developed this theory using the assumptions proposed by Kirchhoff.
4) Reissner–Mindlin theory (Onate, 2013): This is an extension of Kirchhoff–Love theory and is suitable for moderately thick plates. The theory is also called *first-order shear deformation theory of plates*, as it accounts for linear transverse shear effects for thin and moderately thick plates. This method is suitable for vibration analysis of plates.
5) High-order composite theory (Cho and Parmerter, 1993): Useful for analysing the interlaminate shear effects of layered composites.
6) Exact theories (Ladeveze, 2002): Based on 3D elasticity theory that accounts for additional effects compared to 2D theories.

Membrane shell and von Karman theories are geometrically nonlinear. In the other models (geometrically linear), the governing equations refer to the initially flat shape of the plate. High-order composite and exact theories are mainly applied in detailed analysis for point loading, local stress calculation, near-edge responses, openings in plates and stress factor calculation. Other types of nonlinearities (non-geometrical) such as material nonlinearities and nonlinearities due to special considerations such as boundary conditions, delamination, composite fracture and cracking, can be included in all the mentioned theories.

Kirchhoff is the simplest acceptable theory for plates which covers a wide range of applications (and is reasonably accurate for engineering purposes). Below mainly covers explanation of this theory for homogeneous isotropic plates with uniform thickness (Bhaskar and Varadan, 2013). This method is widely applied for static analysis of plates in different structural engineering problems.

Lagrange, Poisson and Kirchhoff had efforts for the development of thin plate theory. Kirchhoff finalized the mathematical formulation, and hence the model is often called *Kirchhoff plate theory*. The following assumptions are made for the Kirchhoff plate model (Timoshenko and Woinowsky-Krieger, 1959):

- The plate is thin, and thickness compared to other dimensions is small. However, the thickness is not very small, and the lateral deflections should be much smaller compared to thickness.
- Ideally, the thickness is uniform; slow variations are allowed (as long as it does not influence the assumption that the effects of 3D stresses are neglected).
- The plate is symmetric with respect to the mid-surface.
- Transverse loads are distributed over the plate surface area larger than the thickness (this avoid local stresses). The Kirchhoff theory for point/line loads calculates the deflections and stresses with acceptable accuracy. However, detailed stress analysis is required.
- Extension of the mid-surface should be limited (care is needed regarding the boundary conditions to avoid significant extensions).

Thin plate theory is applicable when the ratio of the thickness to the characteristic length of the plate is between 0.01 and 0.1. Based on Kirchhoff theory, the equation of motion for small deflections can be presented as:

$$D\left(\frac{\partial^4 w(x,y,t)}{\partial x^4} + 2\frac{\partial^4 w(x,y,t)}{\partial x^2 \partial y^2} + \frac{\partial^4 w(x,y,t)}{\partial y^4}\right) = \rho h \frac{\partial^2 w(x,y,t)}{\partial t^2} \qquad \text{Equation 5.110}$$

where $w(x,y,t)$ is the deflection of the plate, ρ is the density, h is the plate thickness and D is the plate flexural rigidity. The left-hand side of the equation can be presented by biharmonic-operator. The flexural rigidity of plates is defined by:

$$D = \frac{Eh^3}{12(1-\vartheta)} \qquad \text{Equation 5.111}$$

If the thickness is very small, the flexural rigidity approaches zero. So, the structural behaviour changes from thin plate to membrane. Membranes can only resist tension, and under compression, they have unstable behaviour called *wrinkling* (Bloom and Coffin, 2000).

References

Adams, R. A., C. Essex. (2010). *Calculus*. Upper Saddle River, NJ: Pearson.

Akin, J. (2009). *Buckling Analysis*. Houston, TX: Curricular Linux Environment at Rice (CLEAR), Rice University.

Andreassen, M. J. (2012). *Distortional Mechanics of Thin-Walled Structural Elements*. Kongens Lyngby, Denmark: DTU.

Bathe, K. J. (2009). *Finite Element Analysis of Solids & Fluids I*. Cambridge, MA: MIT OpenCourseWare.

Beer, F. P., E. R. Johnston Jr., J. T. Dewolf, D. F. Mazurek. (2012). *Mechanics of Materials*, 6th ed. New York: McGraw-Hill.

Bhaskar, K., T. K. Varadan. (2013). *Plates: Theories and Applications*. Chichester: John Wiley & Sons.

Bloom, F., D. Coffin. (2000). *Handbook of Thin Plate Buckling and Postbuckling*. Boca Raton, FL: Chapman and Hall/CRC.

Bucciarelli, L. L. Jr. (2002a). *Deflections Due to Bending*. Cambridge, MA: MIT Press. Retrieved from http://web.mit.edu/emech/dontindex-build/full-text/emechbk_8.pdf

Bucciarelli, L. L. Jr. (2002b). *Engineering Mechanics of Solids*. Retrieved from http://web.mit.edu/: http://web.mit.edu/emech/dontindex-build/

Bucciarelli, L. L. Jr. (2002c). *Stresses: Beams in Bending*. Cambridge, MA: MIT Press.

Budynas, R. G., K. J. Nisbett. (2014). *Shigley's Mechanical Engineering Design*. New York: McGraw-Hill.

Chakrabarty, J. (2010). *Problems in Plane Stress (in Applied Plasticity)*. New York: Springer Science + Business Media.

Cho, M., R. Parmerter. (1993). Efficient higher order composite plate theory for general lamination configurations. *American Institute of Aeronautics and Astronautics (AIAA) Journal*, 1299–1306.

colorado.edu. (2015a). *Kirchhoff Plates: Field Equations*. Retrieved from http://www.colorado.edu/engineering/CAS/courses.d/AFEM.d/AFEM.Ch20.d/AFEM.Ch20.pdf

colorado.edu. (2015b). *Stability Of Structures: Additional Topics*. Department of Aerospace Engineering Sciences, University of Colorado at Boulder. Retrieved from http://www.colorado.edu/engineering/CAS/courses.d/Structures.d/IAST.Lect26.d/IAST.Lect26.pdf

Crandall, S. H., N. C. Dahl, T. J. Lardner. (1972). *An Introduction to the Mechanics of Solids*. New York: McGraw-Hill.

Damkilde, L. (2000). *Stress and Stiffness Analysis of Beam-Sections*. Kongens Lyngby, Denmark: DTU.

Department of Aerospace Engineering Sciences. (2014). *Lecture 8: Torsion Of Open Thin Wall (OTW) Sections*. Boulder: University of Colorado.

Domokos, G., R. Holmes, B. Royce. (1997). Constrained Euler buckling. *Nonlinear Science*, 281–314.

European Committee for Standardization. (2005). *Eurocode 3: Design of Steel Structures*. NS-EN 1993-1-1:2005. Brussels: European Committee for Standardization.

Felippa, C. A. (2015). *The Plane Stress Problem*. Boulder, CO: Center for Aerospace Structures (CAS), Department of Aerospace Engineering Sciences.

Gruttmann, F., W. Wagner. (2001). *Shear Correction Factors in Timoshenko's Beam Theory for Arbitrary Shaped Cross-Sections*. Karlsruhe, Germany: Universitat Karlsruhe (TH), Institut fur Baustatik.

Han, S. M., H. Benaroya, T. Wei. (1999). Dynamics of transversely vibrating beams using four engineering theories. *Journal of Sound and Vibration*, 225(5), 935–988. Retrieved from http://www.idealibrary.com

Hartsuijker, C., H. Welleman. (2008). *Introduction into Continuum Mechanics*. Delft, Netherlands: Civil Engineering Department, TU-Delft.

Haukaas, T. (2012). *Warping Torsion*. University of British Columbia. Retrieved from www.inrisk.ubc.ca

Hibbeler, R. C. (2013). *Mechanics of Materials*. Upper Saddle River, NJ: Prentice Hall International.

Hughes, A. F., D. C. Iles, A. S. Malik. (2011). *Design of Steel Beams in Torsion*. London: SCI, The Steel Construction Institute.

Iowa State University of Science and Technology. (2011). *Prandtl Stress Function*. Retrieved from http://www.public.iastate.edu/: http://www.public.iastate.edu/~e_m.424/Torsion%20Prandtl%20Stress%20F.pdf

Jennings, A. (2004). *Structures: From Theory to Practice*. Boca Raton, FL: CRC Press, Taylor & Francis.

Keeports, D. (2006). The common forces: conservative or nonconservative? *Physics Education*, 219–222.

Kelly, P. (2015). *Mechanics Lecture Notes*. Retrieved from http://homepages.engineering.auckland.ac.nz/~pkel015/SolidMechanicsBooks/index.html: http://www.engineering.auckland.ac.nz/en.html

Ladeveze, P. (2002). The exact theory of plate bending. *Journal of Elasticity*, 37–71.

Lagace, P. A. (2009). *The Column and Buckling*. Cambridge, MA: Department of Aeronautics and Astronautics, Massachusetts Institute of Technology (MIT).

Le Dret, H., A. Raoult. (1996). The membrane shell model in nonlinear elasticity: a variational asymptotic derivation. *Journal of Nonlinear Science*, 6, 59–84.

Lund, J. (2014). *Buckling of Cylindrical Members with Respect to Axial Loads*. Trondheim, Norway: Norwegian University of Science and Technology.

MARINTEK. (2014). *RIFLEX-Program Documentation*, version 4.4. Trondheim, Norway: MARINTEK.

Mase, G.M. (1970). *Schaum's Outline of Theory and Problems of Continuum Mechanics*. New York: McGraw-Hill.

Megson, T. (1996). *Structural and Stress Analysis*. Amsterdam: Elsevier.

Neukamm, S., I. Velcic. (2010). *Derivation of the Homogenized von Karman Plate Theory from 3D Elasticity*. Retrieved from http://www.bcamath.org/: http://www.bcamath.org/documentos_public/archivos/publicaciones/VKhomogenization2.pdf

Noels, L. (2013–2014). *Aircraft Structures Beams – Torsion & Section Idealization*. Computational & Multiscale Mechanics of Materials – CM3. Retrieved from http://www.ltas-cm3.ulg.ac.be/

Onate, E. (2013). *Structural Analysis with the Finite Element Method Linear Statics*. New York: Springer.

Peirce, A. (2014). *Separation of Variables and Fourier Series*. Vancouver: University of British Columbia.

Preumont, A. (2013). *Twelve Lectures on Structural Dynamics*. Berlin: Springer.

Rao, S. S. (1995). *Mechanical Vibrations*. Reading, MA: Addison-Wesley.

Reddy, J. N. (2006). *Theory and Analysis of Elastic Plates and Shells*. Boca Raton, FL: CRC Press.

Reismann, H., P. S. Pawlik. (1974). *Elastokinetics*. St. Paul, MN: West Publishing.

Roylance, D. (2000a). *The Equilibrium Equations*. Cambridge, MA: Department of Materials Science and Engineering, Massachusetts Institute of Technology.

Roylance, D. (2000b). *Stresses in Beams*. Cambridge, MA: Department of Materials Science and Engineering, Massachusetts Institute of Technology.

SIMULIA. (2015). *ABAQUS Analysis User's Manual, 6.2.3 Eigenvalue Buckling Prediction*. Retrieved from http://www.intrinsys.com/software/simulia?gclid=CjwKEAjw1MSvBRDj2IyP-o7PygsSJAC_6zod5nO9Uvgi4vM1tEvDxrYhPFbklOGQxAXtIoq_4uXgUBoCLLrw_wcB.

Stephen, N. G., M. Levinson. (1979). A second order beam theory. *Journal of Sound and Vibration*, 293–305.

Thomson, W. (1993). *Theory of Vibration with Applications*. USA: Prentice-Hall.

Timoshenko, S., J. N. Goodier. (1951). *Theory of Elasticity*. New York: Mcgraw-Hill.

Timoshenko, S., S. Woinowsky-Krieger. (1959). *Theory of Plates and Shells*. New York: McGraw-Iill.

Tiwari, R. (2010). *The Continuous and Finite Element Transverse Vibration Analyses of Simple Rotor Systems*. Guwahati, India: IIT Guwahati.

Traill-Nash, R. W., A. R. Collar. (1953). The elects of shear flexibility and rotatory inertia on the bending vibrations of beams. *Quarterly Journal of Mechanics and Applied Mathematics*, 6, 186–213.

University of Colorado at Boulder. (2014). *Beam Deflections: Second-Order Method*. Introduction to Aerospace Structures (ASEN 3112). USA: Department of Aerospace Engineering Sciences, University of Colorado at Boulder. Retrieved July 25, 2015, from http://www.colorado.edu/engineering/CAS/courses.d/Structures.d/Home.html.

web.aeromech.usyd.edu.au. (2015). *Bending Moments and Shear Force Diagrams for Beams*. Retrieved from Aerospace, Mechanical & Mechatronic Engineering, University of Sydney: http://web.aeromech.usyd.edu.au/AMME2301/Documents/

Weaver, W. Jr., J. M. Gere. (1980). *Analysis of Framed Structures*. New York: Van Nostrand Reinhold.

Wolf, L., M. S. Kazimi and N. E. Todreas. (2003). *Introduction to Structural Mechanics*. Cambridge, MA: Department of Nuclear Engineering, Massachusetts Institute of Technology.

Yu, W. (2012). *Beam Models*. Logan, UT: Department of Mechanical and Aerospace Engineering, Utah State University.

6

Numerical Methods in Offshore Structural Mechanics

6.1 Structural Dynamics

This chapter presents numerical methods that are used for the dynamic analysis of structures in offshore engineering. It is very common for offshore structures to be subjected to dynamic loads (e.g. waves and wind). The term *dynamic loads* means that the magnitude, direction and/or position of these loads vary with time. In fact, all the possible loads in nature are dynamic since they have to be applied involving a time variation of the applied load. Additionally, offshore structures have the following characteristics: (a) dynamic loadings, (b) structures that are moving, (c) flexible structures, (d) structures consisting of rigid interconnected modules, (e) suddenly applied loads like gusts of wind or slamming water loads and (f) intense nonlinear fluid–structure interaction phenomena. Structural dynamic effects are important, dominate the response and should be accounted for in the design of offshore structures.

For the solution of typical structural dynamic problems, inertia and damping forces to the elastic resistance forces and the time dependency of all force quantities should be included. Based on the dynamic analysis of an offshore structure, determination of the time variation of the deflection in any point of the structure can be achieved; with the use of the deflection, the time variation of the stresses can be calculated, and the offshore structure will be assessed for its structural integrity and compliance with relevant regulations. It is expected that the resulting stresses and deflections also will be time varying (dynamic). Note that for the case in which a load is applied dynamically, the deflection of the structure depends upon the load but also upon the inertial loads that oppose the acceleration that produces it. As a result, the corresponding internal loads of the structure in any possible section must equilibrate the external loads and the inertial loads.

Usually, offshore engineering uses the basic assumption that the mass is concentrated at discrete points; as a result, the related analytical problem is simplified, and it is necessary to define the displacements and accelerations only at these discrete points where the mass is concentrated. A very basic definition in structural dynamics is the number of degrees of freedom of the structure (e.g. number of displacement components) that should be considered in order to represent appropriately the effects of all the possible and significant inertial and external loads.

It is very common to simplify the study of the response of offshore structures or components of these structures so that they behave like single degree of freedom (SDOF) or multiple degrees of freedom (MDOF) structures and systems. For the case of

Offshore Mechanics: Structural and Fluid Dynamics for Recent Applications, First Edition.
Madjid Karimirad, Constantine Michailides and Ali Nematbakhsh.
© 2018 John Wiley & Sons Ltd. Published 2018 by John Wiley & Sons Ltd.

an SDOF, the oscillatory response of the structure is completely described by one displacement variable; this may be a simplification, but in a lot of applications with MDOF structures, the principle of modal decomposition is used where the MDOF system is reduced to a set of uncoupled SDOF systems.

The basic condition that should be valid for offshore engineering applications is the condition for dynamic equilibrium that states that the total forces are in equilibrium with the inertia forces. This is derived by the use of Newton's second law of motion that mathematically can be expressed as in Equation 6.1 with the differential equation:

$$\mathbf{p}(t) = \frac{d}{dt}\left(m\frac{d\mathbf{v}}{dt}\right) \qquad \text{Equation 6.1}$$

where $\mathbf{p}(t)$ is the load vector and $\mathbf{v}(t)$ is the position vector of the mass m of the SDOF. For applications where the mass does not vary with time, Equation 6.2 holds true:

$$\mathbf{p}(t) = m\frac{d^2\mathbf{v}}{dt} \equiv m\ddot{\mathbf{v}}(t) \qquad \text{Equation 6.2}$$

As a result, the force is equal to the product of the mass and the acceleration; in other words, this is the inertial force that resists the acceleration of the mass.

$$\mathbf{p}(t) - m\ddot{\mathbf{v}}(t) = 0 \qquad \text{Equation 6.3}$$

This condition is known as the d'Alembert principle, and the equation of motion is expressed as an equation of dynamic equilibrium.

Figure 6.1 depicts an idealized mathematical model of an SDOF system. The mass is considered to be lumped at one point and is moved only in one horizontal direction. The SDOF system consists of: (a) a mass element m; (b) a massless spring element with stiffness k that identifies the presence of elastic restoring force and potential energy of the system; (c) a damping element c (e.g. dashpot), representing frictional characteristics of energy loss or dissipation of energy in the system for any possible reason (e.g. radiation damping); and (d) an excitation force p(t), representing the external force acting on the system or the summation of possible external forces acting on the system. The position of the mass is defined with the displacement coordinate u(t). With the use of the d'Alembert principle and by directly expressing the equilibrium of all forces acting on the mass, the equation of motion that is an expression of the equilibrium of these forces is as follows (Zienkiewicz, 2006):

$$f_I(t) + f_D(t) + f_S(t) = p(t) \qquad \text{Equation 6.4}$$

or:

$$m\ddot{u}(t) + c\dot{u}(t) + ku(t) = p(t) \qquad \text{Equation 6.5}$$

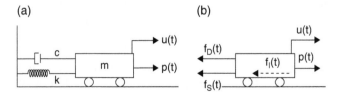

Figure 6.1 A SDOF system (a) and a free body diagram of the system (b).

where the inertia force $f_I(t)$ is the product of the mass and acceleration, the damping force $f_D(t)$ is the product of the damping constant and the velocity, and the spring force $f_S(t)$ is the product of the spring constant and the displacement (Figure 6.1).

It is very important to study initially the case where $p(t)$ equals zero and there is no damping ($c = 0$). For this case, and when an initial displacement is given to the mass as excitation, the mass afterwards is released to oscillate without any further external force. The equation of motion for the free vibration of the system is as follows:

$$m\ddot{u}(t)+ku(t)=0 \qquad \text{Equation 6.6}$$

or:

$$\ddot{u}(t)+\frac{k}{m}u(t)=0 \qquad \text{Equation 6.7}$$

The solution of this equation is:

$$u(t)=\frac{\dot{u}(t)}{\omega}\sin\omega t+u_0\cos\omega t \qquad \text{Equation 6.8}$$

where u_o is the initial displacement and $\omega=\sqrt{k/m}$ is the natural circular frequency of the system's oscillation. As can be seen in Equation 6.8, the free vibration is harmonic since $u(t)$ varies sinusoidally with t. The parameter ω (rad/sec) provides very critical information about the frequency of the system that the resonance will occur; consequently, when a cyclic loading is applied to the system, the appropriate selection of the m and k values can lead to a system that will have resonance for frequencies far from the frequency of the excitation. With the use of simple mathematical calculations, the natural period T and natural frequency f of the system are defined as follows:

$$T=2\pi\sqrt{\frac{m}{k}} \quad \text{and} \quad f=\frac{1}{2\pi}\sqrt{\frac{k}{m}} \qquad \text{Equation 6.9}$$

Note that the natural frequency of the system does not depend upon the external loading but only depends upon the characteristics of the system.

For the case that an external force P with a constant magnitude is applied to the same system as described previously ($c = 0$), then Equation 6.7 becomes:

$$\ddot{u}(t)+\frac{k}{m}u(t)=\frac{P}{m} \qquad \text{Equation 6.10}$$

and the solution of this equation is:

$$u(t)=\frac{P}{k}(1-\cos\omega t) \qquad \text{Equation 6.11}$$

By comparing Equations 6.8 and 6.11, it can be noted that the maximum displacement 2P/k is two times the displacement that the system would have if the force was applied statically, and also that the axis of vibration has been shifted by P/k.

For the case of damped systems with zero external loading and with damping that can be approximated with linear models (e.g. viscous damping that is proportional to the velocity of the system), the equation of motion is as follows:

$$m\ddot{u}(t)+c\dot{u}(t)+ku(t)=0 \qquad \text{Equation 6.12}$$

This equation is a second-order, homogeneous, ordinary differential equation (ODE) and can be solved with the use of the characteristic equation method. Based on this method, three possible cases exist:

1) $c^2 - 4mk < 0$: The motion of the system is called *underdamped*, and the motion corresponds to an oscillation with an exponential decay in its amplitude.
2) $c^2 - 4mk = 0$: The motion of the system is called *critically damped*, and the motion corresponds to a simple decaying.
3) $c^2 - 4mk > 0$: The motion of the system is called *overdamped*, and the motion corresponds to an oscillation with a simple exponential decay.

Before continuing, the three different cases – the critical damping β, the damping ratio ξ and the damped frequency ω_d – of vibration of the system are defined as follows:

$$\beta = 2m\sqrt{k/m} = 2m\omega \qquad \text{Equation 6.13}$$

$$\xi = c/\beta \qquad \text{Equation 6.14}$$

$$\omega_d = \sqrt{1-\xi^2}\,\omega \qquad \text{Equation 6.15}$$

For the case of underdamped motion, the equation of the displacement of the mass is:

$$u(t) = e^{-\xi\omega t}\left[u_o \cos(\omega_d t) + \frac{\dot{u}_o + \xi\omega u_o}{\omega_d}\sin(\omega_d t)\right] \qquad \text{Equation 6.16}$$

The damping ratio ξ can be defined with the use of the logarithmic decrement δ:

$$\delta = \ln\frac{x_1}{x_2} = \frac{2\pi\xi}{\sqrt{1-\xi^2}} \qquad \text{Equation 6.17}$$

The critical damping β is in a way the minimum damping that results in a non-periodic motion. For the case of critically damped motion, the equation of the displacement of the mass is:

$$u(t) = e^{-\omega t}\left[u_o + (\dot{u}_o + \omega u_o)t\right] \qquad \text{Equation 6.18}$$

For the case of overdamped motion, the equation of the displacement of the mass is non-periodic, as shown here:

$$u(t) = \frac{u_o\omega\left[\xi + \sqrt{\xi^2-1}\right] + \dot{u}_o}{2\omega\sqrt{\xi^2-1}}e^{\left[-\xi+\sqrt{\xi^2-1}\right]\omega t} + \frac{-u_o\omega\left[\xi - \sqrt{\xi^2-1}\right] - \dot{u}_o}{2\omega\sqrt{\xi^2-1}}e^{\left[-\xi-\sqrt{\xi^2-1}\right]\omega t}$$

$$\text{Equation 6.19}$$

or, if we consider that the SDOF is a heavily damped system with $\xi \gg 1$:

$$u(t) = u_o + \frac{\dot{u}_o}{2\xi\omega}\left(1 - e^{-2\xi\omega t}\right) \qquad \text{Equation 6.20}$$

For the case of MDOF systems, the same process is followed as for the case of SDOF systems. Usually, the number of degrees of freedom is equal to the number of independent types of motion (e.g. translational or rotational) that define the behaviour of the MDOF

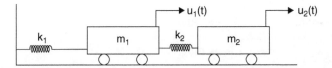

Figure 6.2 An undamped two degrees of freedom MDOF system.

system. Herein, we will study two different cases of MDOF systems (Biggs, 1964). The first one deals with an MDOF with two degrees of freedom (Figure 6.2), consisting of two masses m_1 and m_2 that are connected with a spring k_2, while the mass m_1 is restrained with the spring k_1.

The equation of motion of the coupled, undamped, two degrees of freedom MDOF for the case where no external forces are presented is:

$$m_1\ddot{u}_1 + k_1 u_1 - k_2\left(u_2 - u_1\right) = 0 \hspace{3cm} \text{Equation 6.21}$$

$$m_2\ddot{u}_2 + k_2\left(u_2 - u_1\right) = 0 \hspace{3cm} \text{Equation 6.22}$$

The two displacements that define the vibration of the system are harmonic and in phase, and they can be expressed with Equations 6.23 and 6.24:

$$u_1 = \alpha_1\sin\omega\left(t + \alpha\right) \hspace{3cm} \text{Equation 6.23}$$

$$u_2 = \alpha_2\sin\omega\left(t + \alpha\right) \hspace{3cm} \text{Equation 6.24}$$

By substituting these equations into Equations 6.21 and 6.22, the following system of equations is obtained:

$$\left(-m_1\omega^2 + k_1 + k_2\right)\alpha_1 + \left(-k_2\right)\alpha_2 = 0 \hspace{3cm} \text{Equation 6.25}$$

$$\left(-k_2\right)\alpha_1 + \left(-m_2\omega^2 + k_2\right)\alpha_2 = 0 \hspace{3cm} \text{Equation 6.26}$$

The determinant of the coefficients of the system of equations must be equal to zero. This condition provides Equation 6.27:

$$\left(\omega^2\right)^2 - \omega^2\left(\frac{k_1 + k_2}{m_1} + \frac{k_2}{m_2}\right) + \frac{k_1 k_2}{m_1 m_2} = 0 \hspace{3cm} \text{Equation 6.27}$$

By solving Equation 6.27, the natural circular frequencies of the two first normal modes ω_1 and ω_2 are evaluated. By substituting separately the ω_1 and ω_2 in Equations 6.25 and 6.26, the shape of each mode can be evaluated, and as a result the modes of the MDOF system can be graphically represented.

For the case in which those two linear dampers c_1 and c_2, and two external forces $p_1(t)$ and $p_2(t)$, are added to the two degrees of freedom MDOF system (Figure 6.3), the dynamic equation of motion for this MDOF system is:

$$m_1\ddot{u}_1 + c_1\dot{u}_1 + c_2\left(\dot{u}_1 - \dot{u}_2\right) + k_2\left(u_1 - u_2\right) + k_1 u_1 = P_1\left(t\right) \hspace{2cm} \text{Equation 6.28}$$

$$m_2\ddot{u}_2 + c_2\left(\dot{u}_2 - \dot{u}_1\right) + k_2\left(u_2 - u_1\right) = P_2\left(t\right) \hspace{2cm} \text{Equation 6.29}$$

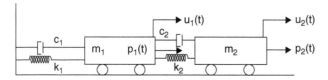

Figure 6.3 A damped two degrees MDOF system with external loadings.

In a matrix form, the system of the two equations can be written as:

$$\mathbf{m\ddot{u} + c\dot{u} + ku = P(t)}$$

Equation 6.30

In order to estimate the natural frequencies of the MDOF system and by neglecting the damping and the external force vectors, Equation 6.31 has to be solved:

$$\left(-\omega^2\mathbf{m} + \mathbf{k}\right)\mathbf{u}_0 = 0$$

Equation 6.31

The natural circular frequencies as well as the mode shapes can be evaluated with Equation 6.32:

$$\det\left(-\omega^2\mathbf{m} + \mathbf{k}\right) = 0$$

Equation 6.32

With exactly the same process, all possible MDOF systems can be solved.

6.2 Stress Analysis

For offshore structures and in order to study the internal effects of stress and strain in a solid body when this body is subjected to any possible kind of external loadings (e.g. wave, wind or current), fundamental theories that exist in the mechanics of materials area can be used. External loadings can be applied in offshore structures as distributed or concentrated surface loadings, or as body loadings that possibly act throughout the volume of the body. In general, the stress at a specific point of the offshore structure is associated with the structural integrity and strength of the material from which the offshore structure is made, while strain is associated with the elastic or plastic deformations of the structure.

For any possible deformable offshore structure and by the use of the equations of equilibrium in all directions and by solving the equation of motion of the body in either the frequency or time domain, the resultant internal loadings can be estimated. The internal loadings act at any possible specific region within the body (e.g. cross section). In general, four different types of internal loadings can be defined, namely, normal forces that act perpendicular to the cross-sectional area, shear forces that lie in the plane of the cross-sectional area, torsion moments about an axis perpendicular to the cross-sectional area and bending moments about an axis lying within the cross-sectional area. The numerical calculation of the stresses is of high importance since stresses are connected with fatigue problems of offshore structures and systems as well as with the phenomena of crack initiation and growth. Note that the induced stresses at any point of an offshore structure are of the dynamic type (e.g. variation of the amplitude over time).

To numerically estimate the stresses, the Lagrangian description is used (Malvern, 1969). The strains are numerically estimated in terms of the Green strain tensor $\mathbf{E}$, which can be calculated with the use of Equation 6.33:

$$dS_n^2 - dS_0^2 = 2 \times d\mathbf{X} \times \mathbf{E} \times d\mathbf{X} \qquad \text{Equation 6.33}$$

where dS_n and dS_0 are the length for a representative distance of two characteristic points for two different time steps of the analysis. The rectangular components of $\mathbf{E}$ are referred to as the *base vectors* and are expressed as follows:

$$E_{ij} = \frac{1}{2}\left(\frac{\partial u_i}{\partial X_j} + \frac{\partial u_j}{\partial X_i} + \frac{\partial u_m}{\partial X_i}\frac{\partial u_m}{\partial X_j} \right) \qquad \text{Equation 6.34}$$

where the components of the displacement vector $\mathbf{u}$ are known. Note that E is a symmetric tensor consisting of both linear and quadratic terms; the quadratic contributions are of importance for only large displacements and stability problems. For example, and for the case of a beam element, the Green strain (Lubliner, 2008) is expressed as:

$$E_{ij} = \frac{1}{2L_0^2}\left(\Delta x^2 + \Delta y^2 + \Delta z^2 - L_0^2 \right) \qquad \text{Equation 6.35}$$

where Δx, Δy and Δz are the difference of the x, y and z coordinates, respectively, of the two nodes of the beam element; and L_0 is the initial length of the element.

For the case of dynamic stresses and for a cyclic loading, the following quantities can be defined: (a) the maximum stress during the loading cycle S_{max}, (b) the minimum stress during the loading cycle S_{min}, (c) the stress range $\Delta S = S_{max} - S_{min}$ and (d) the stress ratio $S = S_{min}/S_{max}$. These quantities are related as follows:

$$S_m = \frac{\Delta S}{2}\left(\frac{1+R}{1-R} \right) \qquad \text{Equation 6.36}$$

The fatigue of the material is connected with the stresses and is normally presented in a stress–life diagram that is named the *SN diagram*; it is the stress range ΔS as a function of the cycles to failure (N). Fatigue analysis is highly recommended by regulations (e.g. DNV-RP-C203; see Det Norske Veritas, 2011). Usually, and since the cycles of failures have a span of several decades in cycles, it is plotted in a log-log format. Two different ranges of fatigue life exist: the high-cycle range of fatigue and the low-cycle range. The high-cycle range of fatigue is greater than 10^5 (e.g. common value), and in this range the behaviour is usually elastic. The low-cycle range of fatigue is less than 10^5 (e.g. common value), and in this range the behaviour is usually plastic. As a result, the stress range cannot express the condition of the element, and a strain range is needed. Usually, the offshore structures are designed in the high-cycle range.

6.3 Time-Domain and Frequency-Domain Analysis

Dynamic analysis is in principle necessary for the analysis of offshore structures and systems. Moreover, the wave load contains energy in the range of the eigenfrequencies of the structures and systems. To perform the dynamic analysis and calculate the performance of offshore structures and systems by examining different response quantities, the equation of motion should be formulated appropriately and be solved numerically in the time or frequency domain. It is straightforward that there is not any specific rule if a time-domain or frequency-domain analysis is required; it depends upon the characteristics of the physical problem. Factors that may influence the use of either the time domain or the frequency domain are: (a) the frequency dependence of properties of the structure/system (e.g. mass, stiffness, damping and loadings), (b) the loading type (e.g. transient or accidental) and (c) the nonlinear effect of loading or structural response (e.g. slamming or quadratic damping).

Time-domain dynamic analysis is very often used for the analysis of offshore structures and systems. During this type of analysis, the equation of motion is solved numerically for every examined time step irrespective of its size in real time (e.g. 0.005 sec). The time step size and the integration algorithm for the solution of the equation of motion should be selected after appropriate sensitivity studies. The selection of these quantities depends upon the computational efficiency, the numerical stability and the convergence of the solver. The equation of motion results directly from appropriate use of Newton's second law (Newton, 1729). This section presents how the formulation of the time-domain dynamic analysis of floating structures is developed mathematically.

Initially, we assume that the floating structure is at rest at time $t = t_0$. For the case that the displacement Δx is applied to a body with a constant velocity V for a short time interval Δt, the water particles will move too. A velocity potential Φ, proportional to the velocity V, is defined as follows:

$$\Phi(x,y,x,t) = \Psi(x,y,z)V(t) \quad t_0 < t < t_0 + \Delta t \qquad \text{Equation 6.37}$$

where Ψ is the normalized velocity potential. It must be noted that the assumption that the flow is assumed as potential is used. After the displacement has occurred, the water particles are still moving; and because:

$$\Delta(x) = V \times \Delta(t) \qquad \text{Equation 6.38}$$

the motions of the fluid are proportional to the displacement Δx:

$$\Phi(x,y,x,t) = x(x,y,z,t)\Delta(x) \quad t > t_0 + \Delta t \qquad \text{Equation 6.39}$$

where x again is the normalized velocity potential. As can be seen from Equation 6.39, the displacement Δx influences the motions of the fluid during the time interval $(t_0, t_0 + \Delta t)$ and, most importantly, during the later time intervals. The same happens also for the motions of the fluid for the time interval $(t_0, t_0 + \Delta t)$ and are influenced by the motions of the structure before this interval; as a result, it can be concluded that a "memory effect" exists when a structure is moving in a fluid (Dhanak, 2016). If we generalize this effect on any floating object that performs a time-dependent varying motion, the total velocity potential $\Phi(t)$ during any possible time interval $(t_m, t_m + \Delta t)$ can be calculated as follows:

$$\Phi(t) = V_m \Psi + \sum_{k=1}^{m} \left[x(t_{m-k}, t_{m-k} + \Delta t)V_k \Delta t \right] \qquad \text{Equation 6.40}$$

where m is the number of time steps, $t_m = t_0 + m\Delta t$, $t_{m-k} = t_0 + (m-k)\Delta t$; V_m is the velocity during the time interval $(t_m, t_m + \Delta t)$; V_k is the velocity during the time interval

$(t_{m-k}, t_{m-k} + \Delta t)$; Ψ is the normalized velocity potential in the time interval $(t_m, t_m + \Delta t)$; and x is the normalized velocity potential in the time interval $(t_{m-k}, t_{m-k} + \Delta t)$. For the case where Δt tends to zero, the total velocity potential $\Phi(t)$ is estimated as follows:

$$\Phi(t) = \dot{x}(t)\Psi + \int_{-\infty}^{t} x(t-\tau)\dot{x}(t)d\tau \qquad \text{Equation 6.41}$$

where $\dot{x}(t)$ is the velocity of the structure at time τ. The hydrodynamic force F, which acts on the wetted surface S of the floating body, can be estimated with the integration of the pressure p in the fluid that follows from the very well-known linearized Bernoulli equation:

$$p = -\rho\frac{\partial\Phi}{\partial t} \qquad \text{Equation 6.42}$$

$$F = -\iint_S p n dS = \left(\rho\iint_S \Psi n dS\right)\ddot{x}(t) + \int_{-\infty}^{t}\left(\rho\iint_S \frac{\partial x(t-\tau)}{\partial t} n dS\right)\dot{x}(\tau)d\tau \qquad \text{Equation 6.43}$$

or:

$$F = A\ddot{x}(t) + \int_{-\infty}^{t} B(t-\tau)\dot{x}(\tau)d\tau \qquad \text{Equation 6.44}$$

with:

$$A = \rho\iint_S \Psi n dS \qquad \text{Equation 6.45}$$

$$B(t) = \rho\iint_S \frac{\partial x(t-\tau)}{\partial t} n dS \qquad \text{Equation 6.46}$$

With the use of Newton's second law, the equation of motion of the floating body in the time domain is as follows:

$$(M+A)\ddot{x}(t) + \int_{-\infty}^{t} B(t-\tau)\dot{x}(\tau)d\tau + Cx(t) = X(t) \qquad \text{Equation 6.47}$$

where $x(t), \dot{x}(\tau), \ddot{x}(t)$ are the displacement, velocity and acceleration of any translational or rotational degree of freedom; M is the structural mass; A is the added mass of the fluid; B(t) is the retardation function for any possible translational or rotational degree of freedom; C is the hydrostatic stiffness; and X is any possible external loading (e.g. wave loading). In Equation 6.47, and by replacing the term τ with the term $t - \tau$ and by changing the integration boundaries, the equation of motion in the time domain takes the following form (Cummins, 1962):

$$(M+A)\ddot{x}(t) + \int_{0}^{\infty} B(\tau)\dot{x}(t-\tau)d\tau + Cx(t) = X(t) \qquad \text{Equation 6.48}$$

Note that the added mass is frequency dependent, and usually the assumption of using the added mass that corresponds to infinite frequency is utilized when solving Equation 6.48.

Usually, the retardation function and the added mass that corresponds to infinite frequency are calculated with the use of the Kramer–Kronig relationship (King, 2009):

$$B(t) = \frac{2}{\pi} \int_0^\infty C_B(\omega) \cos\omega t \, d\omega \qquad\qquad \text{Equation 6.49}$$

$$A^\infty = A(\omega) + \frac{1}{\omega} \int_0^\infty B(t) \sin\omega t \, dt \qquad\qquad \text{Equation 6.50}$$

It is stressed that the calculation of the retardation function and the added mass in infinite frequency is not straightforward due to the semi-infinite integral and also due to the possibility of instability for the numerical calculation of the coefficients. It is very critical to stress the fact that the retardation function and the convolution integral must be iterated in time; this numerical process is highly computationally demanding. Different research groups have researched replacing the convolution terms with other approximate methods (Taghipour, 2008). These methods can be categorized into those that replace the frequency-dependent added mass and radiation damping with constant values and those that replace the convolution integral with a state-space formulation.

For the solution of the equation of motion in time-domain dynamic analysis of offshore structures and systems, methods with numerical time integration are commonly used (Hilber, 1977). Basic assumptions of these methods are that the dynamic response is determined not as a continuous function but as values in discrete points in time and also that the result of the time integration contains both the homogeneous and the particular solution. Note that the solutions have a limited accuracy that depends upon the calculation method and the length of the time step in the integration. Common types of error that should be evaluated are the negative numerical damping, positive numerical damping, period error and unstable integration. In general, all the previous methods are using the Newmark methods (Newmark, 1959); alternatively, the Runge–Kutta method can be used.

It is very common that the integration methods (Butcher, 2008) are using: (a) constant initial acceleration, (b) constant average acceleration and (c) linear acceleration. The velocity and displacement are defined as:

$$d\dot{u} = \ddot{u}dt \quad \text{and} \quad du = \dot{u}dt \qquad\qquad \text{Equation 6.51}$$

For the case that the acceleration is changing over the time step h, then Equation 6.51 can be integrated directly as follows:

$$\dot{u}(t) = \dot{u}_1 + \int_0^\tau \ddot{u}(\tau)d\tau \qquad\qquad \text{Equation 6.52}$$

$$u(t) = u_1 + \int_0^\tau \dot{u}(\tau)d\tau \qquad\qquad \text{Equation 6.53}$$

For the case in which constant initial acceleration is used (e.g. the conditional stable method), the acceleration is considered to be constant during the time step h and equal

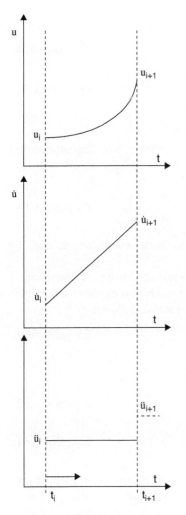

Figure 6.4 Constant initial acceleration for numerical integration.

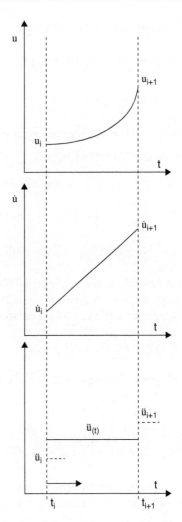

Figure 6.5 Constant average acceleration for numerical integration.

to the acceleration at the beginning of the time step $\ddot{u}(t) = \ddot{u}_1$ (Figure 6.4). As a result, Equations 6.54, 6.55 and 6.56 can be solved for every time step:

$$u_{i+1} = u_i + h\dot{u}_i + \frac{h^2}{2}\ddot{u}_i \qquad\qquad \text{Equation 6.54}$$

$$\dot{u}_{i+1} = \dot{u}_i + h\ddot{u}_i \qquad\qquad \text{Equation 6.55}$$

$$\ddot{u}_i = \frac{1}{m}\left(P_i - c\dot{u}_i - ku_i\right) \qquad\qquad \text{Equation 6.56}$$

For the case that the average value during the time step is used, the method is implicit and is named the *constant average acceleration* (Figure 6.5):

$$u(\tau) = \frac{1}{2}\left(\ddot{u}_1 + \ddot{u}_{1+1}\right) \qquad\qquad \text{Equation 6.57}$$

After integration, Equations 6.58 and 6.59 hold:

$$u_{i+1} = u_i + h\dot{u}_i + \frac{h^2}{4}(\ddot{u}_i + \ddot{u}_{i+1})$$

Equation 6.58

$$\dot{u}_{i+1} = \dot{u}_i + \frac{1}{2}h(\ddot{u}_i + \ddot{u}_{i+1})$$

Equation 6.59

With the use of the equilibrium equation for every time step:

$$m\ddot{u}_{i+1} + c\dot{u}_{i+1} + ku_{i+1} = p_{i+1}$$

Equation 6.60

for Equations 6.58 and 6.59, the following relation holds:

$$m\ddot{u}_i + \left(\frac{4}{h}m + c\right)\dot{u}_i + p_{i+1} + \left(\frac{4}{h^2}m + \frac{2}{h}c\right)u_i = \left(\frac{4}{h^2}m + \frac{2}{h}c + k\right)u_{i+1}$$

Equation 6.61

which can be solved for every possible time step if the initial conditions u_0 and $\dot{u}_0$ are known.

For the case where the acceleration for the time step is assumed to be the linear combined value between the two time limits of the step, the method is explicit; it is called *linear acceleration* (Figure 6.6). The acceleration is calculated as follows:

$$u(\tau) = \ddot{u}_1 + \frac{\tau}{h}(\ddot{u}_{1+1} - \ddot{u}_1)$$

Equation 6.62

After integration, Equations 6.63 and 6.64 hold true:

$$u_{i+1} = h\dot{u}_i + \frac{h^2}{3}\ddot{u}_i + \frac{h^2}{6}\ddot{u}_{i+1}$$

Equation 6.63

$$\dot{u}_{i+1} = \dot{u}_i + h\ddot{u}_i + \frac{h}{2}(\ddot{u}_{i+1} - \ddot{u}_i)$$

Equation 6.64

These equations with the equilibrium equation can be used for the estimation of the displacement, velocity and acceleration at any time step as follows:

$$\left(2m + \frac{h}{2}c\right)\ddot{u}_i + \left(\frac{6}{h}m + 2c\right)\dot{u}_i + p_{i+1} + \left(\frac{6}{h^2}m + \frac{3}{h}c\right)u_i = \left(\frac{6}{h^2}m + \frac{3}{h}c + k\right)u_{i+1}$$

Equation 6.65

The Newmark methods are based on the following set of equations:

$$\dot{u}_{i+1} = \dot{u}_i + (1-\gamma)\Delta t\ddot{u}_i + \gamma\Delta t\ddot{u}_{i+1}$$

Equation 6.66

$$u_{i+1} = u_i + \Delta t\dot{u}_i + (0.5-\beta)\Delta t^2\ddot{u}_i + \beta\Delta t^2\ddot{u}_{i+1}$$

Equation 6.67

combined with the equation of dynamic equilibrium:

$$\ddot{u}_{i+1} = m^{-1}(-c\dot{u}_{i+1} - f(u_{i+1},\dot{u}_{i+1}) + p_{i+1})$$

Equation 6.68

where γ and β are weighted numbers that determine how the values at time steps t and t_{i+1} shall be weighted. These parameters also determine the stability and accuracy of the method. Typical values that are used for these two constants are $1/6 \leq \beta \leq 1/2$ and $\gamma = 1/2$; for the case that $\beta = \gamma = 1/2$, the constant average acceleration method is used, while for the case that $\beta = 1/6$ and $\gamma = 1/2$, the linear variation of acceleration is used. With the system of the three equations, it is possible to determine numerically the displacement, velocity and acceleration at the end of the $i^{th} + 1$ time step. The initial conditions u_0 and $\dot{u}_0$ that are given as input combined with the previous equations represent the required information for starting the process.

For the solution of Equations 6.66 ~ 6.68, an iteration process has to be performed; this method of solution is called *implicit*. Meanwhile, for the case of linear systems, an *explicit* method can be alternatively used. Equations 6.66 ~ 6.68 are reformulated in terms of the incremental quantities $\Delta u_i = u_{i+1} - u_i$ as follows:

$$\Delta \dot{u}_i = \Delta t \ddot{u}_i + \gamma \Delta t \Delta \ddot{u}_i \qquad \text{Equation 6.69}$$

$$\Delta u_i = \Delta t \dot{u}_i + \frac{\Delta t^2}{2} \ddot{u}_i + \beta \Delta t^2 \Delta \ddot{u}_i \qquad \text{Equation 6.70}$$

With appropriate substitutions, we have:

$$\Delta \ddot{u}_i = \frac{1}{\beta \Delta t^2} \Delta u_i - \frac{1}{\beta \Delta t} \dot{u}_i - \frac{1}{2\beta} \ddot{u}_i \qquad \text{Equation 6.71}$$

$$\Delta \dot{u}_i = \frac{\gamma}{\beta \Delta t} \Delta u_i - \frac{\gamma}{\beta} \dot{u}_i + \Delta t \left(1 - \frac{\gamma}{2\beta}\right) \ddot{u}_i$$

$$\text{Equation 6.72}$$

and, with the use of the following equation of motion in an incremental form:

$$m \Delta \ddot{u}_i + c \Delta \dot{u}_i + k \Delta u_i = \Delta p_i \qquad \text{Equation 6.73}$$

and, for the case of a linear system, we have the following result:

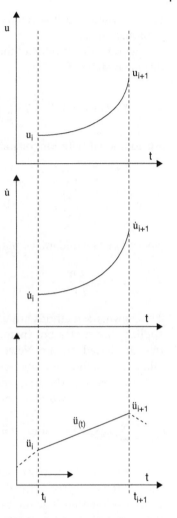

Figure 6.6 Linear acceleration for numerical integration.

$$\left(k + \frac{\gamma}{\beta \Delta t} c + \frac{1}{\beta \Delta t^2} m\right) \Delta u_i = \Delta p_i + \left(\frac{1}{\beta \Delta t} m + \frac{\gamma}{\beta} c\right) \dot{u}_i + \left(\frac{1}{2\beta} m + \Delta t \left(\frac{\gamma}{2\beta} - 1\right) c\right) \ddot{u}_i$$

$$\text{Equation 6.74}$$

With the use of Equation 6.74, the Δu_i is calculated; from Equations 6.71 and 6.72, the $\Delta \dot{u}_i$ is calculated; and from Equation 6.72, the $\Delta \ddot{u}_i$. Based on Equation 6.74, it can be concluded that the constant γ controls the numerical damping in the numerical solution; three different cases can be found: (a) $\gamma = 0.5$ and as a result there is no numerical damping,

(b) $\gamma > 0.5$ and a positive numerical damping exists and (c) $\gamma < 0.5$ and negative numerical damping exists.

Regarding the stability of the Newmark methods, the Newmark methods are unconditionally stable if:

$$\gamma \geq 1/2 \text{ and } \beta \geq \frac{1}{4}\left(\gamma + \frac{1}{2}\right)^2 \qquad \text{Equation 6.75}$$

An additional criterion for stability is expressed with Equation 6.76:

$$\frac{\Delta t}{T_n} \leq \frac{1}{2\pi} \frac{1}{\sqrt{\frac{1}{4}\left(\gamma + \frac{1}{2}\right)^2 - \beta}} \qquad \text{Equation 6.76}$$

where T_n is the undamped natural period of the system:

$$T_n = 2\pi\sqrt{\frac{m}{k}} \qquad \text{Equation 6.77}$$

The Newmark methods are used widely by a huge number of researchers in different applications in offshore engineering. Several researchers proposed improved methods that are based on the Newmark methods; some examples are Wilson's θ-method, Hilber's α-method and the Bossak–Newmark method.

With regard to the frequency-domain dynamic analysis of floating structures and with the use of Newton's second law, the general equation of motion takes the form of:

$$\sum_{j=1}^{6} m_{ij}\ddot{x}_j = F_i \quad \text{for} \quad i = 1,\ldots,6 \qquad \text{Equation 6.78}$$

where m_{ij} is a 6×6 matrix of the mass and inertia of the body, $\ddot{x}_j$ is the acceleration of the body in the jth degree of freedom and F_i is the total of the forces that are acting in the ith degree of freedom. The forces that are acting on a floating body are the summation of the wave excitation forces and the forces acting on an oscillating body in still water, and are related to the added mass, radiation damping and hydrostatic stiffness. The equation of motion in the frequency domain is (Faltinsen, 1990):

$$\sum_{j=1}^{6}\left[-\omega^2\left(M_{ij} + A_{ij}\right) + i\omega B_{ij} + C_{ij}\right]\xi_j = X_i \quad \text{for} \quad i = 1,\ldots,6 \qquad \text{Equation 6.79}$$

where ω is the frequency of the incident wave, M_{ij} are coefficients of structural mass, A_{ij} are coefficients of added mass, B_{ij} are coefficients of radiation damping, C_{ij} are coefficients of the hydrostatic stiffness, ξ_j are the motions' amplitude and X_i are excitation loads induced by the incident waves. Note that for the case of any possible external mass, damping or stiffness coefficients (e.g. mooring lines and viscous structural damping), the relevant coefficients are added to the aforementioned coefficients. The solution of Equation 6.79 can be numerically achieved with the use of simple matrix calculations.

6.4   Multibody Approach

Very frequently, offshore structures and systems are composed of two or more rigid bodies that are interconnected as a total body. Examples are wave energy converters, combined energy systems and multipurpose offshore structures and systems. In general, numerical analysis of multi-body offshore structures and systems in the frequency and time domains can be performed either using multi-interconnected floating rigid bodies with the adoption of hydrodynamic analysis or using one floating body that behaves flexibly according to predefined calculated eigenmodes with the adoption of hydro-elastic analysis (Michailides, 2012); the latter case also requires a structural model for calculation of the eigenmodes of the offshore structure and system.

One example of a multi-body system is the combined *semisubmersible wind energy and flap-type wave energy converter* (SFC) that consists of a braceless semisubmersible floating platform with four cylindrical-shaped columns (one central column and three side columns) and three rectangular-shaped pontoons with large dimensions that connect the side columns to the central column, a 5 MW wind turbine placed on the central column of the semisubmersible platform, three rotating flap-type WECs hinged at the pontoons of the semisubmersible through rigid structural arms and linear power take-off (PTO) mechanisms, and three catenary mooring lines positioned at the three side columns of the semisubmersible. In total, four rigid bodies exist, namely, the semisubmersible platform and the three WECs (Michailides, 2016). For this kind of structure and in order to perform the time-domain numerical analysis, the following coupling terms between the different rigid bodies should be accounted and included: (a) the coupled added mass coefficients at infinite frequency and (b) the coupled retardation functions. For the case that two rigid bodies exist, the inertia terms associated with the added mass are as follows:

$$
\begin{bmatrix} (\mathbf{m} + \mathbf{A}_\infty)_{i,j} & (\mathbf{A}_\infty)_{i,j} \\ (\mathbf{A}_\infty)_{j,i} & (\mathbf{m} + \mathbf{A}_\infty)_{j,i} \end{bmatrix} \begin{bmatrix} \ddot{\mathbf{x}}_i \\ \ddot{\mathbf{x}}_j \end{bmatrix}
\qquad\qquad \text{Equation 6.80}
$$

where i and j denote the two bodies, and $\ddot{\mathbf{x}}_i$ and $\ddot{\mathbf{x}}_j$ are their accelerations in six rigid body degrees of freedom referred to as the *relevant fixed-body coordinate systems*. The coupling terms that are associated with the retardation functions (Wan, 2016) are calculated as follows:

$$
\int_0^\tau \begin{bmatrix} \mathbf{h}(t-\tau)_{i,i} & \mathbf{h}(t-\tau)_{i,j} \\ \mathbf{h}(t-\tau)_{j,i} & \mathbf{h}(t-\tau)_{j,j} \end{bmatrix} \begin{bmatrix} \dot{\mathbf{x}}_i \\ \dot{\mathbf{x}}_j \end{bmatrix} d\tau
\qquad\qquad \text{Equation 6.81}
$$

For the case in which one floating body that behaves flexibly according to predefined calculated eigenmodes is under study, then the multi-body analysis can be achieved with the use of the hydro-elastic analysis and can be performed in either the time or frequency domain. A relevant example is the case of a multipurpose structure that operates both as a floating breakwater and as a wave energy converter (Michailides, 2015). The floating structure is composed of four rigid modules that are interconnected with the use of flexible connectors with specific mechanical characteristics and with the use of PTO mechanisms for producing power from the relative motion of the modules.

The amplitudes of the body's motions ξ_j are obtained from the solution of the linear system of equations:

$$\sum_{j=1}^{N}\left[-\omega^2\left(M_{ij}+A_{ij}\right)+i\omega\left(B_{ij}+B_{ij}^E\right)+\left(C_{ij}+K_{ij}\right)\right]\xi_j = X_i \quad i,j=1,\ldots,N \qquad \text{Equation 6.82}$$

where ω is the incident wave frequency; A_{ij} and B_{ij} are the coefficients of the added mass matrix and radiation damping matrix, respectively; X_i are the exciting forces of the incident wave; C_{ij} are the coefficients of the hydrostatic-gravitational stiffness matrix (Newman, 1994; Senjanovic, 2008); N is the total number of degrees of freedom of the system; and modes for $j = 1,\ldots, 6$ represent the six rigid body modes, namely, surge, sway, heave, roll, pitch and yaw (respectively, ξ_1, ξ_2, ξ_3, ξ_4, ξ_5 and ξ_6), while modes for $j = 7,\ldots,N$ represent the additional generalized modes of the system that correspond to the "dry" or "wet" eigenmodes of the structure. The additional generalized modes are estimated based on analytical solutions or based on finite element method (FEM) models. For the estimation of the eigenmodes, the solution of the free vibration equation of the structure has to be solved:

$$\left(K^{str}-\omega_j^2 M^{str}\right)S_j = \{0\} \qquad \text{Equation 6.83}$$

where K_{str} and M_{str} are the structural stiffness and mass matrices of the structure, respectively; and ω_j and S_j, $j = 1,\ldots,N$, are the eigenfrequency and the eigenmodes of each jth mode, respectively. The solution of the boundary value problem is based on the three-dimensional (3D) panel method utilizing Green's theorem, imposing the appropriate boundary conditions on the free surface, on the sea bottom and on the floating body, and the radiation condition for the outgoing waves (Lee, 2000). The radiation potential of each one of the generalized modes φ_j, $j = 7,\ldots,N$, is subjected to the following boundary condition on the body boundary (Lee, 2002):

$$\frac{\partial\varphi_j}{\partial n} = n_j = u_j n_x + v_j n_y + w_j n_z \qquad \text{Equation 6.84}$$

where $\mathbf{n} = (n_x, n_y$ and $n_z)$ is the unit normal vector and u_j, v_j and w_j are the components of the displacement vector of mode j, $j = 7,\ldots,N$, in the x, y and z directions, respectively, as calculated in the 3D structural model. For the implementation of hydro-elastic analysis, commercial programs like WAMIT (2008) can be used for the solution of both the radiation and the diffraction problem. The response of the floating structure in each mode is expressed in terms of the response amplitude operator (RAO):

$$RAO_j = \frac{|\xi_j|}{A}, \text{ with } j = 1,\ldots,N \qquad \text{Equation 6.85}$$

where $|\xi_j|$ is the amplitude of the quantity ξ_j. The hydro-elastic displacement vector (structural deformations) of every i^{th} point of the structure is proportional to the displacement vector of the eigenmodes S_j and to the calculated RAO_j:

$$U_F^i = \sum_{j=1}^{N} S_j RAO_j, \quad j = 1,\ldots,N \qquad \text{Equation 6.86}$$

With the use of the displacement vector and the FEM model of the offshore structure, critical quantities (e.g. stress and strain) are estimated and the structure's integrity is assessed.

6.5   Finite Element Method

As for the analysis of any possible structure, the FEM can be used and applied for the case of offshore structures. Fundamental theory for the development of the FEM can be found in Bathe (1996). Usually, the structure is discretized into many elements; as an example, these elements can be beam elements with two nodes, shell elements with four nodes or more and solid elements with eight nodes or more. Each node of the elements has up to six degrees of freedom, which are three translational degrees and three rotational degrees of freedom. Every element, based on its properties, results in matrices of mass, damping and stiffness. On the other hand, all possible external loads can be added in any node of the structure. The summation of the matrices of mass, damping and stiffness, and of all the external loads in all the nodes of the structure, provides the final matrices for the numerical analysis. Note that the size of the matrices equals the total of the degrees of freedom of all the nodes. The method is straightforward, and the equation of motion in both the time and frequency domain is ready to be solved numerically.

In offshore engineering problems, the FEM is used extensively. For the case of fixed-bottom structures, the FEM is used exactly as for the case of any other structure onshore. It is stressed that with the FEM, a simulation or representation of the real structure is attempted. Therefore, the fidelity of the numerical model compared to the real structure depends upon the user who builds the model and not upon the method itself. Usually, in the beginning, the structure is assembled into numerous elements. Then, the element properties are assigned; in other words, the properties of mass, damping and stiffness are assigned. Appropriate boundary conditions are assigned that correspond to the conditions of the real physical problem. External loads are assigned to the nodes when existing. The solution of the equation of motion in the time or frequency domain is then achieved. For the case in which a rigid or deformable floating body exists, the FEM is applied in an alternative way. The rigid body is considered as a special node; in other words, one node corresponds in one structure (e.g. semisubmersible). For this special node, all the required information is estimated and inserted in the equation of motion. For example, the hydrodynamic coefficients of the added mass, radiation damping and excitation-wave first-order loads are estimated with the use of the boundary element method or any other possible method, and then are given as input for the special node.

6.6   Nonlinear Analysis

It is very common for the dynamic analysis of offshore structures and systems' nonlinear phenomena to be included and solved with the solution of the equation of motion. Nonlinearities can be found both in the loading terms and in the response quantities of the structure (Argyris, 1991). Here, numerical methods that are used for incorporating the nonlinear phenomena during the dynamic analysis are presented.

For the case of an offshore structure with nonlinearities, and for any possible increment Δt of time, the incremental form of the equation of motion is as follows:

$$\left(m\ddot{u}(t+\Delta t)-m\ddot{u}(t)\right)+\left(c\dot{u}(t+\Delta t)-c\dot{u}(t)\right)+\left(ku(t+\Delta t)-ku(t)\right)=\left(\Delta p(t+\Delta t)-\Delta p(t)\right)$$

Equation 6.87

It is stated that the increment in the left side that is caused by inertia, damping and reaction forces is balanced by the increments in external loadings. The numerical solution of the incremental equation of motion is achieved by introducing the tangential damping c_i and stiffness k_i matrices at the beginning of the time increment (Hughes, 1987). The incremental form of the equation of motion is as follows:

$$m\Delta\ddot{u}_i+\left(c+c_i\right)\Delta\dot{u}_i+k_i\Delta u_i=\Delta p_i$$

Equation 6.88

where $k_i=k^T(u_i,\dot{u}_i)$ with $k^T(x,y)=\partial f(x,y)/\partial x$, and $c_i=c^T(u_i,\dot{u}_i)$ with $c^T(x,y)=\partial f(x,y)/\partial y$.

For the case in which the time step Δt is constant, significant errors during the implementation of the numerical analysis can be presented mainly due to the use of tangential damping and stiffness that require knowledge of the u_{i+1} and $\dot{u}_{i+1}$ that are unknown or due to sudden changes in the nonlinearities that are not captured with the constant value of Δt; it is straightforward that the decrease of the time step Δt will improve the solution, but on the other hand the computational cost will greatly increase.

To reduce the errors, an iterative solution method for each time step has to be performed. Analogous to the case of linear dynamic analysis and based on substitutions exactly the same as in Equations 6.50 ~ 6.55, we have:

$$\left(k_i+\frac{\gamma}{\beta\Delta t}(c+c_i)+\frac{1}{\beta\Delta t^2}m\right)\Delta u_i=\Delta p_i+\left(\frac{1}{\beta\Delta t}m+\frac{\gamma}{\beta}(c+c_i)\right)\dot{u}_i$$
$$+\left(\frac{1}{2\beta}m+\Delta t\left(\frac{\gamma}{2\beta}-1\right)(c+c_i)\right)\ddot{u}_i$$

Equation 6.89

Note that the level of nonlinearity between the real and the tangential stiffness is the same. For the first time step t_1, and by solving Equation 6.89, the $\Delta u_{i,1}$, $\Delta\dot{u}_{i,1}$ and $\Delta\ddot{u}_{i,1}$ are calculated. For the estimated solution, the error is estimated. If the error is smaller compared to the threshold that is defined in the beginning of the analysis, the calculated solution is the final solution of the nonlinear dynamic analysis. Contrary to if the error is larger compared to the threshold, then, a corrected load increment is introduced as follows:

Equation 6.90

$$\Delta p_{i,1}=\Delta p_i-\left(m\Delta\ddot{u}_{i,1}+c\Delta\dot{u}_{i,1}+\Delta f_{i,1}\right)$$

which gives the following correction:

$$\Delta\hat{p}_{i,1}=\Delta p_{i,1}+\left(\frac{1}{\beta\Delta t}m+\frac{\gamma}{\beta}(c+c_{i,1})\right)\dot{u}_{i,1}+\left(\frac{1}{2\beta}m+\Delta t\left(\frac{\gamma}{2\beta}-1\right)(c+c_{i,1})\right)\ddot{u}_{i,1}$$

Equation 6.91

where $c_{i,1} = c^T(u_{i,1}, \dot{u}_{i,1})$ with $u_{i,1} = u_i + \Delta u_{i,1}$, $\dot{u}_{i,1} = \dot{u}_i + \Delta \dot{u}_{i,1}$ and $\ddot{u}_{i,1} = \ddot{u}_i + \Delta \ddot{u}_{i,1}$. Equally, the stiffness matrix takes the form:

$$\hat{k}_{i,1} = k_{i,1} + \frac{\gamma}{\beta \Delta t}(c + c_{i,1}) + \frac{1}{\beta \Delta t^2} m \qquad \text{Equation 6.92}$$

where $k_{i,1} = k^T(u_{i,1}, \dot{u}_{i,1})$. By solving Equation 6.93, a corrected solution $\Delta u_{i,2}$ with regard to the first solution is obtained:

$$\hat{k}_{i,1} \Delta u_{i,2} = \Delta \hat{p}_{i,1} \qquad \text{Equation 6.93}$$

and $\Delta \dot{u}_{i,1}$ and $\Delta \ddot{u}_{i,1}$ are calculated, too. Again, the error of this second solution is estimated with the use of Equation 6.94:

$$m\Delta \ddot{u}_{i,2} + c\Delta \dot{u}_{i,2} + \Delta f_{i,2} = \Delta p_{i,1} \qquad \text{Equation 6.94}$$

If the error is again larger compared to the threshold that defines the acceptance (or not) of the solution, a new corrected load increment is introduced. The same process is stopped when the error is smaller compared to the threshold value $\Delta p_{i,2}$, and a new solution is calculated with the use of exactly the same equations.

6.7 Extreme Response Analysis and Prediction

Offshore structures and systems have to be designed to ensure serviceability and survivability over their entire service life. Structural integrity of all parts of offshore structures should be adequate. It is very common that the extreme response quantities must be estimated appropriately for the engineers to perform accurate calculations and examine/ensure the structural integrity of the offshore structures and systems for expected extreme conditions. For estimating and predicting extreme responses, three methods are widely used and proposed by regulations and research institutes. These methods are: (a) the characteristic design wave method, (b) the short-term sea state method and (c) the full long-term method.

The characteristic design wave method is the simplest method among the three. The design wave is considered as a wave or a group of waves that will maximize the response quantities. The response quantity can be any possible internal load but also any possible stress or strain along the offshore structure. It is very common for the 100-year wave height H^{100} to be established and used as the design wave based on measured data for a specific site. H^{100} corresponds to the height of a wave that is exceeded on average once every 100 years, or the wave height that is exceeded in a time period of one year with a probability of 10^{-2}. After the establishment of the H^{100}, correlated wave periods T and wave directions are estimated for this wave. Correlated wave periods are usually suggested in the bibliography; for example, in NORSOKN003 (NORSOK, 2007), it is suggested that the range of the correlated wave periods is in between:

$$\sqrt{6.5 H^{100}} \leq T \leq \sqrt{11 H^{100}} \qquad \text{Equation 6.95}$$

For the specified environmental cases, numerical analysis is performed with appropriate numerical tools and the estimated structural response that is considered as the extreme response is calculated. In general, with the design wave method, the estimation of the response quantities for a prescribed annual exceedance probability (e.g. 10^{-2} or 10^{-3}) is achieved without needing to analyse a big number of different environmental conditions.

For establishing the short-term sea state method, usually a design check for load effects established during an N-year storm of duration (commonly, 3 or 6 hours) is performed; usually, the 100-year storm of a specific duration is examined. Alternatively, the contour line approach can be used for the estimation of the maximum responses. A basic assumption for using the aforementioned methods is that both the environmental loading (e.g. wave) and the response quantities follow a Gaussian process. In general, a process x(t) is considered as Gaussian if the random variable $x(t_i)$, where t_i is a random point in time, is Gaussian distributed. In cases where the responses cannot be considered as Gaussian processes, other simulation methods can be applied (e.g. Monte Carlo). For the case of the N-year storm analysis, the maxima N_{max} of a specific period of time of analysis (e.g. 10 minutes; the process is observed for several time intervals of the same time duration) are recorded. With the use of these maxima values, the distribution of the largest maxima within the period of time can be obtained; mathematically, this distribution can be estimated with the following Rayleigh distribution:

$$f_{X_{max}}(x_{max}) = P(X_{max} < x_{max}) = \left[1 - \exp\left(\frac{-x_{max}^2}{2\sigma_x^2}\right)\right]^N \qquad \text{Equation 6.96}$$

where X_{max} is the largest amplitude among N_{max}, and σ_x is the standard deviation of the Gaussian process. It is clear that a relationship between probability and amplitude for a response quantity in a given sea state is established. Based on the previous formula, it is clear that the extreme distribution will change for a different time T or for a different number of maxima N_{max} in this specific time; for a decrease of the number of maxima N_{max}, the mean value of X_{max} decreases too, while the standard deviation of X_{max} increases. Two characteristic values are of high importance and are used for design purposes: (a) the expected largest value that corresponds to the mean value of the distribution (Equation 6.97), and (b) the most probable largest value that corresponds to the peak of the probability density function (Equation 6.98). These are calculated with the use of Equations 6.98:

$$\overline{X_{max,N}} = \sigma_x\left[\sqrt{2\ln N} + \frac{0.5722}{\sqrt{2\ln N}}\right] \qquad \text{Equation 6.97}$$

$$X_{max,N} = \sigma_x\sqrt{2\ln N} \qquad \text{Equation 6.98}$$

where σ_x is the process standard deviation. Additionally, the standard deviation of the largest amplitude among N can be estimated with:

$$\sigma[X_{max}] = \sigma_x\frac{\pi}{\sqrt{6}}\frac{1}{\sqrt{2\ln N}} \qquad \text{Equation 6.99}$$

Alternatively, and for a short-term approach of the extreme responses of an offshore structure, the environmental contour line approach (Winterstein, 1993) can be performed. Usually, this method is suggested when due to time restrictions, a full long-term extreme method cannot be performed. In general, the contour lines provide reasonable extreme values by concentrating the short-term method in a small certain amount of sea states of the scatter diagram of a specific site. The basic idea behind the contour line method is that any q-probability response quantity can be estimated by studying the short-term response for the sea state that is placed along the q-probability contour line. It is stated that in the Gaussian space, the contour line that corresponds to the annual exceedance probability of q will be circles with a radius of:

$$r = \Phi^{-1}(1 - q/n)$$ Equation 6.100

where Φ is the cumulative distribution function of the Gaussian process, and n is the number of the examined sea states per year (e.g. $n = 2920$ for 3-hour sea states). The contour line can be estimated by the joint environmental distribution of H_s and T_p for a specific site and with the use of the following equation:

$$\frac{q}{n} = \iint f_{H,T_p}(h,t)dtdh$$ Equation 6.101

where $f_{H,T_p}(h,t)$ is the joint long-term distribution of the variation of H_s and T_p represented with a scatter diagram. Once the contour line is developed, several sea states along the contour line (e.g. ten different H_s and T_p, or more) are used as input for the implementation of the numerical analysis of the numerical model. With the use of the response time series for these examined environmental conditions and the maxima of the response quantities, and with the use of Equations $6.96 \sim 6.99$, the extreme response quantities can be predicted based on a short-term method (Naess, 2013).

The full long-term method is a straightforward approach for the estimation of the long-term extreme response and the evaluation of the extreme response's probability distribution. Two types of statistical information are needed: (a) statistics of the sea characteristics for the entire lifetime of the offshore structure, and (b) statistics of the responses for all the possible sea states. With regard to the sea characteristics, it is very common that the long-term distribution of H_s and T_p is usually established as a simultaneous distribution $P(H_s, T_p)$ and presented in a table format as a scatter diagram. The full long-term method calculates the long-term response quantities by directly integrating all environmental parameters and the corresponding short-term response probability functions. In general, the long-term extreme of a response can be found with the use of the short-short-term extremes, with Equation 6.102:

$$F_X^{LT}(y) = \int F_{X|S}^{ST}(y|s)f_S(s)ds$$ Equation 6.102

where F is the cumulative distribution function of response X, s is the environmental condition (Equation 6.103) and $f_S(s)$ is the probability density function of s. Note that for the short term (ST), the probability distribution is valid in only one sea state.

$$\int f_S(s)ds = 1$$ Equation 6.103

For the case in which we have two environmental conditions, H_s and T_p, the long-term extreme of the response will be:

$$F_X^{LT}(y) = \sum_{H_s}\sum_{T_p} F_{X|H_s,T_p}^{ST}(y|H_s,T_p)\frac{\bar{T}_z}{T_z(H_s,T_p)}f_{H_s,T_p}(H_s,T_p)$$ Equation 6.104

where $\overline{T}_z$ is the average zero-upcrossing period for the response quantity X for all the sea states, and T_z is the average zero-upcrossing period for the response quantity X in one specific sea state. Note that the long-term extreme response quantity is independent of the sea state that this maximum is taken from. It is very common for the long-term distribution to be transformed to a histogram that shows the number of response amplitudes of a given value that is expected to occur during the lifetime of a structure. For the case that the wave directionality θ is added to the environmental conditions s, then the long-term extreme of any response can be estimated with Equation 6.105:

$$F_X^{LT}(y) = \sum_\theta \sum_{H_s} \sum_{T_p} F_{X|\theta,H_s,T_p}^{ST}(y|\theta,H_s,T_p) \frac{\overline{T}_z}{T_z(\theta,H_s,T_p)} f_{H_s,T_p}(H_s,T_p) f_\theta(\theta)$$

Equation 6.105

For the case in which the wave spectrum parameters H_s and T_p, and wave direction θ, are statistically dependent, the two last terms of the probability density function in Equation 6.105 are replaced with the probability density function $f_{\theta,H_s,T_p}(\theta,H_s,T_p)$ but with the existence of a 3D scatter diagram.

6.8 Testing and Validation of Offshore Structures

Experimental investigation of offshore structures is a very valuable tool. Typically, the objectives for performing physical model tests are: (a) for performing feasibility studies during the early stage of new concepts and for their proof of concept, (b) for verifying the behaviour and response of concepts by applying design loads to the structure and verifying that it satisfies its structural integrity, (c) for examining operational limits, (d) for performing tests in parts of a complex offshore structure, (e) for determining coefficients (e.g. added mass, drag, linear and quadratic damping, and RAOs) that afterwards will be used in integrated numerical analysis and (f) for validating and developing numerical methods and computational tools.

For performing the physical model tests of offshore structures, a suite of world-class research infrastructures for both scientific and commercial missions exists. Each of them has advantages and disadvantages compared to the rest related to the operational depth and the horizontal dimensions of the corresponding facilities (basin or flume), the characteristics/range of the generated waves, the capability of generating current and wind, the characteristics/range of the generated current and wind, the enabled modelling scale and so on. As a result, none of the laboratories could be identified as the best one.

With regard to the state of the art currently existing worldwide, important and significant large-scale research infrastructures working on the field of offshore structures can be summarized as follows:

- MARINTEK/CeSOS–NTNU (Norway), with one offshore basin and towing tanks focusing on offshore structures and ships research (http://www.sintef.no/en/marintek/)
- MARIN (The Netherlands), with one offshore basin for offshore structures physical modelling and other facilities (e.g. towing tanks etc.) for physical modelling of ships (http://www.marin.nl/web/Facilities-Tools/Basins.htm)

- Deltares (The Netherlands), with four wave basins and four wave flumes (including one large-scale flume) focusing mainly on coastal structures research (http://www.deltares.nl/en/facilities/experimental-facilities)
- DHI (Denmark), with one offshore basin for offshore structures testing and one coastal basin for physical modelling of coastal structures (www.dhigroup.com)
- HR Wallingford Ltd (UK), with six basins and other facilities (e.g. wave flumes) for supporting mainly the physical modelling of coastal structures (http://www.hrwallingford.com/facilities/physical-modelling)
- Coastal Research Centre (FZK, Germany), with one large-scale flume capable of near-prototype physical testing of marine structures (http://www.fzk.uni-hannover.de/406.html?L=1)
- ICTS-CIEM (Spain), with one large-scale flume capable of near-prototype physical testing of marine structures (http://ciemlab.upc.edu/facilities/ciem-1)
- BGO-First (France), which is a multipurpose basin mainly dedicated to studies of maritime structures used in offshore oil industry and coastal engineering (http://www.oceanide.net/BGO_ENG.html)
- LabOceano (Brazil), with one offshore basin (Figure 6.7) for physical modelling of offshore structures and ships (http://www.laboceano.coppe.ufrj.br/index_en.php)
- Offshore Technology Research Center (OTRC, Texas, USA), with one offshore basin focusing on physical modelling of offshore structures and ships (http://otrc.tamu.edu)
- Haynes Coastal Engineering Laboratory of Texas A&M University (USA), with one wave basin supporting physical testing of coastal structures (http://coastal.tamu.edu/home.html)

Figure 6.7 The Ocean Tank of the Ocean Technology Laboratory – LabOceano. *Source*: Courtesy of COPPE/UFRJ, 2003; this file is licensed under the Creative Commons CC0 1.0 Universal Public Domain Dedication.

- Oregon State University's laboratory (OSU, USA), with one basin and one large flume for coastal structure physical modelling and testing (http://wave.oregonstate.edu/Facilities/)
- National Research Council Canada (NRCC, Canada), with one offshore basin and three coastal basins as well as other facilities (e.g. large flume) for physical modelling of offshore and coastal structures (http://www.nrc-cnrc.gc.ca/eng/solutions/facilities/marine_performance_index.html)
- Coastal and Hydraulics Laboratory of the Council for Scientific and Industrial Research (CSIR, South Africa), with two wave basins and one large wave flume supporting coastal structure physical testing (http://www.csir.co.za/Built_environment/Infrastructure_engineering/cepi.html#hydraulics).

The physical models should be exact replicas of the prototype structures. The first and most important aspect in physical model tests of offshore structures is the study of the appropriate similitude laws that should be applied for the physical model testing of offshore structures depending on the type of structure.

For the physical model to be an exact replica of the prototype structure, the following three criteria/similarities should be achieved:

- Geometrical similarity: The physical model and the prototype must have the same shape, and also the linear dimensions of the physical model and prototype must have the same scale ratio. The same requirement has to be satisfied for the environment (e.g. waves, wind and current) in the testing area of the structure.
- Kinematic similarity: The fluid flow must undergo similar motions at both the model and prototype scales; as a result, the velocities in the x and y directions must have the same ratio, so that a circular motion at full scale must be also a circular motion at the model scale.
- Dynamic similarity: The ratios of all forces acting on corresponding fluid particles and boundary surfaces in the physical model and in the prototype should be constant. The following force contributions are important: (a) inertia forces, (b) viscous forces, (c) gravitational forces, (d) pressure forces, (e) compressibility elastic forces in the fluid and (f) surface forces (Heller, 2011).

It is very common to use Froude's similitude law. The Froude number is defined as the ratio of the inertia and gravity force:

$$Fr = \frac{u^2}{gD} \qquad\qquad \text{Equation 6.106}$$

where u is the fluid velocity, g the gravitational acceleration and D a possible characteristic dimension of the structure. For the case in which p symbolizes the prototype, and m the model must satisfy the following relationship based on Froude's law:

$$\frac{u_p^2}{gD_p} = \frac{u_m^2}{gD_m} \qquad\qquad \text{Equation 6.107}$$

Since the geometric similarity must hold true:

$$l_p = \lambda l_m \qquad\qquad \text{Equation 6.108}$$

With the use of Equation 6.107, the relationship for the velocity of the prototype and the model is as follows:

$$u_p = \sqrt{\lambda} u_m \qquad \text{Equation 6.109}$$

The similarity in geometric and kinematic conditions, along with equality in Froude number, will ensure similarity between inertia and gravity forces applied in the model. With the same approximations, the relationships of the prototype and the model for mass, force, acceleration, time and stress/pressure are as follows:

$$m_p = \lambda^3 m_m \qquad \text{Equation 6.110}$$

$$F_p = \lambda^3 F_m \qquad \text{Equation 6.111}$$

$$\dot{u}_p = \dot{u}_m \qquad \text{Equation 6.112}$$

$$t_p = \sqrt{\lambda} t_m \qquad \text{Equation 6.113}$$

$$s_p = \lambda s_m \qquad \text{Equation 6.114}$$

The scaling ratios that are used in testing of marine and offshore structures are tabulated in Table 6.1. Note that it is practically impossible to satisfy all the different scaling laws that are presented in Table 6.1. For example, and for offshore structures, during the tests the surface waves are dominated by the gravitational forces; the same Froude number in model and full scale must be achieved. For a different offshore structure, and if the viscous forces are important, then the equality for the Reynolds number (Re) should in principle be achieved. It is important that for any possible experimental testing of offshore structures, before the construction phase of the physical model a detailed examination of the laws that will be used or not and their effect on the results should be examined, evaluated and reported.

With regard to Re, the equality in Re ensures that viscous forces are correctly scaled and are the same as the prototype structure, assuming that the same fluid is used in the model system. With the use of Froude's law, the effect of Re is not scaled properly, and an error by a factor of $\lambda^{1.5}$ exists. For example, and for a physical model with $\lambda = 50$, the

Table 6.1 Scaling ratios that are used in testing of marine and offshore structures.

Dimensionless number	Force ratio	Definition
Reynolds number	Inertia/viscous	UL/v
Froude number	Inertia/gravity	$U/\sqrt{gL}$
Mach's number	Inertia/elasticity	$U/\sqrt{E/\rho}$
Weber's number	Inertia/surface tension	$U/\sqrt{\sigma/\rho L}$
Strouhall number	–	fD/U
Keulegan–Carpenter number	Drag/inertia	$U_A T/D$
Cauchy number	–	$U^2 L^4/EI$

prototype Re is 354 times larger than the Re of the physical model. For structures where the viscous effects are significant, the Reynolds scaling should be accounted (Chakrabarti, 2005).

Based on Re, an equal ratio between inertia and viscous forces between the physical model and the prototype has to be achieved. The equal ratio between inertia and viscous forces will give:

$$Re = \frac{UL}{v}$$

Equation 6.115

where v (m^2/sec) is the kinematic viscosity. Note that it will be possible that the model flow will be laminar, while the flow that exists in the prototype model is in the turbulent region. Special attention is required prior to the tests for the estimation of the pattern of the flow. Usually, the Froude scaling laws are employed and account for the Reynolds disparity by other means. A simple method that is widely used for achieving a proper Re effect at the boundary layer of the offshore structure is by applying roughness on the surface of the front walls of the model (Chakrabarti, 2005).

With regard to Mach's number M, the equality between inertia and elastic forces will give:

$$M = \frac{U}{\sqrt{E_v / \rho}}$$

Equation 6.116

where E_v is the volume elasticity, and $\sqrt{E_v / \rho}$ is the speed of sound in the water. Note that with Mach's number, the importance of the elasticity of water that may influence the pressure transmission is addressed.

With regard to Weber's number W, the equality between inertia and surface tension forces will give:

$$W = \frac{U}{\sqrt{\sigma / (\rho L)}}$$

Equation 6.117

where σ is the surface tension. Note that with Mach's number, the importance of the elasticity of water that may influence the pressure transmission is addressed.

The Strouhall number is not derived by any possible equality between different forces. In Table 6.1, and with regard to the Strouhall number, the quantity f is the vortex shedding frequency and D is the diameter of the cylinder that the wave forces are applied against.

With regard to the Keulegan–Carpenter number (KC), the equality between inertia and drag forces will give (Keulegan & Carpenter, 1958):

$$KC = \frac{U_A T}{D}$$

Equation 6.118

where U_A is the amplitude of the velocity, and T is the period of oscillation.

With regard to Cauchy's number C, this number is used mainly for the experimental investigation of structures where the hydrodynamic forces are influenced by the elastic deformations of the structure itself; or, in other words, when the hydro-elasticity dominates the response of the structure (e.g. risers, mooring lines, wave energy converters and floating bridges). For these structures, the elasticity of the full-scale structure should be maintained in the physical model, and the elastic deformations have to be similar. In addition to the Froude laws, the Cauchy similitude is required where the stiffness of the model must be connected with the stiffness of the physical model in Equation 6.119:

$$E_f I_f = \lambda^5 \left(E_m I_m \right)$$

Equation 6.119

where E is the modulus of elasticity of the structure, I is the moment of inertia in any possible direction, f depicts the full-scale prototype and m depicts the physical model. It is noted that if all the shape dimensions are scaled according to the Froude laws, then the modulus of elasticity of the structure for the physical model must be $1/\lambda$ times the value of the full-scale prototype, since for the moment of inertia the following relation holds:

$$I_f = \lambda^4 I_m$$

Equation 6.120

To perform reliable physical model tests, controlled environmental conditions should be properly simulated in the laboratory. Environmental conditions that should be modelled in both time and the laboratory's space include waves, wind and current. Usually, the following characteristic environmental cases are examined during the tests of offshore structures: (a) regular unidirectional waves, (b) irregular unidirectional waves, (c) white noise, (d) wave groups, (e) multidirectional waves, (f) winds with or without uniform turbulence in the testing area and (g) currents with uniform or non-uniform profile.

For the experimental testing of offshore structures, the wave loading is the environmental condition that is simulated in all possible tests. The waves in the wave basin are generated with the use of a wave generator system that is based on flap-type or piston-type wave generation. The regular waves are generated in a straightforward way and are given in terms of the wave height and wave period. For the case of flap-type wave generators, the generation of the waves is controlled by the frequency and amplitude of the flap. Based on previous studies, a correlation between flap stroke and wave height with the wavelength and the laboratory's water depth exists (Dean, 1991); this correlation is named the *transfer function of the wave maker*, and the regular waves can be generated using this function and the required control signal to wave generators. It is very common that the elevation of the sea surface can be numerically estimated if we consider the elevation as a stationary Gaussian process with zero mean value. In this case, the random sea surface elevation can be simulated with the summation of a finite number of Fourier components as a function of time. As a result, the surface elevation $\eta(t)$ is represented as:

$$\eta(t) = \sum_{n=1}^{N} \alpha_n \cos\left(\omega_n t + \varepsilon_n \right)$$

Equation 6.121

where α_n is the amplitude, ω_n is the frequency and ε_n is the phase of the different wave components n. Usually, 2000 or more components are used for the experimental testing of offshore structures, to avoid repetition of the wave signal and aliasing. For each of these frequency bands, the Fourier amplitude is obtained from the spectrum density value S(ω) as follows:

$$\alpha_n = \sqrt{2S(\omega_n)\Delta\omega} \qquad \text{Equation 6.122}$$

where $\Delta\omega$ is the frequency interval for each one component n. It is noted that in most cases, the JONSWAP spectrum is used for wave spectra generation.

For the generation of multidirectional waves, an array of flaps on one side of the wave basin are used. The surface elevation is expressed as a function of space and time, and the directional spreading function is applied in addition to the energy density function, as shown here:

$$\eta(x,y,t) = \sum_{n=1}^{N} \alpha_n \cos\left[k_n\left(x\cos\theta_n + y\sin\theta_n\right) - \omega_n t + \varepsilon_n\right] \qquad \text{Equation 6.123}$$

where α_n is the Fourier amplitude of component n:

$$\alpha_n = \sqrt{2S(\omega_n)D(\omega_n,\theta_n)\Delta\omega\Delta\theta} \qquad \text{Equation 6.124}$$

During the experiments, different types of sensors are used for measuring the environmental conditions as well as the response quantities of the structure. Usually, the quantities that are measured during the tests are the environmental conditions (e.g. sea surface elevation in different places, wind force or wind speed), motions of the structure, tension of mooring lines, and strain and stresses in specific points placed on the structure. Special types of sensors are used for measuring quantities in tests oriented by explicit behaviour (e.g. slamming loads into offshore structures). Prior to the tests, all the instruments that are used are calibrated in the laboratory by comparing the recorded response against an expected response under fully controlled experimental conditions.

For experimental investigation of the response of offshore structures and systems, different types of tests are performed that depend upon the purpose of the tests and the type of the structure. If applicable, the following types of tests have to be performed:

- Static draft, trim and heel test: For the case of a floating structure, initially a test for evaluating the static draft, trim and heel is required.
- Inclining test: For determining the metacentric height, and to provide information about the stability of the platform
- Pull-out test: For determining the stiffness and estimating the tension-offset curves of the mooring lines
- Decay test: For determining the natural frequencies of the platform in all six rigid body degrees of freedom
- Hammer test: For determining the natural frequencies of the tower of the wind turbine for the case of the experimental investigation of offshore wind turbines
- Response test: For evaluating the response quantities of different structural responses for regular waves, irregular waves, wind and current
- White noise test: For evaluating the transfer functions (e.g. RAOs) for the responses that are measured
- Second-order slow-drift test: For determining the quadratic transfer functions for the motions of the platform.

6.9 Examples

6.9.1 Example 6.1

Consider a rectangular floating body (FB) with overall dimensions equal to $L = 30.0\,m$ (length), $B = 6.0\,m$ (width) and $H = 1.2\,m$ (height). The total weight of the FB is 120,000 kg, the draft dr is equal to 0.66 m and the water depth is d (Figure 6.8). Please estimate the RAOs for surge ξ_1, heave ξ_3 and pitch ξ_5, for the case of incident waves with wave frequency ω_i, $i = 1 \sim 10$, equal to 0.3, 0.4, 0.7, 0.8, 1.0, 1.25, 1.5, 2.0, 2.5 and 3.0. The FB is considered to be free floating. No external damping exists.

For the solution of Example 6.1, Equation 6.79 has to be solved. This can be achieved with the use of sophisticated software (e.g. WAMIT) that can solve the equation but also, most importantly, estimate the hydrodynamic coefficients, namely, the added mass, the radiation damping and the excitation wave loads. Alternatively, software that can handle simple matrix equations can solve Equation 6.79. For the latter case, the estimation of the hydrodynamic coefficients can be achieved with formulas that are presented in both Chapter 7 and Chapter 6 with user codes, or alternatively with the use of potential theory and a hydrodynamic model. For the case in which a hydrodynamic model is built, an appropriate convergence study for the size of the panels of the hydrodynamic model is required. Usually, a small number of panels is examined initially. Hydrodynamic analysis is performed. Afterwards, a second hydrodynamic model consisting of a larger number of panels is created with the use of sophisticated software or with the use of user codes. This process is repeated until hydrodynamic coefficients will not change a lot. It is strongly recommended, and for not having a big number of panels that may lead to computational efficiency problems, that users generate hydrodynamic models that have refine mesh close to all sharp edges of the structure. For our example, in Figure 6.9, the final mesh of the hydrodynamic model is presented. The hydrodynamic model consists of 5600 panels of unequal dimensions. Based on the hydrodynamic model, the hydrodynamic coefficients are estimated. The RAOs for surge ξ_1, heave ξ_3 and pitch ξ_5, and for the case of incident wave with wave frequency

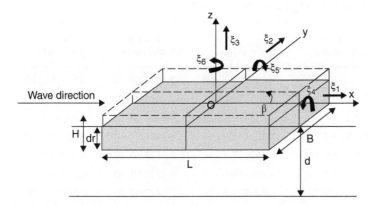

Figure 6.8 Definition and basic parameters of the examined FB.

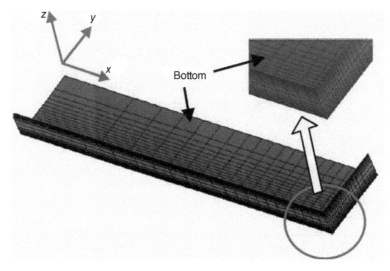

Figure 6.9 Hydrodynamic model of the examined FB.

ω_i, i = 1 ~ 10, are presented in Table 6.2. The structural mass matrix is presented here in relevant units of kilograms:

Structural mass matrix:

$$
\begin{bmatrix}
121770 & 0 & 0 & 0 & -7306 & 0 \\
0 & 121770 & 0 & 7306 & 0 & 0 \\
0 & 0 & 121770 & 0 & 0 & 0 \\
0 & 7306 & 0 & 380361 & 0 & 0 \\
-7306 & 0 & 0 & 0 & 9147801 & 0 \\
0 & 0 & 0 & 0 & 0 & 9498060
\end{bmatrix}
$$

Table 6.2 RAOs of surge, heave and roll motions of the floating body.

Examined frequency (rad/sec)	Amplitude surge (m/m)	Phase surge	Amplitude heave (m/m)	Amplitude pitch (rad/m)
0.3	5.492251E-02	−9.297858E+01	2.102620E-02	3.825566E-03
0.4	7.325104E-02	4.529470E+01	4.389928E-02	1.175001E-02
0.7	6.352055E-02	−1.384740E+02	9.143695E-02	2.937846E-02
0.8	7.821205E-02	8.853335E+01	1.240117E-01	4.913766E-02
1.0	1.816568E-01	−8.920647E+01	2.312770E-01	8.352493E-02
1.25	5.652103E-01	−9.011538E+01	5.272623E-01	8.732279E-02
1.5	9.323410E-01	−9.004243E+01	7.221388E-01	7.591391E-02
2.0	1.156997E+00	−9.002109E+01	7.962372E-01	6.791725E-02
2.5	2.321401E+00	−9.000238E+01	9.396843E-01	4.004760E-02
3.0	3.167019E+00	−9.000117E+01	9.667482E-01	3.016314E-02

6.9.2 Example 6.2

For the combined wind/wave SFC concept, please provide the values of all the presented quantities of Table 6.3 for a physical model that will be built at 1:50 scale. For the quantities, use Froude laws of similitude. Moreover, please provide an equivalent stiffness value for the mooring lines, and select the mooring line characteristics for the scaled model.

Based on Froude's laws of similitude, the factors in Table 6.4 have to be used for the estimation of the physical model values. It is very useful to explain how these factors have been calculated; for the case of the power produced by the PTO of the WECs, the unit that corresponds to the power is equal to N*m*sec/degree, since the factor for force

Table 6.3 Properties of a SFC in full scale.

Property	Full scale	1:50 value
Semisubmersible platform		
Diameter of the centre and outer columns (m)	6.5	0.130
Height of the pontoon (m)	6	0.120
Width of the pontoon (m)	9	0.180
Distance from the centre line of the centre column to the edge of the pontoon (m)	45.5	0.910
Draft (m)	30	0.600
Freeboard (m)	20	0.400
Water depth (m)	200	4.000
Steel mass of semisubmersible (kg) (wind turbine is included)	2,387,620	19.101
Flap-type WECs		
Length of the flap (m)	20	0.400
Height of the flap (m)	7	0.140
Elliptical axis of flap (m)	3.5	0.070
Distance of the upper part of the flap from SWL (m)	2	0.040
Distance of the left part of the flap from central column (m)	15	0.300
Steel mass of each flap (kg)	75,000	0.560
Buoyancy of each flap (kg)	395,000	3.160
Damping coefficient of power take-off mechanism (N*m*sec/degree)	2,250,000	0.051
Average power for one WEC (kW)	150	1.697E-4
Whole system		
Total mass, including ballast (total) (kg)	11,336,400	90.69
Centre of gravity Z (m)	−18.33	−0.367
Centre of buoyancy Z (m)	−21.27	−0.425
Steel mass (kg)	2,917,620	23.34
Ballast mass (kg)	2,290,000	18.32

SFC, Semisubmersible wind energy and flap-type wave energy converter; SWL, safe working load; WEC, wave energy converter.

Table 6.4 Estimated scale factors for use based on Froude's laws.

Variables	Scale factor	Value of scale factor in 1:50
Linear dimensions (length, height, width etc.)	λ	0.020
Mass	λ^3	8.000E-6
Time	$\lambda^{0.5}$	0.141
Force	λ^3	8.000E-6
Torque	λ^4	1.600E-7
Power	$\lambda^{3.5}$	1.131E-6
Angular stiffness	λ^4	1.600E-7
Angular damping	$\lambda^{4.5}$	2.263E-8

is λ^3, for lineal dimensions is λ, for time is $\lambda^{0.5}$ and for angular dimensions is λ; the scale factor for the power is $\lambda^{3.5}$. These factors can be used for the values of the corresponding physical quantities to be calculated. In Table 6.4, the values of the parameters that the physical model of SFC must have for its experimental testing in 1:50 scale are presented.

The estimation of the mooring lines stiffness will be achieved based on a mooring line numerical model. For the mooring line of the SFC, an offset equal to 2 m is given as excitation in both horizontal directions. The tension component in line with the horizontal displacement at the fairlead of the mooring line is measured equal to 1606 kN and 1302 kN for −2 m and 2 m offset, respectively. The tension component in the vertical direction at the fairlead of the mooring line is measured equal to 883 kN and 807 kN for −2 m and 2 m offset, respectively. Based on the estimated aforementioned response values, the equivalent horizontal and vertical mooring line stiffness is 76 kN/m and 19 kN/m, respectively. In scale values, the equivalent horizontal and vertical mooring line stiffness is 30.4 N/m and 7.6 kN/m, respectively. For any material that can provide these two values of stiffness, the mooring line physical modelling is expected to give reasonable results, especially for motions that are in the range of −2 m to 2 m. For larger expected offsets, relevant studies have to be performed.

References

Argyris, J., 1991. *Dynamics of Structures*. Amsterdam: Elsevier Science Publishers.

Bathe, K.-J., 1996. *Finite Element Procedures*. Upper Saddle River, NJ: Prentice Hall.

Biggs, J., 1964. *Introduction to Structural Dynamics*. New York: McGraw-Hill.

Butcher, J., 2008. *Numerical Methods for Ordinary Differential Equations*. Hoboken, NJ: John Wiley & Sons.

Chakrabarti, S., 2005. *Handbook of Offshore Engineering*. Amsterdam: Elsevier.

Cummins, W., 1962. *The Impulse Response Function and Ship Motions*. Bethesda, MD: Navy Department, David Taylor Model Basin.

Dean, R. D. R., 1991. *Water Wave Mechanics for Engineers and Scientists*. Singapore: World Scientific.

Det Norske Veritas, 2011. *Fatigue Design of Offshore Steel Structures*. DNV-RP-C203. Oslo: Det Norske Veritas.

Dhanak, M. X. N., 2016. *Handbook of Ocean Engineering*. New York: Springer.

Faltinsen, O. M., 1990. *Sea Loads on Ships and Offshore Structures*. Cambridge: Cambridge University Press.

Heller, V., 2011. Scale effects in physical hydraulic engineering models. *Journal of Hydraulic Research*, 293–306.

Hilber, H. H., 1977. Improved numerical dissipation for time integration algorithms in structural dynamics. *Earthquake Engineering & Structural Dynamics*, 283–292.

Hughes, T. J. R., 1987. *The Finite Element Method*. Englewood Cliffs, NJ: Prentice Hall.

Keulegan, G. H. & Carpenter, L. H., 1958. Forces on cylinders and plates in an oscillating fluid. *Journal of Research of the National Bureau of Standards*, 60, 423–440.

King, F., 2009. *Hilbert Transforms*. Cambridge: Cambridge University Press.

Lee, C. N. J., 2000. An assessment of hydroelasticity for very large hinged. *Journal of Fluids and Structures*, 957–970.

Lee, C. N. J., 2002. Boundary-element methods in offshore structure. *Journal of Offshore Mechanics and Arctic Engineering*, 81–89.

Lubliner, J., 2008. *Plasticity Theory*, rev. ed. New York: Dover.

Malvern, L., 1969. *Introduction to the Mechanics of a Continuous Medium*. Englewood Cliffs, NJ: Prentice Hall.

Michailides, C. A. D., 2012. Modeling of energy extraction and behavior of a flexible floating breakwater. *Applied Ocean Research*, 77–94.

Michailides, C. A. D., 2015. Optimization of a flexible floating structure for wave energy production and protection effectiveness. *Engineering Structures*, 249–263.

Michailides, C. G., 2016. Experimental study of the functionality of a semisubmersible wind turbine combined with flap-type wave energy converters. *Renewable Energy*, 675–690.

Naess, A. M. T., 2013. *Stochastic Dynamics of Marine Structures*. Cambridge: Cambridge University Press.

Newman, J., 1994. Wave effects on deformable bodies. *Applied Ocean Research*, 47–59.

Newmark, N., 1959. A method of computation for structural dynamics. *Journal of Engineering Mechanics*, 67–94.

Newton, I., 1729. *The Mathematical Principles of Natural Philosophy*. London: n.p.

NORSOK, 2007. *NORSOK Standard N-003: Actions and Action Effects*, 2nd ed. NORSOKN-003. Trondheim, Norway: NORSOK.

Senjanovic, I. M., 2008. Investigation of ship hydroelasticity. *Ocean Engineering*, 523–535.

Taghipour, R. P., 2008. Hybrid frequency–time domain models for dynamic response analysis of marine structures. *Ocean Engineering*, 685–705.

WAMIT, 2008. [Home page]. www.wamit.com

Wan, L. G., 2016. Comparative experimental study of the survivability of a combined wind and wave energy converter in two testing facilities. *Ocean Engineering*, 82–94.

Winterstein, S. U., 1993. Environmental parameters for extreme response: inverse form with omission factors. In: *Proceedings of the ICOSSAR-93*.

Zienkiewicz, O. C., 2006. *The Finite Element Method for Solid and Structural*. Amsterdam: Elsevier.

7

Numerical Methods in Offshore Fluid Mechanics

7.1 Introduction

Analyzing hydrodynamic loads on offshore structures can be performed by analytical models, numerical approaches and/or experimental tests. The numerical models are gradually becoming more and more reliable due to advances in computational tools and resources. Different approaches have been developed for numerical analysis so far, and potential theory–based models and computational fluid dynamic (CFD)-based models are becoming more popular. There are also some models under development, such as Lattice Boltzmann–based models and smooth particle hydrodynamic models, which are now mostly in the research phase. Also, the finite element approach has received some interest for hydrodynamic analysis of offshore structures.

In this chapter, we first briefly describe potential flow theory models, then discuss CFD models in detail. Since the potential flow theory model has been discussed extensively (see e.g. Faltinsen, 1993), the focus here is more on the CFD approach.

7.2 Potential Flow Theory Approach

Chapter 4, on ideal flow passing, very simple structures were studied. For example, to study a uniform flow passing a circular cylinder, the superposition of fundamental solutions (source, sink and uniform flow) has been used to generate the desired flow field. However, real offshore structures are more complicated; therefore, assessing the flow field around complicated structures needs more consideration. Figure 7.1 shows two platforms for a floating wind turbine that has been studied by the potential flow theory approach (Karimirad and Michailides, 2015; Gao *et al.*, 2016). Another problem that arises for studying offshore structures is the imposition of boundary conditions. In the simplified problem of uniform flow passing a cylinder, only the slip boundary condition on the cylinder needs to be satisfied. However, regarding offshore structures, in addition to slip boundary conditions on the surface of the body, free surface boundary conditions on the water's free surface and finite/infinite-depth boundary conditions on the sea floor need to be satisfied.

Modeling complex geometric structures by using potential flow theory can be addressed by, for example, using distribution of sources and sinks and setting the

Offshore Mechanics: Structural and Fluid Dynamics for Recent Applications, First Edition.
Madjid Karimirad, Constantine Michailides and Ali Nematbakhsh.
© 2018 John Wiley & Sons Ltd. Published 2018 by John Wiley & Sons Ltd.

(a) (b)

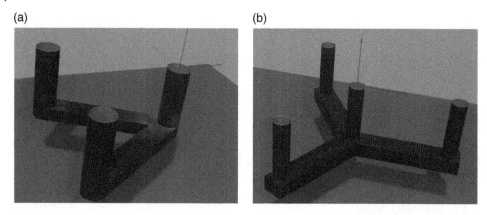

Figure 7.1 Two offshore platforms, discretized for hydrodynamic analysis based on the potential flow theory approach.

(a) (b) (c)

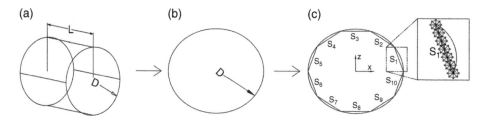

Figure 7.2 (a) Cylinder used to study the force heave motion. It is assumed that L >> D; therefore, it can be represented in the heave direction by (b) a 2D model. The slip boundary condition is imposed by defining the strength of the number of sources on the perimeter of the circle.

strength of these sources and sinks such that the proper boundary condition is imposed. Following Faltinsen (1993), let's assume a cylinder in an unbounded domain that has a known heave motion. The goal is to find the flow field around the cylinder, and corresponding added mass and damping in the heave direction. The cylinder is assumed to be infinitely long, so a two-dimensional (2D) model is enough for describing the heave motion. Although an analytical model for the above-mentioned problem is available, for learning purposes, let's try to study this problem by using sources and sinks. For simplicity, we use only the source, since the sink is just a source with negative value.

The strategy is to distribute the number of sources around the cylinder perimeter (circle), and we try to find the appropriate strength of each source to satisfy the slip boundary condition.

The velocity of the points on the perimeter of the cylinder (see Figure 7.2) can be written in the polar coordinate system as follows:

$$\frac{\partial \phi}{\partial r} = -\cos\theta |\eta_3| \omega \cos(\omega t) \; at \, r = R \; for \; 0 < \theta < 2\pi \qquad \text{Equation 7.1}$$

where η_3 is the amplitude of heave motion, ω is the frequency of the given heave motion and R is the radius of the cylinder. In a 2D model, a source can be defined as follows:

$$\phi_s = \frac{Q}{2\pi}\ln r \qquad\qquad \text{Equation 7.2}$$

where Q is the strength of the source, and r is the distance from the source. The sources are distributed over the perimeter of the cylinder; therefore, the potential field at any point in the domain can be calculated by integration of sources as follows:

$$\phi(x,z,t) = \int_S f(\alpha,\gamma)G(x,z,\alpha,\gamma)ds \qquad\qquad \text{Equation 7.3}$$

$$f(\alpha,\gamma) = Q(\alpha,\gamma)/2\pi$$

$$G(x,z,\alpha,\gamma) = \ln\left(\left(x-\alpha(s)\right)^2 + \left(z-\gamma(s)\right)^2\right)^{1/2}$$

where s is the distance from an arbitrary starting point of a line curve of a cylinder perimeter. Coordinates of points, where the sources are placed, are defined by $(\alpha(s), \gamma(s))$ as follows:

$$x_{PCyl} = \alpha(s)$$
$$x_{PCyl} = \gamma(s) \qquad\qquad \text{Equation 7.4}$$

So, the functions α and γ input a positive value based on the distance of traveling along the perimeter of the cylinder from a starting point. Output is the location point on the perimeter in an x–y coordinate system. We will assume that we are going to use 10 points to discretize the perimeter of the cylinder. Therefore, the cylinder will be approximated by a polygon with 10 sides, and on each side the potential field is assumed constant. Note that although 10 points are used for discretization, the sources are still uniformly distributed along the approximated body surface, and we just assume that it is constant over one-tenth of the cylinder. Therefore, Equation 7.3 can be written in the following form:

$$\phi(x,z,t) = \sum_{i=1}^{10} Q_i \int_{S_i} \frac{1}{2\pi}\ln\left(\left(x-\alpha(s)\right)^2 + \left(z-\gamma(s)\right)^2\right)^{1/2} ds \qquad \text{Equation 7.5}$$

Now we need to set Q_i, where i varies from 1 to 10, such that the slip boundary condition on the body surface (Equation 7.1) can be satisfied. In Equation 7.1, the term $-|\eta_3|\omega\cos(\omega t)$ is constant; therefore, all the unknown potential values can be normalized based on this constant, and we can write:

$$Q_i = -\bar{Q}_i|\eta_3|\omega\cos(\omega t) \qquad\qquad \text{Equation 7.6}$$

$$\phi(x,z,t) = -\bar{\phi}(x,z)|\eta_3|\omega\cos(\omega t)$$

$$\bar{\phi}(x,z) = \sum_{i=1}^{10} \bar{Q}_i \int_{S_i} \frac{1}{2\pi}\ln\left(\left(x-\alpha(s)\right)^2 + \left(z-\gamma(s)\right)^2\right)^{1/2} ds \qquad \text{Equation 7.7}$$

The boundary condition must be satisfied on all 10 points used to approximate the cylinder perimeter. Taking the derivative of Equation 7.7, and writing down the equation for all 10 points, results in the following set of equations:

$$\frac{\partial \bar{\phi}(x_1,z_1)}{\partial r} = \frac{\partial}{\partial r} \left(\frac{\bar{Q}_1}{2\pi} \int_{S_1} \ln\left(\left(x-\alpha(s)\right)^2 + \left(z-\gamma(s)\right)^2\right)^{1/2} ds + .. \right.$$

$$\left. + \frac{\bar{Q}_{10}}{2\pi} \int_{S_{10}} \ln\left(\left(x-\alpha(s)\right)^2 + \left(z-\gamma(s)\right)^2\right)^{1/2} ds \right) = \cos(\theta_1)$$

$$\frac{\partial \bar{\phi}(x_2,z_2)}{\partial r} = \frac{\partial}{\partial r} \left(\frac{\bar{Q}_1}{2\pi} \int_{S_1} \ln\left(\left(x-\alpha(s)\right)^2 + \left(z-\gamma(s)\right)^2\right)^{1/2} ds + .. \right.$$

$$\left. + \frac{\bar{Q}_{10}}{2\pi} \int_{S_{10}} \ln\left(\left(x-\alpha(s)\right)^2 + \left(z-\gamma(s)\right)^2\right)^{1/2} ds \right) = \cos(\theta_2)$$

....

....

$$\frac{\partial \bar{\phi}(x_{10},z_{10})}{\partial r} = \frac{\partial}{\partial r} \left(\frac{\bar{Q}_1}{2\pi} \int_{S_1} \ln\left(\left(z-\alpha(s)\right)^2 + \left(z-\gamma(s)\right)^2\right)^{1/2} ds + .. \right.$$

$$\left. + \frac{\bar{Q}_{10}}{2\pi} \int_{S_{10}} \ln\left(\left(x-\alpha(s)\right)^2 + \left(z-\gamma(s)\right)^2\right)^{1/2} ds \right) = \cos(\theta_{10}) \quad \text{Equation 7.8}$$

Since $\bar{Q}_i$ is assumed constant over the related element, Equation 7.8 can be written as a set of algebraic equations in the following form:

$$\bar{Q}_1 \frac{\partial}{\partial r} \left[\left(\int_{S_1} \ln\left(\left(x-\alpha(s)\right)^2 + \left(z-\gamma(s)\right)^2\right)^{1/2} ds \right) \right]_{x_1,z_1} + ..$$

$$+ \bar{Q}_{10} \frac{\partial}{\partial r} \left[\left(\int_{S_{10}} \ln\left(\left(x-\alpha(s)\right)^2 + \left(z-\gamma(s)\right)^2\right)^{1/2} ds \right) \right]_{x_1,z_1} = 2\pi \cos(\theta_1)$$

$$\bar{Q}_1 \frac{\partial}{\partial r} \left[\left(\int_{S_1} \ln\left(\left(x-\alpha(s)\right)^2 + \left(z-\gamma(s)\right)^2\right)^{1/2} ds \right) \right]_{x_2,z_2} + ..$$

$$+ \bar{Q}_{10} \frac{\partial}{\partial r} \left[\left(\int_{S_{10}} \ln\left(\left(x-\alpha(s)\right)^2 + \left(z-\gamma(s)\right)^2\right)^{1/2} ds \right) \right]_{x_2,z_2} = 2\pi \cos(\theta_2)$$

....

....

$$\bar{Q}_1 \frac{\partial}{\partial r} \left[\left(\int_{S_1} \ln\left(\left(x-\alpha(s)\right)^2 + \left(z-\gamma(s)\right)^2\right)^{1/2} ds \right) \right]_{x_{10},z_{10}} + ..$$

$$+ \bar{Q}_{10} \frac{\partial}{\partial r} \left[\left(\int_{S_{10}} \ln\left(\left(x-\alpha(s)\right)^2 + \left(z-\gamma(s)\right)^2\right)^{1/2} ds \right) \right]_{x_{10},z_{10}} = 2\pi \cos(\theta_{10})$$

$$\text{Equation 7.9}$$

Although imposing boundary conditions in a polar coordinate system is easier, calculating the derivative of integrals in a polar coordinate is not very straightforward. The partial derivative over r can be written as follows:

$$\frac{\partial(\ )}{\partial r} = \frac{\partial(\ )}{\partial x}\frac{\partial x}{\partial r} + \frac{\partial(\ )}{\partial z}\frac{\partial z}{\partial r}$$

$$\begin{cases} x = r\sin\theta \\ z = -r\cos\theta \end{cases}$$

$$\frac{\partial(\ )}{\partial r} = \frac{\partial(\ )}{\partial x}\sin\theta - \frac{\partial(\ )}{\partial z}\cos\theta \qquad\qquad \text{Equation 7.10}$$

Based on the above relations, we rewrite the set of Equation 7.9 in the following form:

$$\bar{Q}_1\left(\sin\theta_1 \frac{\partial}{\partial x}\left[\int_{S_1} f(x,z,s)ds\right]_{x_1,z_1} - \cos\theta_1 \frac{\partial}{\partial z}\left[\int_{S_1} f(x,z,s)ds\right]_{x_1,z_1}\right) + \ldots$$

$$+ \bar{Q}_{10}\left(\sin\theta_{10} \frac{\partial}{\partial x}\left[\int_{S_{10}} f(x,z,s)ds\right]_{x_1,z_1} - \cos\theta_{10} \frac{\partial}{\partial z}\left[\int_{S_{10}} f(x,z,s)ds\right]_{x_1,z_1}\right) = 2\pi\cos(\theta_1)$$

$$\bar{Q}_1\left(\sin\theta_1 \frac{\partial}{\partial x}\left[\int_{S_1} f(x,z,s)ds\right]_{x_2,z_2} - \cos\theta_1 \frac{\partial}{\partial z}\left[\int_{S_1} f(x,z,s)ds\right]_{x_2,z_2}\right) + \ldots$$

$$+ \bar{Q}_{10}\left(\sin\theta_{10} \frac{\partial}{\partial x}\left[\int_{S_{10}} f(x,z,s)ds\right]_{x_2,z_2} - \cos\theta_{10} \frac{\partial}{\partial z}\left[\int_{S_{10}} f(x,z,s)ds\right]_{x_2,z_2}\right) = 2\pi\cos(\theta_2)$$

......

......

$$\bar{Q}_1\left(\sin\theta_1 \frac{\partial}{\partial x}\left[\int_{S_1} f(x,z,s)ds\right]_{x_{10},z_{10}} - \cos\theta_1 \frac{\partial}{\partial z}\left[\int_{S_1} f(x,z,s)ds\right]_{x_{10},z_{10}}\right) + \ldots$$

$$+ \bar{Q}_{10}\left(\sin\theta_{10} \frac{\partial}{\partial x}\left[\int_{S_{10}} f(x,z,s)ds\right]_{x_{10},z_{10}} - \cos\theta_{10} \frac{\partial}{\partial z}\left[\int_{S_{10}} f(x,z,s)ds\right]_{x_{10},z_{10}}\right) = 2\pi\cos(\theta_{10})$$

$$f(x,z,s) = \ln\left(\left(x-\alpha(s)\right)^2 + \left(z-\gamma(s)\right)^2\right)^{1/2} \qquad\qquad \text{Equation 7.11}$$

Either the integration involved in the algebraic set of Equation 7.11 can be computed numerically, or, if $\alpha(s)$ and $\gamma(s)$ are sufficiently simple, it may be computed analytically. Therefore, Equation 7.11 is simplified to a set of equations:

$$\bar{Q}_1 A_{1-1} + \bar{Q}_2 A_{1-2} + \ldots + \bar{Q}_{10} A_{1-10} = B_1$$
$$\bar{Q}_1 A_{1-2} + \bar{Q}_2 A_{2-2} + \ldots + \bar{Q}_{10} A_{2-10} = B_2$$

....

$$\bar{Q}_1 A_{10-1} + \bar{Q}_2 A_{10-2} + \ldots + \bar{Q}_{10} A_{10-10} = B_{10}$$

Equation 7.12

Or, simply: $\bar{Q}_j A_{ij} = B_i$ or $A_{ij} \bar{Q}_j = B_i$, A_{ij} can be interpreted as the influence of source j for imposing boundary condition i in the numerical domain. Finally, we reach a 10×10 matrix with constant coefficients. A linear set of equations with constant coefficients is easy to solve.

After finding $\bar{Q}_i$ from Equation 7.11, we can write the normalized potential function in the general form of Equation 7.7, and hence obtain velocity and pressure in the domain. Using the Bernoulli equation and neglecting the nonlinear term (velocity square), the pressure can be written in the following form:

$$p = -\rho \frac{\partial \phi}{\partial t} = -\rho |\eta_3| \omega^2 \sin(\omega t) \bar{\phi}(x,z)$$

Equation 7.13

where $\bar{\phi}(x,z)$ is known from Equation 7.7. To calculate the added mass in the heave direction, the z component of force induced by hydrodynamic pressure is calculated:

$$p = -\rho \frac{\partial \phi}{\partial t} = -\rho |\eta_3| \omega^2 \sin(\omega t) \bar{\phi}(x,z)$$

$$F_3 = -\int_s p \cos\theta ds = -\int_s \rho |\eta_3| \omega^2 \sin(\omega t) \bar{\phi}(x,z) \cos\theta ds = -\rho |\eta_3| \omega^2 \sin(\omega t) \sum_{i=1}^{10} \int_{s_i} \bar{\phi}(x,z) \cos\theta ds$$

Equation 7.14

Therefore, added mass and damping coefficients in the heave direction are equal to:

$$A_{33} = -\sum_{i=1}^{10} \int_{s_i} \bar{\phi}(x,z) \cos\theta ds$$

$$B_{33} = 0$$

Equation 7.15

Since there is no load proportional to the velocity of heave motion, the damping coefficient in heave is equal to zero in this example. Numerical or analytical models can be used for the integration involved for calculating A_{33}.

7.2.1 Three-dimensional Problem

The above problem was solved for a 2D approximation of a cylinder in the case that there is no boundary condition other than a slip boundary on the body surface. Also, the structure's geometry was very simple. Offshore structures can be more complicated in shape, and often the free surface and finite-depth boundary conditions need to be satisfied. Furthermore, the wave load response of the offshore structure needs to be calculated. So, normally for an offshore structure, the following equation and boundary conditions need to be satisfied. If we assume a fixed large structure in the ocean, the following equation and boundary conditions can be written (Sarpkaya and Isaacson, 1981):

$$\frac{\partial^2 \phi}{\partial x^2} + \frac{\partial^2 \phi}{\partial y^2} + \frac{\partial^2 \phi}{\partial z^2} = 0 \qquad\qquad \text{Equation 7.16}$$

$$\frac{\partial^2 \phi}{\partial t^2} + g\frac{\partial \phi}{\partial z} = 0 \; at \; z = 0 \qquad\qquad \text{Equation 7.17}$$

$$\frac{\partial \phi}{\partial t} + g\xi = 0 \; at \; z = 0 \qquad\qquad \text{Equation 7.18}$$

$$\frac{\partial \phi}{\partial n} = 0 \; at \; the \; surface \; of \; solid \qquad\qquad \text{Equation 7.19}$$

$$\frac{\partial \phi}{\partial z} = 0 \; at \; z = -h \;\; or \; |\nabla\phi| \to 0 \; at \; z \to -\infty \qquad\qquad \text{Equation 7.20}$$

Furthermore, we need to make sure that the ϕ will converge to zero as the waves outgo far from the structure. It can be fair to assume that the waves are harmonic (at least in linear theory approximation), hence the resultant motion also needs to be harmonic. Therefore, the potential function of such a problem can be written as:

$$\phi(x,y,z) = \text{Re}\left\{\phi'(x,y,z)e^{-i\omega t}\right\} \qquad\qquad \text{Equation 7.21}$$

The potential function is possibly a summation of incoming waves plus the effects of the waves scattered from the structure, so it can be decomposed into the following components with the same frequency of oscillation:

$$\phi'(x,y,z) = \phi'_I + \phi'_s \qquad\qquad \text{Equation 7.22}$$

The incident potential is the given potential of wave that we are studying, and it is known from Stoke's flow theory. Following the previous example, we will try to find an appropriate scattering potential function by distribution of sources on the surface of the body. Following the general pattern of source function, the following general formula can be considered as a starting point:

$$\phi'_s = \frac{1}{4\pi}\int_s f(\alpha)G(\alpha,x)ds \qquad\qquad \text{Equation 7.23}$$

Equation 7.23 is very similar to Equation 7.3. The main difference is that the function in Equation 7.23 needs to satisfy not only the Laplace equation (this time in 3D) but also free surface kinematic and dynamic boundary conditions; also, the boundary conditions at the sea floor (Equations 7.17, 7.18 and 7.20) need to be satisfied. Therefore, Equation 7.23 needs some rather complicated corrections with respect to the simple source written in Equation 7.3. John (1950) developed a formula as follows, which is a solution to the Laplace equation and satisfies the mentioned boundary condition in the form of (Sarpkaya and Isaacson, 1981):

$$G(\alpha,x) = \frac{1}{R} + \frac{1}{R'} + 2\int_0^\infty \frac{(\mu+\upsilon)e^{-\mu d}\cosh(\mu(\gamma+d))\cosh(\mu(z+d))}{\mu\sinh(\mu d)-\upsilon\cosh(\upsilon d)}J_0(\mu r)d\mu$$
$$-iC_0\cosh(k(\gamma+d))\cosh(k(z+d))J_0(kr)$$

Equation 7.24

where:

$$R = \left[(x-\alpha)^2+(y-\beta)^2+(x-\gamma)^2\right]^{1/2}$$
$$R' = \left[(x-\alpha)^2+(y-\beta)^2+(z+2d+\gamma)^2\right]^{1/2}$$
$$r = \left[(x-\alpha)^2+(y-\beta)^2\right]^{1/2}$$

Equation 7.25

$$\upsilon = k\tanh(kd)$$
$$C_0 = \frac{2\pi(\upsilon^2-k^2)}{(k^2-\upsilon^2)d+\upsilon}$$

Or $G(\alpha,x)$ can be written as follows:

$$G(\alpha,x) = -iC_0\cosh(k(\gamma+d))\cosh(k(z+d))H_0^{(1)}(kr)$$
$$+4\sum_{i=1}^\infty C_m\cos(\mu_m(\gamma+d)\cos(\mu_m(z+d))k_0(\mu_m r)$$

Equation 7.26

where k_0 is the modified Bessel function (second kind and zero order), and c_m is:

$$c_m = \frac{\mu_m^2+\upsilon^2}{(\mu_m^2+\upsilon^2)d-\upsilon}$$

Equation 7.27

where μ_m is the real root of Equation 7.28:

$$\mu_m\tan(\mu_m d)+\upsilon = 0$$

Equation 7.28

The function $G(\alpha,x)$ is called the *Green function*. Efficiently finding the Green function and integration involved in this equation is an important and time-consuming step in hydrodynamic analysis of offshore structures based on potential flow theory. After finding the Green function for different sections of the surface, an algebraic set of equations similar to Equation 7.12 will be set up to impose the no-through-boundary condition on the surface of the solid. After finding the strengths of the Green function, the potential function can be derived, and added mass and damping of the structure are easy to calculate, very similar to the procedure given for a circular cylinder in heave. Also, the force and moment on the body can be calculated by following similar steps defined in the example of heave motion of a cylinder in unbounded fluid.

7.2.2   Numerical Consideration

The method of distributing the source function on the solid's surface (source function method) to satisfy the slip boundary condition may not give unique results in all the wave frequencies; unphysical results may be obtained in some wave frequencies, which is known as the *irregular frequencies problem*. This problem usually arises for surface-piercing structures. Immersed bodies below the water usually don't have such problems. Fortunately, the problem arises often for incoming waves, which are in the order of characteristic size of the offshore structure. Usually, such waves are not important for design of the offshore structure; however, this still needs to be taken into account. It should be noted that this is a fully numerical problem and doesn't resemble physical behavior. For example, an analytical solution for a potential flow field around a surface-piercing cylinder is available in all the range of waves' frequencies without any issue.

The number of grids that are used to discretize the body surface highly affects the results. The best method for numerical analysis is to systematically increase the number of grids until convergence is reached. Usually, using one-eighth of a wave length (as a characteristic length) can give reasonable results. Although different types of elements can be used for discretizing the domain, quadrilaterals are the most common elements used for discretizing the body surface. A structure's curvatures may need more refined grids for discretization.

Finally, other than source techniques, a combination of sources and dipoles can also be used for discretization; it is slightly more complicated but can handle nonlinear free surface waves. For more information, see Newman (1977) and Faltinsen (1993).

7.3 CFD Approach

Significant efforts have been made for about half a century to develop reliable tools based on potential flow theory. This approach is quite popular since it is low in computational cost, and it is also very reliable if the assumptions required for this approach are satisfied. These assumptions are linear motion of the structure, linear or weakly nonlinear waves, and limited effects of damping.

There is a more robust method based on solving the equation of motion for all the elements of fluid. It is called the Navier–Stokes (NS) equation and is based on momentum conservation. The NS equation is more general, can handle nonlinear motion of the structure and nonlinear waves, and also can consider the effects of damping. However, it comes with considerably higher computational cost (Nematbakhsh *et al.*, 2015). In a 3D problem, the only outer surface of the structure needs to be discretized; therefore, the number of grids is N^2. However, in the CFD model, the whole domain needs to be discretized; therefore, the number of required grid points is N^3. Prior to recent advances in computer processing, CFD model applications were limited to understanding the physics of flow, and the simulations were usually limited to 2D models. Recently, however, they have been used for practical applications. To study hydrodynamics of offshore structures by using CFD techniques, different approaches can be used. The approaches can be divided into immersed boundary methods and body-fitted grid methods. In the immersed boundary methods, an offshore structure, whether fixed or floating, is immersed in the structured grids, while in the body-fitted method, a solid body is considered as a boundary for the structures and usually a re-meshing technique will be used if the structure is moving. In this section, we start with a numerical solution of the NS equation in the simplest form without any offshore structure or free surface involved in the problem. Then, we will study the effects of free surfaces and the presence of offshore structures on the model.

7.3.1 Discretization of the Navier–Stokes Equation on Rectangular Structured Grids

The NS equations, in two dimensions for incompressible single-phase flow, can be written as follows:

$$\rho\left(\frac{\partial u}{\partial t} + u\frac{\partial u}{\partial x} + v\frac{\partial u}{\partial y}\right) = -\frac{\partial p}{\partial x} + \frac{\partial \tau_{xx}}{\partial x} + \frac{\partial \tau_{xy}}{\partial y}$$

$$\rho\left(\frac{\partial v}{\partial t} + u\frac{\partial v}{\partial x} + v\frac{\partial v}{\partial y}\right) = -\frac{\partial p}{\partial y} + \frac{\partial \tau_{yx}}{\partial x} + \frac{\partial \tau_{yy}}{\partial y} + \rho g$$

Equation 7.29

The equations in Equation 7.29 consist of four major terms that need to be discretized: the temporal term, advection term, pressure term and viscous term. We will try to explain the discretization of these four terms by the following example:

Let's assume that we have a cavity full of fluid, and the top of the cavity is sliding with uniform velocity. The other sides are walls that do not let the fluid go through or slip on it (a no-slip boundary condition). The schematic of this domain is shown in Figure 7.3.

The task is to numerically solve the NS equation to obtain the pressure and the flow field in the steady-state condition. The cavity problem is one of the simplest benchmark problems used in CFD (Ghia *et al.*, 1982), and it is usually used for checking the accuracy of the CFD models that are written from scratch. The cavity box is discretized with square grids in here, and the NS equation needs to discretized on those grids (Figure 7.3).

The first term on the left-hand side of Equation 7.29 is the temporal term $\frac{\partial u}{\partial t}$. The easiest approach is to write it in the following form, which is first order in time:

$$\frac{\partial u}{\partial t} = \frac{u^{n+1} - u^n}{\Delta t} + O\left(\Delta t^2\right)$$

Equation 7.30

Higher order discretization techniques come with higher computational cost. Methods such as the predictor-corrector method (which is second order in time) can be used, or high-order Runge–Kutta methods can be used. Usually, for normal application, second-order accuracy will give sufficiently reliable results. The NS equation in the x-direction, using Equation 7.30 for discretizing the temporal, can be written as follows:

$$u^{n+1} = u^n + \Delta t\left(-u\frac{\delta u}{\delta x} - v\frac{\delta u}{\delta y} - \frac{1}{\rho}\frac{\delta p}{\delta x} + \frac{\delta \tau_{xx}}{\delta x} + \frac{\delta \tau_{xy}}{\delta y}\right)$$

Equation 7.31

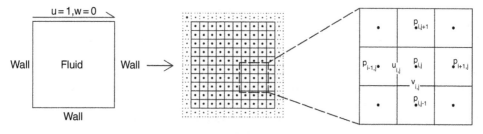

Figure 7.3 Cavity problem and a structured grid used for discretization of the domain.

We can use the predictor-corrector method as follows:

$$u^{n+1^*} = u^n + \Delta t \left(-u \frac{\delta u^n}{\delta x} - v \frac{\delta u^n}{\delta y} - \frac{1}{\rho} \frac{\delta p^n}{\delta x} + \frac{\delta \tau^n_{xx}}{\delta x} + \frac{\delta \tau^n_{xy}}{\delta y} \right)$$

$$u^{n+2^*} = u^{n+1^*} + \Delta t \left(-u \frac{\delta u^{n+1^*}}{\delta x} - v \frac{\delta u^{n+1^*}}{\delta y} - \frac{1}{\rho} \frac{\delta p^{n+1^*}}{\delta x} + \frac{\delta \tau^{n+1^*}_{xx}}{\delta x} + \frac{\delta \tau^{n+1^*}_{xy}}{\delta y} \right)$$

$$u^{n+1} = \frac{u^{n+2^*} + u^n}{2}$$

Equation 7.32

It can be seen that the computer cost is nearly twice that of a first-order method (Equation 7.30). The method is also second order in time. In this method, we use a first-order method in two consecutive time steps and then average the velocities in time. In other words, we first guess the results with first-order methods and then correct our prediction by averaging the results.

7.3.2 Advection Terms

Advection terms can be discretized using the first-order upwind method. Higher order methods are also available, such as the second-order essentially none oscillatory (ENO) method and third-order ENO method (Harten, 1997). Here, the first-order upwind method will be described.

The advection term for the *u* velocity is as follows:

$$u \frac{\partial u}{\partial x} + v \frac{\partial u}{\partial y}$$

Equation 7.33

The upwind method uses a first-order discretization of derivate. The important point about the upwind method is that the direction of the discretization depends on the flow direction and is written as:

$$u \frac{\partial u}{\partial x}\bigg|_{i,j} = \begin{cases} u_{i,j} \dfrac{u_{i,j} - u_{i-1,j}}{h} & if \ u_{i,j} > 0 \\ u_{i,j} \dfrac{u_{i+1,j} - u_{i,j}}{h} & if \ u_{i,j} < 0 \end{cases}$$

Equation 7.34

$$v \frac{\partial u}{\partial y}\bigg|_{i,j} \approx \begin{cases} 0.25\left(v_{i,j} + v_{i,j+1} + v_{i-1,j} + v_{i-1,j+1}\right)\dfrac{u_{i,j} - u_{i,j-1}}{h} & if \ v_{i,j} > 0 \\ 0.25\left(v_{i,j} + v_{i,j+1} + v_{i-1,j} + v_{i-1,j+1}\right)\dfrac{u_{i,j+1} - u_{i,j}}{h} & if \ v_{i,j} < 0 \end{cases}$$

Equation 7.35

The reader may wonder why we do not simply use the central difference method, which is second order in space and more straightforward. The reasons are that advection terms behave like hyperbolic equations, and using central difference will lead to instability of the solution. For more discussions about stability analysis of hyperbolic equations, please see Ferziger and Peric (2012).

7.3.3 Viscous Terms

The viscous term for the NS equation in the x-direction may be written as:

$$\frac{\partial \tau_{xx}}{\partial x} + \frac{\partial \tau_{xy}}{\partial y} \qquad \text{Equation 7.36}$$

where τ_{xx} is normal stress in the x-direction, and τ_{xy} is tangential stress in the x-direction on a plane with a normal vector in the y-direction. The viscous stress can be written as:

$$\frac{\partial \tau_{xx}}{\partial x} = \frac{\partial(2\mu D_{xx})}{\partial x} = \frac{\partial}{\partial x}\left(2\mu \frac{1}{2}\left(\frac{\partial u}{\partial x} + \frac{\partial u}{\partial x}\right)\right) = \frac{\partial}{\partial x}\left(2\mu \frac{\partial u}{\partial x}\right)$$

$$\frac{\partial \tau_{xy}}{\partial y} = \frac{\partial(2\mu D_{xy})}{\partial y} = \frac{\partial}{\partial y}\left(2\mu \frac{1}{2}\left(\frac{\partial u}{\partial y} + \frac{\partial v}{\partial x}\right)\right) = \frac{\partial}{\partial y}\left(\mu\left(\frac{\partial u}{\partial y} + \frac{\partial v}{\partial x}\right)\right)$$

$$\text{Equation 7.37}$$

where D_{xx} and D_{xy} are the rates of strain in the corresponding directions. To discretize the viscosity term, the second-order central difference method will be used.

$$\frac{\partial}{\partial x}\left(2\mu \frac{\partial u}{\partial x}\right) = \frac{\left(2\mu \frac{\partial u}{\partial x}\right)\Big|_{i+1/2,j} - \left(2\mu \frac{\partial u}{\partial x}\right)\Big|_{i-1/2,j}}{h} = \frac{2\mu \frac{u_{i+1,j}-u_{i,j}}{h} - 2\mu \frac{u_{i,j}-u_{i-1,j}}{h}}{h}$$

$$\frac{\partial}{\partial y}\left(\mu\left(\frac{\partial u}{\partial y} + \frac{\partial v}{\partial x}\right)\right) = \frac{\partial}{\partial y}\left(\mu \frac{\partial u}{\partial y}\right) + \frac{\partial}{\partial y}\left(\mu \frac{\partial v}{\partial x}\right) = \frac{\mu \frac{\partial u}{\partial y}\Big|_{i,j+1/2} - \mu \frac{\partial u}{\partial y}\Big|_{i,j-1/2}}{h}$$

$$= \frac{\mu \frac{u_{i,j+1}-u_{i,j}}{h} - \mu \frac{u_{i,j}-u_{i,j-1}}{h}}{h} + \frac{\mu \frac{\partial v}{\partial x}\Big|_{i,j+1/2} - \mu \frac{\partial v}{\partial x}\Big|_{i,j-1/2}}{h}$$

$$= \frac{\mu \frac{v_{i,j+1}-v_{i-1,j+1}}{h} - \mu \frac{v_{i,j}-v_{i-1,j}}{h}}{h} \qquad \text{Equation 7.38}$$

The viscosity terms reduce the momentum of the flow and result in damping of the flow field. Therefore, usually viscous terms, in contrast with advection terms, will not result in instability of the problem and even stabilize the solution.

7.3.4 Pressure Term and Mass Conservation Equation

The pressure term leads to a force in the negative direction of the gradient of the pressure's magnitude. The force direction is in agreement with common sense that flow moves from a high-pressure field to a low-pressure field. The pressure term can be discretized by a second-order central difference method in the following form in the x-direction:

$$\frac{1}{\rho}\frac{\partial p}{\partial x} \approx \frac{1}{\rho}\frac{p_{i,j}-p_{i-1,j}}{h} \qquad \text{Equation 7.39}$$

In contrast with velocity terms in NS equations, the pressure term does not include any temporal term; and, in fact, the NS equations themselves are not closed equations. For a 2D problem, there exist three unknowns (u, v and p) with two equations. A mass conservation equation will be used to close the system of equations. The mass conservation equation for incompressible flows can be written as follows:

$$\nabla \cdot u = \frac{\partial u}{\partial x} + \frac{\partial v}{\partial y} = 0 \ (2D\ flow)$$

Equation 7.40

The mass conservation equation is not directly discretized and solved; instead, it is used "indirectly" to correct the pressure term.

7.3.5 Solving Navier–Stokes Equations

The general strategy for solving NS equations (Equation 7.29) is to initially guess a value for the pressure term and predict the x- and y-directions' velocities. Then, based on the mass conservation equation, correct the initial guess for the pressure term and correct the velocity predictions accordingly.

NS equations are discretized as follows. Based on values assigned to the advection, pressure and viscous terms, initial predictions for velocities are obtained:

$$u^* = u^n + \Delta t \left(-u\frac{\delta u^n}{\delta x} - v\frac{\delta u^n}{\delta y} - \frac{1}{\rho}\frac{\delta p^n}{\delta x} + \frac{\delta \tau_{xx}^n}{\delta x} + \frac{\delta \tau_{xy}^n}{\delta y} \right) = u^n + \Delta t H_x^n$$

$$v^* = v^n + \Delta t \left(-u\frac{\delta v^n}{\delta x} - v\frac{\delta v^n}{\delta y} - \frac{1}{\rho}\frac{\delta p^n}{\delta y} + \frac{\delta \tau_{yx}^n}{\delta x} + \frac{\delta \tau_{yy}^n}{\delta y} \right) = v^n + \Delta t H_y^n$$

Equation 7.41

Since the pressure term is assigned from the previous time step, the resultant velocities are not necessarily divergence free (mass conservation is not satisfied). So, we need to correct the pressure term to obtain velocities that are divergence free. The unknown pressure correction term is denoted by p'. Using p', we can write the following equations:

$$u^{n+1} = u^n + \Delta t H_x^n - \frac{1}{\rho}\frac{\partial p'}{\partial x}$$

$$v^{n+1} = v^n + \Delta t H_y^n - \frac{1}{\rho}\frac{\partial p'}{\partial y}$$

Equation 7.42

Now, we take the derivative with respect to x from the u component of Equation 7.42 and with respect to y from the v component of Equation 7.42. Then, we sum up two equations; in other words, we are taking divergence from the NS equation in vector form.

$$\frac{\partial u^{n+1}}{\partial x} + \frac{\partial v^{n+1}}{\partial y} = \frac{\partial u^n}{\partial x} + \frac{\partial v^n}{\partial y} + \Delta t\left(\frac{\partial H_x^n}{\partial x} + \frac{\partial H_y^n}{\partial y}\right) - \left(\frac{\partial}{\partial x}\left(\frac{1}{\rho}\frac{\partial p'}{\partial x}\right)\right) - \left(\frac{\partial}{\partial y}\left(\frac{1}{\rho}\frac{\partial p'}{\partial y}\right)\right)$$

Equation 7.43

We recall that the resultant velocities at time step $n+1$ should conserve the mass (Equation 7.40). Also, the velocities at time step n conserve the mass. Therefore, Equation 7.44 reduces to:

$$\left(\frac{\partial}{\partial x}\left(\frac{1}{\rho}\frac{\partial p'}{\partial x}\right)\right) + \left(\frac{\partial}{\partial y}\left(\frac{1}{\rho}\frac{\partial p'}{\partial y}\right)\right) = \Delta t\left(\frac{\partial H_x^n}{\partial x} + \frac{\partial H_y^n}{\partial y}\right)$$

Equation 7.44

Since the density is assumed constant, the equation can be written as:

$$\frac{\partial^2 p'}{\partial x^2} + \frac{\partial^2 p'}{\partial y^2} = \frac{\Delta t}{\rho}\left(\frac{\partial H_x^n}{\partial x} + \frac{\partial H_y^n}{\partial y}\right)$$

Equation 7.45

Equation 7.45 is the standard type of Poisson equation, which is an elliptical type of equation. This is usually the most time-consuming equation needed to be solved in each time step when updating the fluid velocities in CFD methods. Discretization and solving this equation will be described in Section 7.3.6, and for now we suppose that Equation 7.45 is solved and p' is found. After finding p' from the Poisson equation, the velocities will be updated as follows:

$$u^{n+1} = u^* - \frac{1}{\rho} \frac{p'_{i,j} - p'_{i-1,j}}{h}$$

$$v^{n+1} = v^* - \frac{1}{\rho} \frac{p'_{i,j} - p'_{i,j-1}}{h}$$

Equation 7.46

The obtained velocities from Equation 7.46 are divergence free. Note that since we have used velocities from previous time steps for advection and viscous terms, the time steps should be relatively small to keep the solution stable. In the above-mentioned method, other than the temporal term, all the velocities are approximated by the previous time step; therefore, an explicit equation for updating the velocity is available. That is the reason why that method is called the *explicit method*. Implicit methods for solving NS equations are also available. In the implicit methods, the values for the advection and viscous terms are studied at time step $n + 1$; therefore, the equation is more stable and larger time steps can be taken. But an explicit equation for velocity at time step $n + 1$ is not available anymore. More complicated methods, such as the SIMPLE, SIMPLEC, SIMPLER or fractional step method, can be used. For more information on using implicit methods for solving NS equations, see Ferziger and Peric (2012).

7.3.6 Poisson Equation

Due to the importance of solving the Poisson equation in CFD methods, this subsection will specifically discuss it. The Poisson equation is a well-known equation in mathematics, and it arises not only in fluid mechanics but also in other branches of physics such as electrostatic and heat transfer. Since solving this equation is computationally expensive, much research has been devoted to reduce the time to solve this equation, and some advanced methods, such as the biconjugate gradient methods, multigrid method and Krylov method, have been developed. There are some advanced libraries that particularly deal with efficiently solving Poisson-type equations (Falgout and Yang, 2002). Usually, for 2D problems, simple approaches are enough to obtain a Poisson equation solution. Here, a simple iterative method will be discussed.

Equation 7.45 can be discretized as follows:

$$\frac{\partial^2 p}{\partial x^2} + \frac{\partial^2 p}{\partial y^2} = \alpha$$

$$\frac{\left.\frac{\partial p}{\partial x}\right|_{i+1/2,j} - \left.\frac{\partial p}{\partial x}\right|_{i-1/2,j}}{h} + \frac{\left.\frac{\partial p}{\partial y}\right|_{i,j+1/2} - \left.\frac{\partial p}{\partial y}\right|_{i,j-1/2}}{h} = \alpha$$

Equation 7.47

$$\frac{\frac{p_{i+1,j} - p_{i,j}}{h} - \frac{p_{i,j} - p_{i-1,j}}{h}}{h} + \frac{\frac{p_{i,j+1} - p_{i,j}}{h} - \frac{p_{i,j} - p_{i,j-1}}{h}}{h} = \alpha$$

$$\frac{1}{h^2}\left(p_{i+1,j} + p_{i-1,j} + p_{i,j+1} + p_{i,j-1} - 4p_{i,j}\right) = \alpha$$

It can be seen that the Poisson equation cannot be solved for each element separately in the numerical domain. For ease of demonstration, let's write Equation 7.47 as:

$$p_{i,j-1} + p_{i-1,j} - 4p_{i,j} + p_{i+1,j} + p_{i,j+1} = h^2\alpha_{i,j} = \beta_{i,j} \qquad \text{Equation 7.48}$$

Writing Equation 7.48 for all the grids in the domain leads to:

$$p_{1,0} + p_{0,1} - 4p_{1,1} + p_{21} + p_{1,2} = \beta_{1,1}$$
$$p_{2,0} + p_{1,1} - 4p_{2,1} + p_{3,1} + p_{2,2} = \beta_{2,1}$$

.....

.....

$$p_{n,0} + p_{n-1,1} - 4p_{n,1} + p_{n+1,1} + p_{n,2} = \beta_{n,1}$$
$$p_{1,1} + p_{0,2} - 4p_{1,2} + p_{22} + p_{1,3} = \beta_{1,2} \qquad \text{Equation 7.49}$$

.....

.....

.....

$$p_{n,n-1} + p_{n-1,n} - 4p_{n,n} + p_{n+1,n} + p_{n,n+1} = \beta_{n,n}$$

Equation 7.49 in matrix form can be written as follows:

$$
\begin{bmatrix}
-4\;1........1... \\
1-4\;1........1... \\
0\;1-4\;1.........1.. \\
0\;0\;1-4\;1........1... \\
.... \\
..................1-4\;1.........1.. \\
.... \\
1...........................1-4\;1.........1... \\
0\;1..........................1-4\;1.........1....................................... \\
0\;0\;1.........................1-4\;1.........1.................................... \\
.... \\
.... \\
.................1.......................1-4\;1.........1......................... \\
.................1.........................1-4\;1.........1..................... \\
.................1...........................1-4\;1.........1................. \\
.... \\
.... \\
.... \\
.... \\
.... \\
....1.........................1-4
\end{bmatrix}
\begin{bmatrix}
p_{11} \\ p_{21} \\ p_{3,1} \\ p_{41} \\ \\ p_{n,1} \\ p_{1,2} \\ p_{2,2} \\ p_{3,2} \\ .... \\ ..... \\ .... \\ p_{i,j} \\ p_{i,+1j} \\ p_{i,+2j} \\ \\ \\ \\ \\ p_{in,n}
\end{bmatrix}
=
\begin{bmatrix}
\beta_{11} \\ \beta_{21} \\ \beta_{3,1} \\ \beta_{41} \\ \\ \beta_{n,1} \\ \beta_{1,2} \\ \beta_{2,2} \\ \beta_{3,2} \\ .... \\ ..... \\ .... \\ \beta_{i,j} \\ \beta_{i,+1j} \\ \beta_{i,+2j} \\ \\ \\ \\ \\ p_{n,n}
\end{bmatrix}
$$

Equation 7.50

Equation 7.50 is obtained for uniform grids with uniform fluid density. For non-uniform grids and multiphase flow problems, the coefficient of pressure terms (the left matrix in Equation 7.50) will be a function of the cell size and density of the fluid. Therefore, in a more general format, Equation 7.50 can be written as the coefficients of the pressure term in the south, west, east and north of the central pressure term (A_S, A_W, A_E, A_N, A_P) as follows:

$$
\begin{bmatrix}
A_P & A_E & \dots\dots & A_N & \dots\dots\dots\dots\dots\dots\dots\dots\dots\dots\dots\dots\dots\dots\dots \\
A_W & A_P & A_E & \dots\dots & A_N & \dots\dots\dots\dots\dots\dots\dots\dots\dots\dots\dots\dots \\
0 & A_W & A_P & A_E & \dots\dots & A_N & \dots\dots\dots\dots\dots\dots\dots\dots\dots \\
0\ 0 & A_W & A_P & A_E & \dots\dots & A_N & \dots\dots\dots\dots\dots\dots\dots \\
\multicolumn{1}{c}{} & \dots\dots\dots\dots\dots\dots\dots\dots\dots\dots\dots \\
\multicolumn{1}{c}{} & A_W & A_P & A_E & \dots\dots & A_N & \dots\dots\dots \\
\multicolumn{1}{c}{} & \dots\dots\dots\dots\dots\dots\dots\dots\dots\dots \\
A_S & \dots\dots & A_W & A_P & A_E & \dots\dots & A_N & \dots\dots \\
0 & A_S & \dots\dots & A_W & A_P & A_E & \dots\dots & A_N & \dots \\
0\ 0 & A_S & \dots\dots & A_W & A_P & A_E & \dots\dots & A_N \\
\multicolumn{1}{c}{} & \dots\dots\dots\dots\dots\dots\dots \\
\multicolumn{1}{c}{} & \dots\dots\dots\dots\dots\dots\dots \\
\multicolumn{1}{c}{} & A_S & \dots\dots & A_W & A_P & A_E & \dots\dots & A_N \\
\multicolumn{1}{c}{} & A_S & \dots\dots & A_W & A_P & A_E & \dots\dots & A_N \\
\multicolumn{1}{c}{} & A_S & \dots\dots & A_W & A_P & A_E & \dots\dots \\
\multicolumn{1}{c}{} & \dots\dots\dots\dots\dots\dots \\
\multicolumn{1}{c}{} & \dots\dots\dots\dots\dots\dots \\
\multicolumn{1}{c}{} & \dots\dots\dots\dots\dots\dots \\
\multicolumn{1}{c}{} & A_S & \dots\dots & A_W & A_P
\end{bmatrix}
\begin{bmatrix}
p_{11} \\ p_{21} \\ p_{3,1} \\ p_{41} \\ \\ p_{n,1} \\ p_{1,2} \\ p_{2,2} \\ p_{3,2} \\ \dots \\ \dots \\ \dots \\ p_{i,j} \\ p_{i,+1j} \\ p_{i,+2j} \\ \\ \\ \\ p_{n,n}
\end{bmatrix}
=
\begin{bmatrix}
\beta_{11} \\ \beta_{21} \\ \beta_{3,1} \\ \beta_{41} \\ \\ \beta_{n,1} \\ \beta_{1,2} \\ \beta_{2,2} \\ \beta_{3,2} \\ \dots \\ \dots \\ \dots \\ \beta_{i,j} \\ \beta_{i,+1j} \\ \beta_{i,+2j} \\ \\ \\ \\ \beta_{n,n}
\end{bmatrix}
$$

Equation 7.51

Equation 7.51 can be written in the form of $AP = \beta$, which is a linear algebraic equation. A is an $n \times n$ matrix, and the solution to this equation is $P = A^{-1}\beta$.

Finding A^{-1} is not trivial, due to the large size of matrix A. A simple iterative method can be used to solve matrix Equation 7.51; it is called the *Jacobi method*, in which the values for each point are guessed as follows:

$$X_P^{n+1} = \frac{\beta_P - A_S X_S^n - A_W X_W^n - A_N X_N^n - A_E X_E^n}{A_P}$$

Equation 7.52

This equation needs to be solved for all the unknown values (pressure) in the numerical domain. After sweeping all the points, only one iteration is completed, and we need to use the new obtained values for new guessing. We continue this guessing until two successive iterations lead to lower differences of the threshold that we are interested in.

Although this method is very robust, it is extremely slow; a simple but efficient improvement is to use the updated guesses as they are solved. Hence, Equation 7.52 can be written as:

$$P_P^{n+1} = \frac{B_p - A_S P_S^{n+1} - A_W P_W^{n+1} - A_N P_N^n - A_E P_E^n}{A_P}$$ Equation 7.53

This method is called the *Gauss–Seidel method*, and it reaches convergence twice as fast as the simple Jacobi method. Another improvement is to use the *overrelaxation method*, which is based on relying on the fact that the direction of variation of the values of the variable P in each time step is probably correct, and we may accelerate reaching the solution by the following formula:

$$P_P^{n+1} = R_{Rex} \frac{B_p - A_S P_S^{n+1} - A_W P_W^{n+1} - A_N P_N^n - A_E P_E^n}{A_P} + (1 - R_{Rex}) P_P^n$$ Equation 7.54

where R_{Rex} is the relaxation factor and should be greater than one for increase in the convergence speed. This method is called the *successive overrelaxation (SOR) method*. Finding an optimized value for the overrelaxation factor is not trivial; however, a value between 1.2 and 1.3 can be chosen as the starting point. Note that using $R_{Rex} = 1$ will change the SOR method back to the Gauss–Seidel method.

After becoming familiar with the Poisson solution procedure, for our particular problem (cavity problem), we need to apply proper boundary conditions to the Poisson equation. The boundary conditions need to be imposed to the four sides of the cavity. The boundary condition at the walls can be determined by simplifying the NS equations very close to the wall. Since no slip boundary condition is imposed on the right- and left-side walls, $u = 0$, $v = 0$, and derivatives of v remain constant and equal to zero. Similar comment can be made for velocity in the y-direction except at the top velocity of $u = 1$. Therefore, for the cavity problem, the NS equations can be simplified at the walls as follows:

$$\frac{\partial p}{\partial x} = \frac{\mu}{\rho} \frac{\partial^2 u}{\partial x^2} \quad at \ x = 0, x = L$$ Equation 7.55

$$\frac{\partial p}{\partial y} = \frac{\mu}{\rho} \frac{\partial^2 u}{\partial y^2} \quad at \ y = 0, y = L$$ Equation 7.56

Now the Poisson equation with the above-given boundary conditions can be numerically solved. The boundary condition just leads to some changes in the coefficients of matrix A in Equation 7.51. The results of the u velocity at the midline of the cavity are shown in Figure 7.4a for a Reynolds number equal to 100. Also, the iso-pressure contours are shown in Figure 7.4b.

Extension of numerical discretization of the NS equation on rectangular structured grids to three dimensions is straightforward and only requires some more discretization for added terms in the z direction. The cavity problem, although a purely numerically designed problem, is a very valuable test for checking the accuracy of the original codes written for solving the NS equations.

7.3.7 The Effects of Free Surface

Free surface waves in the CFD models can be studied as an interface between two phases of flow (air and water). In other words, they can be studied as a multiphase flow

(a)   (b)

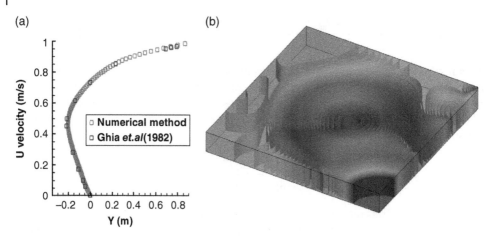

Figure 7.4 Cavity problem for a Reynolds number equal to 100: (a) shows the results of comparison with another numerical model; and (b) iso-pressure contours in the cavity box.

problem. Multiphase flow problems in CFD have a long history; they have been used for different problems such as studying droplet impact on the solid surface, modeling bubbles in the gas chambers and studying jet flow in engines and even oil spills in the ocean. Here, the multiphase models will be used to study hydrodynamic loads on offshore structures. To handle these wide ranges of problems, significant effort has been made starting mainly from the late 1970s, and different approaches were proposed.

The approaches that have been developed for multiphase flow are the volume of fluid (VOF) method (Hirt and Nichols, 1981), level set method (Osher and Sethian, 1988), front-tracking method (Unverdi and Tryggvason, 1992), phase field method (Steinbach et al., 1996) and a constraint interpolation polynomial (CIP) approach (Yabe et al., 2001). Although all of the methods are applicable for modeling wave free surfaces in offshore engineering, the first two are more popular due to their simplicity and robustness for this particular application, and they will be discussed here. For more information about numerical modeling of multiphase flow, please see Tryggvason *et al.* (2011).

7.3.8 Volume of Fluid Method

The VOF method can be considered the first method proposed for modeling two-phase flow problems. It is based on assigning a marker function to each grid in the numerical domain: for example, assigning a value equal to zero to water and one to air. Regarding the cells that are partly filled with water, the value will be based on the filled volume ratio. Figure 7.5 shows assigning marker function to different cells in the numerical domain.

We are trying to develop an equation to track this marker function in time. We note that a particle considered as one type of fluid (e.g. water) remains the same in time. The same goes for the other type of fluid (air). Therefore, a Lagrangian derivative (total derivative) of this marker function is equal to zero. If we name the marker value C, Equation 7.57 is valid:

$$\frac{Dc}{Dt} = 0$$

Equation 7.57

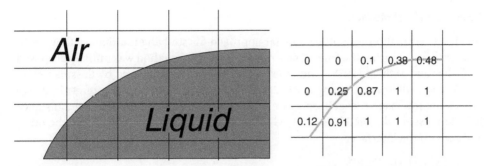

0	0	0.1	0.38	0.48
0	0.25	0.87	1	1
0.12	0.91	1	1	1

Figure 7.5 Marker function used in the VOF method to distinguish between different fluids.

This equation is called the *transport equation* and can be written in the Eulerian form as follows (for 2D problems):

$$\frac{\partial c}{\partial t} + u\frac{\partial c}{\partial x} + v\frac{\partial c}{\partial y} = 0 \qquad \text{Equation 7.58}$$

To track the interface of two-phase flow in the model (water free surface), in addition to solving the NS equation, Equation 7.58 needs to be solved and results in updated values for the cells in the numerical domain.

After solving Equation 7.58, the new values for cells will be obtained. Suppose that for one cell, the value is updated from to 0.4 to 0.5. There are different lines with different slopes, which can give $C = 0.5$. To find the right slope, we can find the normal direction of the interface slope, by calculating the gradient of the marker function:

$$n = \frac{\nabla C}{|\nabla C|} \qquad \text{Equation 7.59}$$

The reason is that the maximum variation of the marker function happened normal to the interface of the fluids. Based on the value and slope of the marker function, the new position of the marker function can be plotted in the numerical grid (Gueyffier *et al.*, 1999).

Another important note regarding the VOF method is that the marker function is a discontinuous function. It means that the slopes at the cells' interface break from one value to another. This makes the discretization of Equation 7.58 not very straightforward. To have a conservative approach in each time step, the advection terms in Equation 7.58 need to calculated based on the position of the cut cell. For more details on discretizing the advection terms in the VOF method, see Gueyffier *et al.* (1999). After updating the interface position, finally the properties of the cells will be updated for the new time step as follows:

$$\rho = C\rho_W + (1-C)\rho_A$$
$$\mu = C\mu_W + (1-C)\mu_A, \qquad \text{Equation 7.60}$$

where ρ_W and ρ_A are densities of water and air, and μ_W and μ_A are the dynamic viscosity of them. The VOF method can conserve the mass very well, but it is not the best method for modeling the interface's curvature due to discontinuity of the marker function. Also, discretization of the VOF method is not as straightforward as that of the level set method, which will be described in details in Section 7.3.9.

7.3.9 Level Set Method

The level set method is another popular approach for studying multiphase flow problems such as capturing the free surface of the waves. The method was initially proposed by Osher and Sethian (1988), and afterward considerably improved by Sussman *et al.* (1994) and Fedkiw *et al.* (Kang *et al.*, 2000). The method gets lots of interest in image processing for shape recognition as well. This method is based on defining a distance function, usually denoted by $\phi(\vec{x},t)$, which is positive at one side, negative at the other and zero on the interface.

$$\begin{cases} \phi(\vec{x},t) > 0 \ \ Fluid\,1 \\ \phi(\vec{x},t) < 0 \ \ Fluid\,2 \\ \phi(\vec{x},t) = 0 \ \ interface \end{cases}$$

Equation 7.61

The absolute value of the distance function is equal to minimum distance from the interface. Tracking the distance function in time can result in capturing the interface of two phases of flow. Figure 7.6 shows the concept of the level set function for free surface of a wave.

As with the VOF method, the particle on the interface should remain on the interface. Therefore, the transport equation is valid for the points on the interface.

$$\frac{D\phi}{Dt} = \frac{\partial \phi}{\partial t} + u\frac{\partial \phi}{\partial x} + v\frac{\partial \phi}{\partial y} = 0 \ \ at \ \phi = 0$$

Equation 7.62

Equation 7.62, however, is not valid for other points in the numerical domain. But we need to have information about the level set function of other points to discretize and solve Equation 7.62. To handle this problem, we solve Equation 7.62 for the whole domain and then try to rearrange the level set function of other points, except $\phi = 0$ to the distance function. Therefore, another equation called the reinitialization equation is used after solving the level set equation (Equation 7.62) to rearrange the level set function in other points of the domain. The reinitialization equation is in the following form:

$$\phi_t + S(\phi_0)\left(|\nabla \phi| - 1\right) = 0$$

$$S(\phi_0) = \frac{\phi_0}{\sqrt{\phi_0^2 + (\Delta x)^2}}$$

Equation 7.63

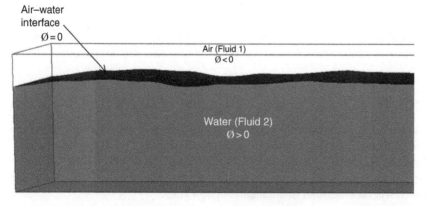

Figure 7.6 Definition of the level set function for free surface flow.

Equation 7.63 will be solved in a few fictitious time steps to reach a steady-state condition. Hence, the level set function returns back to the distance function. Note that $S(\phi_0)$ is based on results obtained from Equation 7.62. It can be seen that by moving toward the interface, $S(\phi_0)$ reaches zero. This is the location where Equation 7.62 is valid.

When the level set function is updated by Equation 7.62 and Equation 7.63, it will be used to update the properties in the numerical domain (very similar to the VOF method; see Equation 7.60) as follows:

$$\rho(x,t) = \rho_w + f(\phi)\left(\rho_a - \rho_w\right)$$
$$\mu(x,t) = \mu_w + f(\phi)\left(\mu_a - \mu_w\right)$$

Equation 7.64

where $f(\phi)$ is a function for smoothly varying from one phase to another. Sharp transition of a fluid's properties from one to the other, especially for high-density-ratio fluids (air–water), will result in creation of non-physical spurious velocity at the interface. Different kinds of smoothing function can be used. An example of the smoothing function is:

$$f(\phi) = 0.5 + 0.5\frac{\phi^3 + 1.5\varepsilon^2\phi}{\left(\phi^2 + \varepsilon^2\right)^{1.5}}$$

Equation 7.65

Equation 7.65 gives a smoothing function between 0 and 1, and in the limit where ϕ is much larger than ε, the function become equal to 1 for positive ϕ and 0 for negative ones. Some cutoff values might also be used to limit the transition region. For example, we can set values less than 0.001 and greater than 0.999 to be equal to 0 and 1, accordingly.

Generally speaking, to solve a multiphase flow problem with the level set method, the NS equation is solved and the velocities are updated. The updated velocities are given as an input to the level set function (Equation 7.62) to update the latter. The updated level set function is corrected by the reinitialization equation (Equation 7.63) and is used in the next time step to update the fluid's properties, density and viscosity (Equation 7.64). In Section 7.3.10, we will discuss details of discretization of the level set function.

7.3.10 Discretization of Level Set Function

The level set equation (Equation 7.62) includes a temporal term and advection terms. The temporal term can be discretized by the second-order predictor corrector (mentioned in this chapter) for discretizing NS equations (Equation 7.32).

The advection terms can also be discretized in the same way as advection terms of NS, by using a first-order upwind method. However, the level set method already suffers from the fact that it is strictly correct only on the interface. Therefore, if low-order methods such as the first-order upwind method are used, we may lose considerable accuracy of the interface position prediction. Therefore, it is highly advised to use higher order methods for discretizing the advection terms. In this regard, families of methods called *ENO methods* can be used (Harten, 1997). ENO methods can be efficiently used to obtain second-, third- and fifth-order approaches. The basic idea of all of them is to use the Newton interpolation polynomial, and if we need higher order methods, we will try to use more terms in this polynomial approximation.

The Newton polynomial theorem is as follows: suppose that we have function $\phi(x)$, and for this function we know $n + 1$ distinct points from #0 to #n. The approximation of the polynomial to the power n of the function $\phi(x)$ can be written as follows (Klee, 2000):

$$\phi(x) \approx b_0 + b_1(x - x_{\#0}) + b_2(x - x_{\#0})(x - x_{\#1}) + \ldots\ldots\ldots + b_n(x - x_{\#0})(x - x_{\#1})\ldots\ldots(x - x_{\#n-1})$$

Equation 7.66

where it can be shown that:

$$b_0 = \phi[x_{\#0}] = \phi(x_0)$$

$$b_1 = \phi[x_{\#1}, x_{\#0}] = \frac{\phi[x_{\#1}] - \phi[x_{\#0}]}{x_{\#1} - x_{\#0}} = \frac{\phi(x_{\#1}) - \phi(x_{\#0})}{x_{\#1} - x_{\#0}}$$

$$b_2 = \phi[x_{\#2}, x_{\#1}, x_{\#0}] = \frac{\phi[x_{\#2}, x_{\#1}] - \phi[x_{\#1}, x_{\#0}]}{x_{\#2} - x_{\#0}} = \frac{\dfrac{\phi[x_{\#2}] - \phi[x_{\#1}]}{x_{\#2} - x_{\#1}} - \dfrac{\phi[x_{\#1}] - \phi[x_{\#0}]}{x_{\#1} - x_{\#0}}}{x_{\#2} - x_{\#0}}$$

$$= \frac{\dfrac{\phi(x_{\#2}) - \phi(x_{\#1})}{x_{\#2} - x_{\#1}} - \dfrac{\phi(x_{\#1}) - \phi(x_{\#0})}{x_{\#1} - x_{\#0}}}{x_{\#2} - x_{\#0}}$$

. . . .

. . . .

$$b_n = \phi[x_{\#n}, \ldots, x_{\#1}, x_{\#0}] = \frac{\phi[x_{\#n}, \ldots, x_{\#1}] - \phi[x_{\#n-1}, \ldots, x_{\#0}]}{x_{\#n} - x_{\#0}}$$

$$= \frac{\dfrac{\phi[x_{\#n}, \ldots, x_{\#2}] - \phi[x_{\#n-1}, \ldots, x_{\#1}]}{x_{\#n} - x_{\#1}} - \dfrac{\phi[x_{\#n-1}, \ldots, x_{\#1}] - \phi[x_{\#n-2}, \ldots, x_{\#0}]}{x_{\#n-1} - x_{\#0}}}{x_{\#n} - x_{\#0}}$$

$$= \ldots .$$

Equation 7.67

We are using the "#" symbol to make clear that no specific ordering for $x_{\#0}, x_{\#1}, \ldots\ldots, x_{\#n}$ is required. Let's say we are interested in using the Newtonian polynomial of power two for the approximate potential function $\phi(x)$:

$$\phi(x) = b_0 + b_1(x - x_{\#0}) + b_2(x - x_{\#0})(x - x_{\#1}) \qquad \text{Equation 7.68}$$

In theory, any points in the numerical grid can be used for approximation; naturally, we are going to select the nodes closer to the interested point x_i. Since we are dealing with a hyperbolic function in advection terms, we are going to choose the polynomial points wisely to prevent instability. Taking the derivative with respect to x from Equation 7.68 results in:

$$\phi_x(x_i) = b_1 + b_2(2x_i - (x_{\#0} + x_{\#1})) \qquad \text{Equation 7.69}$$

where ϕ_x is the derivative with respect to x. Since we are interested in computing the derivative of ϕ at x_i, it is natural to choose $x_{\#0} = x_i$. The second point $x_{\#1}$ can be chosen from either x_{i+1} or x_{i-1} as follows:

$$\phi_x(x_i) = b_1$$

$$x_{\#0} = x_i$$

$$\begin{cases} \text{if select } x_{\#1} = x_{i-1} \rightarrow b_1 = \dfrac{x_{i-1} - x_i}{-h} = \dfrac{x_i - x_{i-1}}{h} = D^-\phi(x_i) \\[2mm] \text{if select } x_{\#1} = x_{i+1} \rightarrow b_1 = \dfrac{x_{i+1} - x_i}{h} = D^+\phi(x_i) \end{cases}$$

Equation 7.70

Note that the Newtonian polynomial is just used to fit a polynomial passing through the given points. We are only trying to use this polynomial for our discretization in advection terms. So it can be seen that if we use the second point on the left side of the first point, the derivative of the Newtonian polynomial resembles the backward first-order approximation, denoted by $D^-\phi(x_i)$; and, choosing from the right side, it resembles the forward first-order one, denoted by $D^+\phi(x_i)$. Based on the upwind method, we can use the second point in the Newtonian interpretation of $\phi_x(x_i)$:

$$
\begin{cases}
\text{if } u > 0 \quad \phi_x(x_i) = b_1 = \dfrac{x_i - x_{i-1}}{h} = D^-\phi(x_i) \\[2mm]
\text{if } u < 0 \quad \phi_x(x_i) = b_1 = \dfrac{x_{i+1} - x_i}{h} = D^+\phi(x_i)
\end{cases}
$$

Equation 7.71

Regarding the second approximation, we will select b_2, which leads to minimum oscillation from the first-order approximation, because we are interested in a more stable solution. So, we can compare two reasonable choices: one is to select the next point for the interpolation at the right side of the other first two points, and the other choice is to select from the left side.

$$
\phi_x(x_i) = b_1 + b_2\left(2x_i - (x_{\#0} + x_{\#1})\right)
$$

$$
\text{if } D^-\phi(x_i)
\begin{cases}
\text{if select } x_{\#2} = x_{i+1} \rightarrow b_2 = \dfrac{\dfrac{\phi(x_{\#2}) - \phi(x_{\#1})}{x_{\#2} - x_{\#1}} - \dfrac{\phi(x_{\#1}) - \phi(x_{\#0})}{x_{\#1} - x_{\#0}}}{x_{\#2} - x_{\#0}} \\[4mm]
\qquad = \dfrac{\phi(x_{i+1}) - 2\phi(x_i) + \phi(x_{i-1})}{2h^2} = D^2\phi(x_i) \\[4mm]
\text{if select } x_{\#2} = x_{i-2} \rightarrow b_2 = \dfrac{\dfrac{\phi(x_{\#2}) - \phi(x_{\#1})}{x_{\#2} - x_{\#1}} - \dfrac{\phi(x_{\#1}) - \phi(x_{\#0})}{x_{\#1} - x_{\#0}}}{x_{\#2} - x_{\#0}} \\[4mm]
\qquad = \dfrac{\phi(x_i) - 2\phi(x_{i-1}) + \phi(x_{i-2})}{2h^2} = D^2\phi(x_{i-1})
\end{cases}
$$

$$
\text{if } D^+\phi(x_i)
\begin{cases}
\text{if select } x_{\#2} = x_{i+2} \rightarrow b_2 = \dfrac{\dfrac{\phi(x_{\#2}) - \phi(x_{\#1})}{x_{\#2} - x_{\#1}} - \dfrac{\phi(x_{\#1}) - \phi(x_{\#0})}{x_{\#1} - x_{\#0}}}{x_{\#2} - x_{\#0}} \\[4mm]
\qquad = \dfrac{\phi(x_{i+2}) - 2\phi(x_{i+1}) + \phi(x_i)}{2h^2} = D^2\phi(x_{i+1}) \\[4mm]
\text{if select } x_{\#2} = x_{i-1} \rightarrow b_2 = \dfrac{\dfrac{\phi(x_{\#2}) - \phi(x_{\#1})}{x_{\#2} - x_{\#1}} - \dfrac{\phi(x_{\#1}) - \phi(x_{\#0})}{x_{\#1} - x_{\#0}}}{x_{\#2} - x_{\#0}} \\[4mm]
\qquad = \dfrac{\phi(x_{i+1}) - 2\phi(x_i) + \phi(x_{i-1})}{2h^2} = D^2\phi(x_i)
\end{cases}
$$

Equation 7.72

Choosing any of these terms will increase the accuracy of discretization to second order. However, in our application (level set method), the stability might be different. To have the most stable solution, we will use the one that leads to less variation to have minimum oscillation in results (ENO). Therefore, the second-order approximation can be written as follows:

$$if\ u > 0\ \phi_x(x_i) = D^-\phi(x_i) + \min\left(\left|D^2\phi(x_i)\right|, \left|D^2\phi(x_{i-1})\right|\right)\left(2x_i - \left(x_{\#0} + x_{\#1}\right)\right)$$
$$if\ u < 0\ \phi_x(x_i) = D^+\phi(x_i) + \min\left(\left|D^2\phi(x_{i+1})\right|, \left|D^2\phi(x_i)\right|\right)\left(2x_i - \left(x_{\#0} + x_{\#1}\right)\right)$$

Equation 7.73

The same method will be used to discretize the advection term in the y-direction for Equation 7.62. The method can be expanded to higher order terms of the Newtonian polynomial to reach higher order approximation, but usually for modeling free surface waves, the second-order method gives reasonable results. For studying the higher discretization methods for the level set function, see Osher and Fedkiw (2006).

7.3.11 Discretization of Reinitialization Equation

After discretizing the level set function, the reinitialization equation (Equation 7.63) needs to be discretized. This equation can be written in the nearly same form of the level set equation, as follows (Kang et al., 2000):

$$\frac{\partial \phi}{\partial t} + \left(\frac{s(\phi_0)\phi_x}{\sqrt{\phi_x^2 + \phi_y^2}}\right)\phi_x + \left(\frac{s(\phi_0)\phi_y}{\sqrt{\phi_x^2 + \phi_y^2}}\right)\phi_y = s(\phi_0)$$

Equation 7.74

Now, based on the sign of the parenthesis in front of the spatial derivatives, Equation 7.74 can be discretized (similar to the ENO method). Since the denominator of the parenthesis in Equation 7.74 is always greater than zero, the nominators can be used as an indicator of the flow direction:

$$\begin{cases} if\ s(\phi_0)\phi_x^- > 0\ \&\ s(\phi_0)\phi_x^+ > 0 \rightarrow \phi_x \approx \phi_x^- \\ if\ s(\phi_0)\phi_x^- < 0\ \&\ s(\phi_0)\phi_x^+ < 0 \rightarrow \phi_x \approx \phi_x^+ \\ if\ s(\phi_0)\phi_x^- < 0\ \&\ s(\phi_0)\phi_x^+ > 0 \rightarrow \phi_x \approx 0 \\ if\ s(\phi_0)\phi_x^- > 0\ \&\ s(\phi_0)\phi_x^+ < 0 \rightarrow \begin{cases} if\ S > 0\ \rightarrow \phi_x \approx \phi_x^- \\ if\ S < 0\ \rightarrow \phi_x \approx \phi_x^+ \end{cases} \end{cases}$$

Equation 7.75

where S is defined as:

$$S = \frac{s(\phi_0)\left(\left|\phi_x^+\right| - \left|\phi_x^-\right|\right)}{\phi_x^+ - \phi_x^-}$$

Equation 7.76

where S is an attempt to estimate the direction of derivatives inside the cell. The same discretization method will be used in the y-direction, and the second-order predictor-corrector method will be used for the temporal term. A time step equaling 0.5 of the characteristic size of grid points at the interface can be used to satisfy the CFL condition

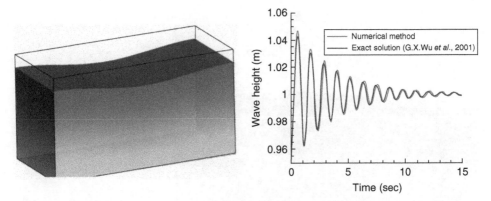

Figure 7.7 Modeling a standing wave with the level set method and comparing the numerical results with an analytical solution.

(Kang *et al.*, 2000) and match Equation 7.74 in time. Three to five fictitious time steps are usually used to reinitialize the level set function to the distance function. Considering that the reinitialization equation is designed to move with the unit velocity in the normal direction to the interface, this number of time steps can transport the level set function 1.5 to 2.5 grid points toward the interface. The time step size and number of fictitious time steps may be increased or decreased based on the complexity of the interface. Usually, sharper and more complex interface problems require a higher number of fictitious time steps to compensate the deviation of the level set function from the distance function.

Level set and reinitialization equations are usually solved in the whole domain. If we are solving these equations for a numerical wave tank, we need boundary conditions for different sides of the flume. Extrapolation is a reasonable approach to obtain the level set function values at the boundary points on different sides, since it also nearly follows the distance function concept on which the level set method is constructed.

In Figure 7.7, the results of the wave elevation of a standing wave with high viscosity are computed by the level set method (Nematbakhsh *et al.*, 2013) and compared with an exact analytical solution provided by Wu *et al.* (2001), and very good agreement is obtained.

7.3.12   Studying Solid–Fluid Interaction

One of the important missions of CFD models is to calculate the fluid's loads on solid objects. This solid object can be an airplane, a car or an offshore structure. Solid–fluid interaction problems in the classical CFD approach can be studied by two general approaches. The first is the body-fitted grid approach, and the second one is the immersed boundary method. In the body-fitted grid approach, the grid points are constructed around the structure; hence, the solid point is a boundary for the grids. Therefore, the boundary conditions necessary for introducing a solid body in the fluid domain, such as a no-slip boundary condition, can be easily imposed. However, if the structure starts to move, the grid point needs to be redefined, and the values need to be transferred from old grids to the new grid points, unless the solid-body motion is

Figure 7.8 Immersed boundary grid points used to study hydrodynamic loads on a wind turbine.

extremely small. Some other approaches, such as the overset grid method (Meakin, 1994), also exist in which two sets of grid points are available, one on the background and another body-fitted grid that overcomes some problems of the simple body-fitted method, but the information still needs to be interpolated between the sets of grid points.

Another approach is the immersed boundary approach in which the solid body is immersed in the numerical domain. Therefore, the structure is free to move, and no re-meshing technique or interpolation is required. Also, the grid generation is very straightforward. However, the main challenge in this approach is imposing the boundary condition at the solid–fluid interface, since the grid point locations are not necessary at the locations where the boundary condition needs to be imposed. Figure 7.8 shows the immersed boundary method used for hydrodynamic analysis of an offshore wind turbine (Nematbakhsh et al., 2015).

In this section, initially the immersed boundary method will be described, and then the body-fitted grid approach will be discussed in detail.

7.3.13 Immersed Boundary Methods

The immersed boundary method is a popular approach for studying solid–fluid interaction since the grid generation and discretization of the numerical scheme are straightforward. Also, faster and more robust solvers can be used. In the standard immersed boundary method, very similar to the VOF method, a marker function will be used to distinguish between the cells occupied by the solid and fluid cells. The marker function can be defined as follows:

$$\begin{cases} C = 1 \, Solid \, cells \\ C = 0 \, Fluid \, cells \end{cases}$$

Equation 7.77

Sharp variation of cells from solid to fluid usually leads to unphysical velocities at the solid–fluid interface. Therefore, a smoothing function can be employed to avoid such problems. For example, the following smoothing function can be used, very similar to the smoothing function used in Equation 7.65. For example, if the considered solid is a circle in 2D flow, the smooth marker function can be written as follows:

$$C(X)=0.5+0.5\frac{\left(R-|X|\right)^{3}+1.5\varepsilon^{2}\left(R-|X|\right)}{\left(\left(R-|X|\right)^{2}+\varepsilon^{2}\right)^{1.5}}$$

Equation 7.78

where X is the vector from the center of a circle to any point in the domain, R is the radius of the cylinder and ε is a variable defining the smoothing length.

In the immersed boundary method, initially we consider the whole domain as fluid, and NS equations are solved for the whole domain, including the cells that are occupied by the solid cells; then, additional constraints are imposed on cells that are occupied by the solid. The NS equation can be written as follows for the immersed boundary method:

$$\rho\left(\frac{\partial u}{\partial t}+u\frac{\partial u}{\partial x}+v\frac{\partial u}{\partial y}\right)=-\frac{\partial p}{\partial x}+\frac{\partial \tau_{xx}}{\partial x}+\frac{\partial \tau_{xy}}{\partial y}+f_{x}$$

$$\rho\left(\frac{\partial v}{\partial t}+u\frac{\partial v}{\partial x}+v\frac{\partial v}{\partial y}\right)=-\frac{\partial p}{\partial y}+\frac{\partial \tau_{yx}}{\partial x}+\frac{\partial \tau_{yy}}{\partial y}+\rho g+f_{y}$$

Equation 7.79

This is the standard NS equation, plus a force term on the right-hand side that is due to the presence of the solid in the numerical domain.

After solving the NS equation, without a forcing term (f_x and f_y), an approximation of the velocities is obtained. The additional constraints that need to be imposed on solid cells are: solid body motion for the solid cells, and a no-penetration boundary condition at the solid–fluid interface. The solid body motion, if the solid body is fixed or has a predefined motion, will be given explicitly to the cells occupied by the solid.

If the solid body is free to float, like floating wind turbines, the solid body linear and angular velocities need to be calculated. This can be done by integration of linear and angular velocities of the cells occupied by the solid, as follows:

$$m_{s}u_{S_{cg}}=\int_{\Omega}C_{s}u\rho d\Omega$$

$$I_{S_{cg}}\omega_{s}=\int_{\Omega}C_{s}\left(r\times u\right)\rho d\Omega$$

$$I_{S_{cg}}=\begin{bmatrix}I_{xx}-I_{xy}-I_{xz}\\-I_{yx}\,I_{yy}-I_{yz}\\-I_{zx}-I_{zy}\,I_{zz}\end{bmatrix}$$

Equation 7.80

where m_s is the solid mass, $u_{S_{cg}}$ is the solid velocity at the center of gravity and $I_{S_{cg}}$ is the mass moment of inertia of the solid with respect to the center of gravity. Solving Equation 7.80, $u_{S_{cg}}$ and ω_s are obtained and used to correct the velocity of cells occupied by the solid. The cells' velocities will be corrected by using the following formula:

$$u_{Cor}=u_{Ini}+C_{s}\left(u_{S_{cg}}+\left(\omega_{s}\times r\right)-u_{Ini}\right)$$

Equation 7.81

The differences between the prediction of solid cell velocities by NS equations (Equation 7.79) and the velocities explicitly given to the solid cells are, in fact, the $f_x \Delta t / \rho$ and $f_y \Delta t / \rho$ on the right-hand side of Equation 7.79. Now that the velocities are updated, the location of the solid also can be updated by calculating two points of the solid as follows:

$$X_s^{n+1} = X_s^n + \left(u_{S_{og}}^{n+1} + \left(\omega_s^{n+1} \times r \right) \right) \Delta t \qquad \text{Equation 7.82}$$

Since the solid is assumed to be rigid, based on these two points, the location of the other points of the solid can be determined.

$$C_s^{n+1} = SolidFinder\left(X_s^{n+1} \right) \qquad \text{Equation 7.83}$$

The solid finder can be a simple function for a cylinder or circle, but it may be more complicated for other structures. Different mathematical algorithms can be used to determine if a point in the domain belongs to inside or outside of the solid (Ogayar *et al.*, 2005).

After velocities in the whole domain and solid position are updated, we need to check if a no-penetration boundary condition is imposed on the solid cells. It can be shown that if divergence of the whole domain is equal to zero and the solid body has rigid motion, the no-penetration boundary condition is satisfied. However, since the velocities are modified by using Equation 7.82, the corrected velocities may have finite divergence. We can solve the Poisson equation again and make the velocities divergence free; however, this leads to slight variation of solid cell velocities from imposed solid-body motion. A couple of iterations can be performed to satisfy both constraints with reasonable accuracy. Some studies (Fadlun *et al.*, 2000) have shown that this finite divergence does not highly affect the physical results. After updating the velocities, the properties of the cells can be updated in the numerical domain using the following formula, which is very similar to the VOF and level set methods:

$$\rho^{n+1} = c_s^{n+1} \rho_s + \left(1 - c_s^{n+1} \right) \rho_F$$
$$\mu^{n+1} = c_s^{n+1} \mu_s + \left(1 - c_s^{n+1} \right) \mu_F \qquad \text{Equation 7.84}$$

After updating the fluid's properties, we can march to the next step. Figure 7.9 shows hydrodynamic analysis of a tension leg platform (TLP) floating wind turbine by using the immersed boundary method to study solid–fluid interaction.

In Section 7.3.14, studying solid–fluid interaction in body-fitted grid methods will be explained.

7.3.14   Discretization of the NS Equation in a Mapped Coordinate System

To study solid–fluid interaction in body-fitted coordinates, we need to first learn the discretization of NS equations in the mapped coordinate system. In this approach, the numerical grid points are not simple blocks anymore, and they can follow the form of the boundary of the domain. Figure 7.10a shows a simple model of the boundary-fitted grid approach.

The advantage of a boundary-fitted coordinate system is that it can be used for most types of boundaries in the numerical domain. This boundary can be a complex shape of a pipe or an offshore structure in the numerical domain.

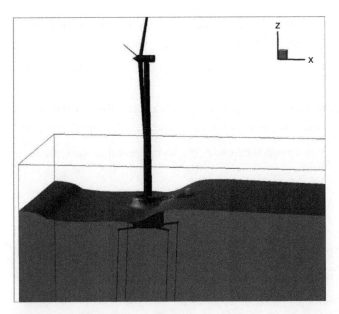

Figure 7.9 CFD simulation of a floating wind turbine on a rectangular structured grid by the level set and immersed boundary methods.

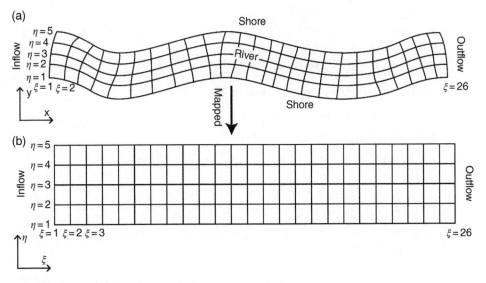

Figure 7.10 Coordinate system for solving the NS equations is mapped from (a) x–y to (b) $\xi - \eta$.

To solve the NS equations in this coordinate system, it is much simpler to map the coordinate system into a simple rectangular structured grid point (Figure 7.10b). Note that the physical coordinate system is x–y, and we need to write the NS equation in the x–y coordinate system. To map the x–y coordinate to ξ–η, we may use the derivative chain law. As an example, suppose that we want to discretize the velocity u: the derivative in the x-direction can be written as follows:

$$\frac{\partial u}{\partial x} = \frac{\partial u}{\partial \xi}\frac{\partial \xi}{\partial x} + \frac{\partial u}{\partial \eta}\frac{\partial \eta}{\partial x}$$

Equation 7.85

The expression in Equation 7.85 is true but didn't completely address the problem, since we still have terms like $\frac{\partial \eta}{\partial x}$ and the numerical domain is discretized in the $\xi-\eta$ coordinate system. So, instead of $\frac{\partial \eta}{\partial x}$, we need terms like $\frac{\partial x}{\partial \eta}$. To address this issue, we may write the following relation:

$$\frac{\partial u}{\partial \xi} = \frac{\partial u}{\partial x}\frac{\partial x}{\partial \xi} + \frac{\partial u}{\partial y}\frac{\partial y}{\partial \xi}$$
$$\frac{\partial u}{\partial \eta} = \frac{\partial u}{\partial x}\frac{\partial x}{\partial \eta} + \frac{\partial u}{\partial y}\frac{\partial y}{\partial \eta}$$

Equation 7.86

We want to obtain an expression for $\frac{\partial u}{\partial y}$ without having any derivative with respect to x or y. To write $\frac{\partial u}{\partial y}$ in the desirable way, we need to get rid of the first term on the right-hand side of both equations (Equation 7.86). Multiplying the first equation by $\frac{\partial x}{\partial \eta}$ and the second equation by $\frac{\partial x}{\partial \xi}$ results in:

$$\frac{\partial x}{\partial \eta}\frac{\partial u}{\partial \xi} = \frac{\partial x}{\partial \eta}\frac{\partial u}{\partial x}\frac{\partial x}{\partial \xi} + \frac{\partial x}{\partial \eta}\frac{\partial u}{\partial y}\frac{\partial y}{\partial \xi}$$
$$\frac{\partial x}{\partial \xi}\frac{\partial u}{\partial \eta} = \frac{\partial x}{\partial \xi}\frac{\partial u}{\partial x}\frac{\partial x}{\partial \eta} + \frac{\partial x}{\partial \xi}\frac{\partial u}{\partial y}\frac{\partial y}{\partial \eta}$$

Equation 7.87

Subtracting two equations written in Equation 7.87:

$$\frac{\partial x}{\partial \eta}\frac{\partial u}{\partial \xi} - \frac{\partial x}{\partial \xi}\frac{\partial u}{\partial \eta} = \frac{\partial u}{\partial y}\left(\frac{\partial x}{\partial \eta}\frac{\partial y}{\partial \xi} - \frac{\partial x}{\partial \xi}\frac{\partial y}{\partial \eta}\right)$$

Equation 7.88

Rearranging Equation 7.87, $\frac{\partial u}{\partial y}$ can be written in the following form:

$$\frac{\partial u}{\partial y} = \frac{1}{J}\left(\frac{\partial u}{\partial \eta}\frac{\partial x}{\partial \xi} - \frac{\partial u}{\partial \xi}\frac{\partial x}{\partial \eta}\right)$$
$$J = \frac{\partial x}{\partial \xi}\frac{\partial y}{\partial \eta} - \frac{\partial x}{\partial \eta}\frac{\partial y}{\partial \xi}$$

Equation 7.89

where J is called the Jacobian transformation. A similar expression can be obtained for $\frac{\partial u}{\partial x}$ by getting rid of the second term on the right-hand side of Equation 7.87, as follows:

$$\frac{\partial u}{\partial x} = \frac{1}{J}\left(\frac{\partial u}{\partial \xi}\frac{\partial y}{\partial \eta} - \frac{\partial u}{\partial \eta}\frac{\partial y}{\partial \xi}\right)$$

Equation 7.90

Each term of the NS equations will be rewritten in a similar format, and the final NS equations on the boundary-fitted structured grid will be obtained. For example, for the advection term, transferring the coordinate system to $\xi - \eta$ results in:

$$\frac{\partial uu}{\partial x} + \frac{\partial uv}{\partial y} = \frac{1}{J}\left(\frac{\partial}{\partial \xi}(uU) + \frac{\partial}{\partial \eta}(uV)\right)$$

Equation 7.91

where U and V are the following:

$$U = uy_\eta - vx_\eta$$
$$V = vx_\xi - uy_\xi$$

Equation 7.92

where U and V in Equation 7.92 can in fact be interpreted as the velocities in the ξ–η coordinate system; this is called *contravariant velocities*. These velocities are shown in Figure 7.11. The reason is as follows:

The unit vector in the η direction, denoted by j', can be written as follows in the x–y coordinate system:

$$j' = \cos\theta\, i + \sin\theta\, j$$
$$\cos\theta = \frac{\partial x}{\partial \eta} = x_\eta$$
$$\sin\theta = \frac{\partial y}{\partial \eta} = y_\eta$$
$$j' = x_\eta\, i + y_\eta\, j = (x_\eta, y_\eta)$$

Equation 7.93

We can conclude from Equation 7.93 that the normal vector to j' is $(-y_\eta, x_\eta)$, which is the unit vector in the ξ direction. To find out the component of the velocity vector in a certain direction, we may use the inner product. The inner product of the velocity vector times the unit vector in ξ is written as:

$$(u,v).(-y_\eta, x_\eta) = -uy_\eta + vx_\eta$$

Equation 7.94

which is equal to U (Figure 7.11). Similar arguments can be made for V, and physical interpretation is given to V in Equation 7.92. Revisiting the advection term in Equation 7.92, the advection term represents the amount of momentum coming in and out of control volumes in the $\xi - \eta$ coordinate system.

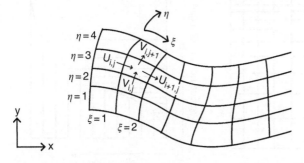

Figure 7.11 Contravariant velocities in the $\xi - \eta$ coordinate system.

The same method of transformation can be used for other terms in the NS equations and in mapped coordinate systems. NS and mass conservation equations can be written as follows:

$$\frac{\partial u}{\partial t} + \frac{1}{J}\left(\frac{\partial Uu}{\partial \xi} + \frac{\partial Vu}{\partial \eta}\right) = -\frac{1}{J\rho}\left(y_\eta\frac{\partial p}{\partial \xi} - y_\xi\frac{\partial p}{\partial \eta}\right)$$
$$+ \frac{\mu}{\rho}\left(\frac{\partial}{\partial \xi}\left(q_1 u_\xi - q_2 u_\eta\right) + \frac{\partial}{\partial \eta}\left(q_3 u_\eta - q_2 u_\xi\right)\right)$$

$$\frac{\partial v}{\partial t} + \frac{1}{J}\left(\frac{\partial Uv}{\partial \xi} + \frac{\partial Vv}{\partial \eta}\right) = -\frac{1}{J\rho}\left(x_\xi\frac{\partial p}{\partial \eta} - x_\eta\frac{\partial p}{\partial \xi}\right)$$
$$+ \frac{\mu}{\rho}\left(\frac{\partial}{\partial \xi}\left(q_1 v_\xi - q_2 v_\eta\right) + \frac{\partial}{\partial \eta}\left(q_3 v_\eta - q_2 v_\xi\right)\right)$$

$$q_1 = x_\eta^2 + y_\eta^2$$
$$q_2 = x_\xi x_\eta + y_\xi y_\eta$$
$$q_3 = x_\xi^2 + y_\xi^2$$

$$\frac{\partial u}{\partial \xi} + \frac{\partial v}{\partial \eta} = 0 \qquad\qquad \text{Equation 7.95}$$

These equations can be discretized in a $\xi - \eta$ coordinate system, and the resultant velocity can be calculated. For more information regarding a solution of the NS equation in a mapped coordinate system, see Ferziger and Peric (2012).

7.3.15   Grid Generation in a Mapped Coordinate System: Stretched Grid

Generating grids in a mapped coordinate system can be as simple as using a stretched grid in the numerical domain. Therefore, as before, rectangular structured grids are used, but the grids are clustered at certain locations and coarsen at some other locations. In other words, the x–y coordinate system will be mapped to $\xi - \eta$, which here are only stretched. Different mapping functions can be used; for example, the two mapped functions in Equation 7.96 will map a uniform grid to non-uniform grids as shown in Figure 7.12.

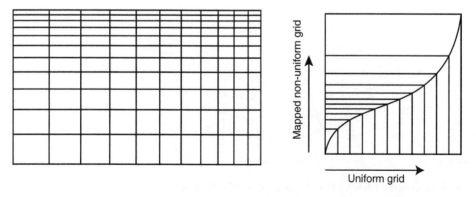

Figure 7.12 Simple mapping function to map a uniform grid to a non-uniform stretched grid.

$$x = \xi^2$$
$$y = \sqrt{\eta}$$

Equation 7.96

The numerical domain may be discretized to different subdomains, and each subdomain can be clustered or stretched at certain locations. For example, in the numerical wave tank shown in Figure 7.13, in the horizontal direction we would like to have a uniform, relatively fine grid from the wave generator to the body; a fine, uniform grid at the solid body location; and gradually coarsened grids in the downstream to damp the wave. So, in the horizontal direction, we can create four subdomains, and for each of them one mapping function can be specified.

More complex mapped functions can be used to cluster certain locations in the numerical domain with different intensities. See Thompson *et al.* (1998) for more details.

7.3.16 Grid Generation in the Mapped Coordinate System: Body-Fitted Grids

A mapped coordinate system can be used to study the flow field around a structure. In this method, grids are constructed with a structured order around the solid body. The grids are then opened from one line and mapped to a rectangular grid domain (see Figure 7.14).

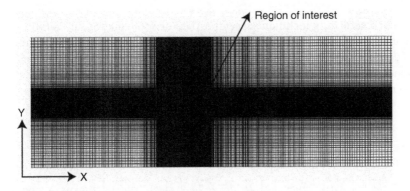

Figure 7.13 Stretching grids used to discretize a 2D numerical wave tank.

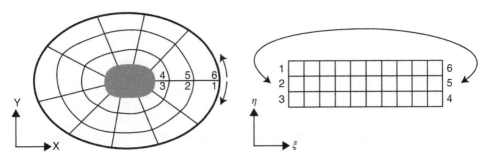

Figure 7.14 Mapped coordinate system for grid generation around a structure.

One of the efficient methods to generate grids around the structure is to solve the Laplace equation in that region. The Laplace equation can resemble solving a steady-state heat equation in the numerical domain. After solving the heat equation, the iso-temperature contour lines are smoothly constructed around the structure, and they can be used as grids around the structure. The heat equation will be solved in both ξ and η directions. The Laplace (heat) equation can be written as:

$$\xi_{xx} + \xi_{yy} = 0$$
$$\eta_{xx} + \eta_{yy} = 0$$

Equation 7.97

This equation looks simple, but derivatives are with respect to x and y. However, we don't have the grids in the domain, and in fact we are solving this equation to obtain the location of the grids. As we did in deriving the NS equation, we need to use the derivative chain rule to change the equation from having derivatives with respect to x and y, to having derivatives with respect to ξ and η, which is the mapped coordinate system. Using the same techniques, the Laplace equation can be written as follows:

$$ax_{\xi\xi} - 2bx_{\xi\eta} + cx_{\eta\eta} = 0$$
$$ay_{\xi\xi} - 2by_{\xi\eta} + cy_{\eta\eta} = 0$$
$$a = x_\eta^2 + y_\eta^2$$
$$b = x_\xi x_\eta + y_\xi y_\eta$$
$$c = x_\xi^2 + y_\xi^2$$

Equation 7.98

Now, the same as in discretizing techniques used for the Poisson equation, an iterative method will be used for solving Equation 7.98. The main difference here is that the coefficients of Equation 7.98 are not constant and should be updated in each iteration until a convergence solution is obtained. For example, for the mapped equation in the x-direction, the equation can be discretized in the following form (supposing a uniform grid is desired):

$$a = \left(\frac{x_{i,j+1}^n - x_{i,j-1}^{n+1}}{2\Delta\eta}\right)^2 + \left(\frac{y_{i,j+1}^n - y_{i,j-1}^{n+1}}{2\Delta\eta}\right)^2$$

$$b = \left(\frac{x_{i,j+1}^n - y_{i,j-1}^{n+1}}{2\Delta\xi}\right)\left(\frac{x_{i,j+1}^n - x_{i,j-1}^{n+1}}{2\Delta\eta}\right) + \left(\frac{y_{i+1,j}^n - y_{i-1,j}^{n+1}}{2\Delta\xi}\right)\left(\frac{y_{i,j+1}^n - y_{i,j-1}^{n+1}}{2\Delta\eta}\right)$$

Equation 7.99

$$c = \left(\frac{x_{i+1,j}^n - x_{i-1,j}^{n+1}}{2\Delta\xi}\right)^2 + \left(\frac{y_{i+1,j}^n - y_{i-1,j}^{n+1}}{2\Delta\xi}\right)^2$$

$$a\left(\frac{x_{i+1,j}^n - 2x_{i,j}^{n+1} + x_{i-1,j}^{n+1}}{(\Delta\xi)^2}\right) - 2b\left(\frac{x_{i+1,j+1}^n - x_{i+1,j-1}^n - x_{i-1,j+1}^n + x_{i-1,j-1}^{s+1}}{4\Delta\xi\Delta\eta}\right)^2 + c\left(\frac{x_{i,j+1}^s - 2x_{i,j}^{s+1} + x_{i,j-1}^{s+1}}{(\Delta\eta)^2}\right) = 0$$

Equation 7.100

Solving for the unknown variable at the new iteration will lead to the following:

$$x_{i,j}^{s} = \frac{\alpha\left(x_{i+1,j}^{n} + x_{i-1,j}^{n+1}\right) + \beta\left(x_{i+1,j+1}^{n} - x_{i+1,j-1}^{n} - x_{i-1,j+1}^{n} + x_{i-1,j-1}^{s+1}\right) + \gamma\left(x_{i,j+1}^{s} + x_{i,j-1}^{s+1}\right)}{2(\alpha + \gamma)}$$

$$\alpha = \frac{a}{\left(\Delta\xi\right)^{2}}, \beta = \frac{-b}{2\Delta\xi\Delta\eta}, \gamma = \frac{c}{\left(\Delta\eta\right)^{2}}$$

Equation 7.101

Equation 7.101 can be solved in an iterative manner until the difference between two successive iterations becomes less than a certain desired threshold. The generated grid around half a cylinder by solving the heat equation (elliptical grid) can be seen in Figure 7.15.

7.3.17   Body-Fitted Grid Generation by Using Unstructured Grids

Sometimes, the geometry of the offshore structure is so complicated that it is hard to create a structured grid around it. In these cases, an unstructured grid will be essential. The difference between structured and unstructured grids is that in the structured grid, there is an ordered layout of the grids so that, by knowing the mapping function, all the information about the grid is available; for the unstructured grid, the neighboring and order of grids should be explicitly described. Examples of structured and unstructured grids are shown in Figure 7.16.

To solve the NS equation by using an unstructured grid, usually a control volume method is used, and conservation of mass and momentum (NS) is written on the control volume. The information may be stored on the vortices of the triangles, and the control volume can be the intersection of lines plotted normal to the triangle lines (Figure 7.16c). For more details about using unstructured grids for solving NS equations, see Ferziger and Peric (2012).

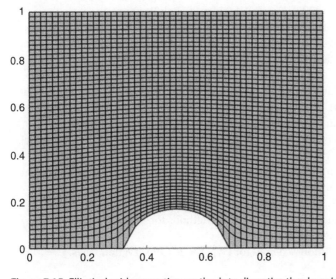

Figure 7.15 Elliptical grid generation methods to discretize the domain around a half a cylinder.

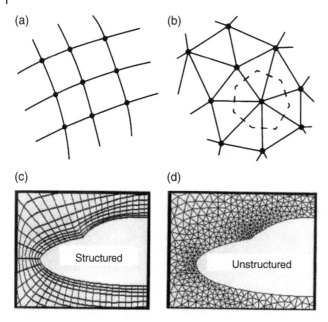

Figure 7.16 Comparison of structured and unstructured grid. (a) Structured grid, (b) unstructured grid, (c) example of structured grid, (d) example of unstructured grid. *Source*: Graphs adopted from Nasa.gov.

References

Fadlun, E.A., Verzicco, R., Orlandi, P., and Mohd-Yusof, J. (2000). Combined Immersed-Boundary Finite-Difference Methods for Three-Dimensional Complex Flow Simulations. J. Comput. Phys. 161, 35–60.

Falgout, R.D., and Yang, U.M. (2002). hypre: A Library of High Performance Preconditioners. In Computational Science – *ICCS 2002*, P.M.A. Sloot, A.G. Hoekstra, C.J.K. Tan, and J.J. Dongarra, eds. (Springer Berlin Heidelberg), pp. 632–641.

Faltinsen, O. (1993). Sea Loads on Ships and Offshore Structures (Cambridge University Press).

Ferziger, J.H., and Peric, M. (2012). Computational Methods for Fluid Dynamics (Springer Science & Business Media).

Gao, Z., Moan, T., Wan, L., and Michailides, C. (2016). Comparative numerical and experimental study of two combined wind and wave energy concepts. J. Ocean Eng. Sci. 1, 36–51.

Ghia, U., Ghia, K.N., and Shin, C.T. (1982). High resolutions for incompressible flow using the Navier–Stokes equations and a multigrid method. J. Comput. Phys. 48, 387–411.

Gueyffier, D., Li, J., Nadim, A., Scardovelli, R., and Zaleski, S. (1999). Volume-of-Fluid Interface Tracking with Smoothed Surface Stress Methods for Three-Dimensional Flows. J. Comput. Phys. 152, 423–456.

Harten, A. (1997). High Resolution Schemes for Hyperbolic Conservation Laws. J. Comput. Phys. 135, 260–278.

Hirt, C.W., and Nichols, B.D. (1981). Volume of fluid (VOF) method for the dynamics of free boundaries. J. Comput. Phys. 39, 201–225.

John, F. (1950). On the motion of floating bodies II. Simple harmonic motions. Commun. Pure Appl. Math. 3, 45–101.

Kang, M., Fedkiw, R.P., and Liu, X.-D. (2000). A Boundary Condition Capturing Method for Multiphase Incompressible Flow. J. Sci. Comput. 15, 323–360.

Karimirad, M., and Michailides, C. (2015). V-shaped semisubmersible offshore wind turbine: An alternative concept for offshore wind technology. Renew. Energy 83, 126–143.

Klee, H. (2000). Newton divided-difference interpolating polynomials. Lecture notes, University of Waterloo.

Meakin, R. (1994). On the spatial and temporal accuracy of overset grid methods for moving body problems. In 12th Applied Aerodynamics Conference (American Institute of Aeronautics and Astronautics).

Nematbakhsh, A., Olinger, D.J., and Tryggvason, G. (2013). A nonlinear computational model of floating wind turbines. J. Fluids Eng. 135, 121103.

Nematbakhsh, A., Bachynski, E.E., Gao, Z., and Moan, T. (2015). Comparison of wave load effects on a TLP wind turbine by using computational fluid dynamics and potential flow theory approaches. Appl. Ocean Res. 53, 142–154.

Newman, J.N. (1977). Marine Hydrodynamics (MIT Press).

Ogayar, C.J., Segura, R.J., and Feito, F.R. (2005). Point in solid strategies. Comput. Graph. 29, 616–624.

Osher, S., and Fedkiw, R. (2006). Level Set Methods and Dynamic Implicit Surfaces (Springer Science & Business Media).

Osher, S., and Sethian, J.A. (1988). Fronts propagating with curvature-dependent speed: Algorithms based on Hamilton-Jacobi formulations. J. Comput. Phys. 79, 12–49.

Sarpkaya, T., & Isaacson, M. (1981). Mechanics of wave forces on offshore structures (Van Nostrand Reinhold Company).

Steinbach, I., Pezzolla, F., Nestler, B., Seeßelberg, M., Prieler, R., Schmitz, G.J., and Rezende, J.L.L. (1996). A phase field concept for multiphase systems. Phys. Nonlinear Phenom. 94, 135–147.

Sussman, M., Smereka, P., and Osher, S. (1994). A Level Set Approach for Computing Solutions to Incompressible Two-Phase Flow. J. Comput. Phys. 114, 146–159.

Thompson, J.F., Soni, B.K., and Weatherill, N.P. (1998). Handbook of Grid Generation (CRC Press).

Tryggvason, G., Scardovelli, R., and Zaleski, S. (2011). Direct Numerical Simulations of Gas–Liquid Multiphase Flows (Cambridge University Press).

Unverdi, S.O., and Tryggvason, G. (1992). A front-tracking method for viscous, incompressible, multi-fluid flows. J. Comput. Phys. 100, 25–37.

Wu, G.X., Taylor, R.E., and Greaves, D.M. (2001). The effect of viscosity on the transient free-surface waves in a two-dimensional tank. J. Eng. Math. 40, 77–90.

Yabe, T., Xiao, F., and Utsumi, T. (2001). The Constrained Interpolation Profile Method for Multiphase Analysis. J. Comput. Phys. 169, 556–593.

8

Mooring and Foundation Analysis

8.1 Mooring Considerations

Floating structures and systems are used in ocean areas by: the oil and gas industry for their different operational phases (e.g. drilling, well intervention, production and storage), offshore renewable energy systems (e.g. wind, wave, tidal and combined systems), the aquaculture industry and transportation-related structures (e.g. floating bridges). During environmental actions (e.g. waves, wind and current), a freely floating structure exhibits offsets that may lead to the offshore structure's overall disaster in order to keep the offshore structure within limits of excursions. An appropriate station-keeping system must be used for the safety of personnel, protection of the environment and stability of the offshore structure. Usually, the station-keeping system is attached to the offshore floating structure. According to ISO 19901-7 (ISO, 2013), the functions of the station-keeping system are to restrict the horizontal excursions of the floating structure within prescribed limits, and to provide means of active or passive directional control when the structure's orientation is important for safety or operational considerations.

Mooring systems are widely used in the offshore industry in order to limit the horizontal excursions of a floating structure from its desired position, provide sufficient restraint to keep surface or subsurface equipment on position, and minimize the combined effects of environmental (wind, current and wave) loads on the offshore floating structure so the structure can fulfil the required tasks. The mooring systems of offshore floating structures (including oil and gas platforms, floating production systems, storage and offloading vessels, offshore renewable energy structures and systems, and auxiliary equipment) are critical elements, and a thorough understanding of their long-term durability is essential to guarantee the survivability of these structures and the structural integrity of the mooring lines. The consequences of a mooring system failure could result in loss of life, environmental disaster or interruption of operations of offshore structures. Therefore, the safety of mooring systems is crucial for successful and viable marine and offshore operations.

In general, a station-keeping system can be passive (e.g. a mooring system), active (e.g. a dynamic positioning system) or combined active and passive (e.g. a thrusters-assisted mooring system), depending upon the main principles of its operation and the way that the system provides the required restoring forces (Barltrop, 1998). Most of the mooring systems for offshore floating structures are passive systems. Previous decades' dynamic positioning systems are used in connection with mooring systems so that the

Offshore Mechanics: Structural and Fluid Dynamics for Recent Applications, First Edition.
Madjid Karimirad, Constantine Michailides and Ali Nematbakhsh.
© 2018 John Wiley & Sons Ltd. Published 2018 by John Wiley & Sons Ltd.

environmental loads that are exerted to the mooring systems are reduced. This is achieved by moving the vessel to new positions when this is needed, or by reducing the quasi-static offset of the floating structure and system. The usual functional requirements that are addressed during mooring system design include offset limitations, lifetime assessment before replacement, installability, maintenance and positioning ability. In this chapter, we will present very basic information on passive station-keeping systems. Different possible mooring system configurations are presented in Figure 8.1. In general, any mooring system is made of one or a number of mooring lines that in their upper ends are attached to different points of the offshore floating structure and in their lower ends are anchored at the seabed with the use of appropriate anchor systems. The mooring lines usually are made of steel chain links, steel wire ropes and synthetic fibre ropes. Usually, multicomponent mooring lines that are composed of two or more line lengths with different materials are used to sustain abrasion at the fairlead and friction on the sea bottom with chain links and to decrease the weight of the whole line using ropes in between. For the design of the mooring system, the tension forces in the cables depend upon the mooring line weight, the material properties (e.g. modulus of elasticity) and the configuration of the mooring system.

Different types of mooring systems that are used widely for the station keeping of offshore structures and systems are:

- Catenary mooring lines
- Taut mooring lines
- Tension leg mooring lines (tendons).

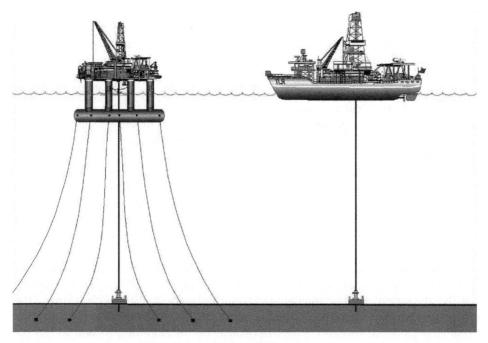

Figure 8.1 Different types of mooring lines. *Source:* Courtesy of Minerals Management Service. Gulf of Mexico OCS Region 2010; this work is in the public domain.

A mooring system with catenary mooring lines is the oldest and most commonly used system in offshore engineering. The restoring force of this system is induced by the lifting and lowering of the weight of the mooring line, and it yields to a hard spring system. Usually, the force increases more than directly proportionally to the offset of the floating platform. Very commonly, an appropriate number of mooring lines (which is defined with the numerical analysis of mooring lines) is placed around the floating structure in a symmetric plane layout to keep it in the desired location (Jeon, 2013). The mooring lines are designed so that a large part of the mooring line lays on the seabed and, as a result, the anchor is kept in position. Systems with taut mooring lines use lines with low net submerged weight that use their elastic stretch to restore force. A very common type of taut mooring line is based on synthetic fibre material. Usually, the synthetic fibre lines are deployed for deep-water offshore engineering applications; for these conditions, the main advantages of their use compared to the catenary mooring lines is that they are considerably lighter, are very flexible, can absorb imposed dynamic motions through extension and do not induce excessive dynamic tension in the platform. Their main disadvantage is that they require very sophisticated numerical models for their required numerical analysis. Tension leg mooring lines (tendons) are used mainly for the station keeping of the tension leg platforms (TLP)-type offshore structures. In general, the buoyancy of the floating platform is greater than the platform's weight, and consequently the net downward force is supplied by the vertical tendons; essentially, the tendons provide the total restraint against the vertical motion of the platform.

In most cases, the mooring system consists of the mooring line (chain, wire, rope or any possible combination), the anchors or piles at the seabed or soil, the fairlead at the platform (e.g. bending shoe types or sheaves), the winches, the power supplies and the rigging (e.g. stoppers, blocks and shackles). Appropriate designs of the mooring systems and of all the different components depend upon several factors like the mooring line loads, the expected environmental loads (e.g. wave, current, wind and earthquake), the water depth, the size and shape of the platform that will be kept at the station and the allowable horizontal motions of the platform (API, 2015).

The environmental loading mechanisms that act on a moored offshore floating structure usually result in time-varying motions in six rigid-body degrees of freedom of the floating platform. As a result, the environmental loads induce first-order motions for the wave frequencies as well as drift motions for low frequencies (Nakamura, 1991). Loading results to the excitation of horizontal motions of the platform that may lead to the increase of the mooring line tension. On the other hand, the mooring system contributes to the damping forces based on different mechanisms. The mooring system contributes to the hydrodynamic drag damping due to the following: transverse drag force that represents energy dissipation, vortex-induced vibration mainly for wire mooring lines, seabed interaction damping through the soil friction and mooring-line in-plane relative motion, and viscous linear material damping. It has been shown that the contribution of the damping that corresponds to the mooring system compared to other contributing factors (e.g. viscous damping of a platform's motions) in some circumstances is very high (Matsumoto, 1991); for the case of a 120,000 DWT (dead weight tons) tanker in 200 m water depth, the contribution of the damping from the mooring lines corresponds to 80% of the total damping, while the contribution of the viscous and wave drift provides the remaining 20%.

In offshore structures, the mooring lines are made by steel chain links, steel wire ropes and synthetic fibre ropes. The most popular material that is used for the case of semi-submersible platforms is the steel chain links. Very often, chains are used in combination with wire ropes. The advantage of the different strengths of the different materials can provide an optimal performance of the mooring system for different water depths. An outline of the three materials is that the chain systems use links that are heavy with high breaking strength and high elasticity; moreover, the chains have no bending effects and are mainly used in shallow waters. For combined mooring systems, chain segments are used close to the fairlead and close to the bottom. Regarding the wires, they are lighter than chains, with slight bending effects, and are used in combined systems in deep waters to reduce the vertical dead loads. Finally, the synthetic fibres are almost neutrally buoyant, they are highly extensible and potentially they can be used in deep waters.

Catenary mooring lines are constructed with the use of stud or stud-less links (Figure 8.2). Stud-link chains have been used for mooring systems of offshore structures in relatively shallow waters. Compared to stud links, the stud-less links reduce the weight per unit of strength and increase the chain's fatigue life. The steel chains for mooring applications are constructed usually based on different steel grades, namely R3, R3S, R4, R4S and R5 (Det Norske Veritas, 2013b). Table 8.1 presents details of the mechanical properties of steel used in the various specifications of chains commonly

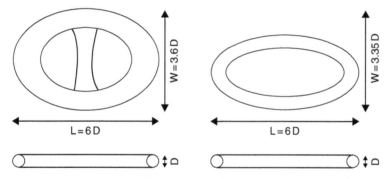

Figure 8.2 Stud and stud-less mooring chain links.

Table 8.1 Details of the mechanical properties of steel used in offshore applications.

Steel grade	Yield stress (N/mm^2)	Tensile stress (N/mm^2)	Elongation (%)	Reduction of area (%)
R3	410	690	17	50[a]
R3S	490	770	15	50[a]
R4	580	860	12	50[b]
R4S	700	960	12	50[b]
R5	760	1000	12	50[b]

a. For cast accessories, the minimum value shall be 40%.
b. For cast accessories, the minimum value shall be 35%.

used in different offshore applications. With regard to the weight and stiffness properties of the chains, Equations 8.1 and 8.2 can be used to estimate these parameters:

$$w_s = 0.1875D^2 N/m \qquad\qquad \text{Equation 8.1}$$

$$AE = 90,000D^2 N \qquad\qquad \text{Equation 8.2}$$

where w_s is the submerged weight per unit length of the mooring line, D is the chain diameter in millimetres and AE is the axial stiffness per unit length of the mooring line.

Another very important parameter for the design of the mooring lines is the breaking strength. The catalogue break strength (CBS) is used by the manufacturers. CBS can be given with Equation 8.3:

$$CBS = c(44 - 0.08D)D^2 N \qquad\qquad \text{Equation 8.3}$$

where c is a material factor. Moreover, with the use of DNV-OS-E302 (Det Norske Veritas, 2013b) formulas for the break strength of the mooring chains are proposed and presented in Table 8.2. As it is expected chains that are constructed with material R5 will provide the largest possible breaking and proof load.

Regarding the hydrodynamic loads that should be accounted in the numerical analysis of mooring lines, drag and inertia coefficients have been proposed so far for use along with the Morison equation. For the inertia coefficient, a value equal to 2.4 has been proposed so far by Larson (1990), while for the drag coefficient, values between 2.4 for stud-less chain and 3.0 have been proposed depending on the dimensions and material of chains.

Steel wire ropes consist of individual wires that are knitted so that they form a "strand." Six-strand rope is the most common type of multistrand rope that is used in offshore applications. Usually, 12, 24, 37 or more wires per strand are used for a mooring line's rope. The material that is used for mooring line wires has a yield strength that is very high and is in between $1770 N/mm^2$ and $1860 N/mm^2$. In Figure 8.3, components of steel wire ropes that are used in offshore engineering are presented.

Table 8.2 Proposed formulas for the proof load and breaking load of mooring lines.

Quantity	Grade R3	Grade R3S	Grade R4	Grade R4S	Grade R5
Proof load, stud link (kN)	$0.0156d^2$ (44–0.08d)	$0.0180d^2$ (44–0.08d)	$0.0216d^2$ (44–0.08d)	$0.0240d^2$ (44–0.08d)	$0.0251d^2$ (44–0.08d)
Proof load, stud-less link (kN)	$0.0156d^2$ (44–0.08d)	$0.0174d^2$ (44–0.08d)	$0.0192d^2$ (44–0.08d)	$0.0213d^2$ (44–0.08d)	$0.0223d^2$ (44–0.08d)
Breaking load (kN)	$0.0223d^2$ (44–0.08d)	$0.0249d^2$ (44–0.08d)	$0.0274d^2$ (44–0.08d)	$0.0304d^2$ (44–0.08d)	$0.0320d^2$ (44–0.08d)
Weight, stud link (kg/m)	$0.0219d^2$				
Five link lengths (mm)	Minimum 22d and maximum 22.55d				

d: Chain nominal diameter.
Source: Det Norske Veritas (2013b).

With regard to the weight and stiffness properties of the wire ropes, empirical Equations 8.4 and 8.5 can be used to estimate these parameters for the case of six-strand rope:

$$w_s = 0.034d^2 \, N/m \qquad\qquad \text{Equation 8.4}$$

$$AE = 45,000d^2 \, N \qquad\qquad \text{Equation 8.5}$$

where d is the nominal diameter of the rope; and, for the case of spiral strand:

$$w_s = 0.043d^2 \, N/m \qquad\qquad \text{Equation 8.6}$$

$$AE = 90,000d^2 \, N \qquad\qquad \text{Equation 8.7}$$

The breaking strength can be given with Equations 8.8 and 8.9 for the case of a six-strand rope with material with yield strength $1770 \, N/mm^2$ and $1860 \, N/mm^2$, respectively:

$$Breaking load = 525d^2 \, N \qquad\qquad \text{Equation 8.8}$$

$$Breaking load = 600d^2 \, N \qquad\qquad \text{Equation 8.9}$$

For the case of a spiral strand with material with yield strength $1570 \, N/mm^2$, the breaking load is given by the following empirical formula:

$$Breaking load = 900d^2 \, N \qquad\qquad \text{Equation 8.10}$$

Regarding the hydrodynamic loads that should be accounted in the numerical analysis of mooring lines, for the inertia coefficient a value equal to 2.0 has been proposed by Larsen (1990), while for the drag coefficient, values between 1.5 and 1.8 have been proposed by different researchers.

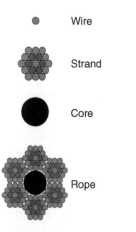

• Wire

Strand

Core

Rope

Figure 8.3 Components of steel wire rope that are used in offshore engineering. *Source:* Courtesy of Tachymètre 2010; this file is licensed under the Creative Commons Attribution-Share Alike 3.0 Unported license.

Mooring lines are usually under cyclic loading and experience fatigue damage in their components. The fatigue damage in mooring lines may lead to the entire fail of the offshore structure. For the design of mooring lines, the fatigue life should be estimated and fatigue analysis should be developed to estimate the damage effect of each load cycle and to determine whether structural failure occurs due to the cumulative damage over a specific period of interest. Fatigue analysis can be carried out by fracture mechanics or an SN approach. SN design curves are proposed in different regulations for the design of mooring lines of different types. Figure 8.4 presents SN design curve types for the design of mooring lines (Det Norske Veritas, 2013a).

Different types of synthetic fibre rope exist, and (a) polyester, (b) aramid and (c) high-molecular-weight polyethylene (HMPE) dominate compared to the others. The development of these synthetic-type mooring lines is important. With regard to the ropes' weight properties, Equations 8.11 through 8.13 can be used to estimate the weight for polyester, aramid and HMPE, respectively:

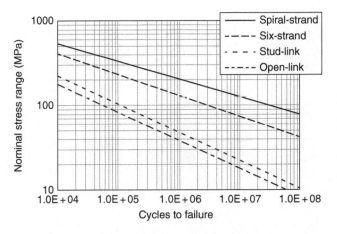

Figure 8.4 S-N design curves for mooring lines of different types.

$$w_{s,p} = 0.0067d^2 N / m \qquad\qquad \text{Equation 8.11}$$

$$w_{s,a} = 0.00565d^2 N / m \qquad\qquad \text{Equation 8.12}$$

$$w_{s,HMPE} = 0.0062d^2 N / m \qquad\qquad \text{Equation 8.13}$$

The breaking strength can be given with Equations 8.14 through 8.16 for polyester, aramid and HMPE, respectively:

$$Breaking load = 250d^2 N \qquad\qquad \text{Equation 8.14}$$

$$Breaking load = 450d^2 N \qquad\qquad \text{Equation 8.15}$$

$$Breaking load = 575d^2 N \qquad\qquad \text{Equation 8.16}$$

Three parameters can be considered as key players for the design of a mooring system: the submerged weight, the breaking strength and the elasticity of the mooring lines.

Usually, the mooring lines are placed in a spread configuration or in a turret. The symmetric spread of mooring lines (Figure 8.5a) is the simplest in terms of design and installation, but sometimes it may not lead to optimum lifetime response and performance. The reasons for this may be related to the wave excitation directionality, subsea spatial layout and space restrictions; under these circumstances, an unsymmetrical spread configuration is selected. On the other hand, single-point mooring of offshore structures can be achieved with the use of internal or external turrets (Figure 8.5b).

8.1.1 Catenary Moorings

The estimation of the tension forces of the mooring lines is fundamental for the coupled numerical analysis of any floating structure. The tension of the mooring lines has to be numerically estimated for every time step for the case of a time-domain numerical analysis and for every examined wave frequency/period for the case of a

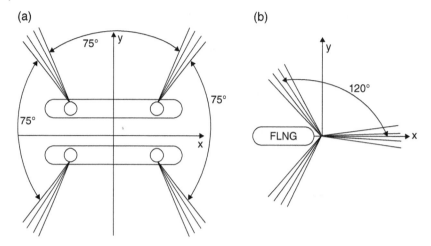

Figure 8.5 Spread symmetric mooring line layout (Figure 8.5a) and a turret mooring system (Figure 8.5b).

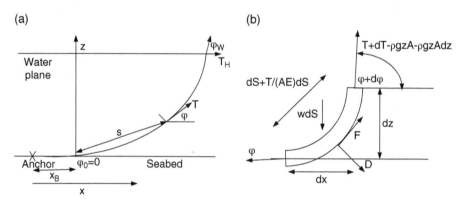

Figure 8.6 A catenary side 2D view (a) of a mooring line and forces that act on an element of the mooring line (b).

frequency-domain numerical analysis. The mooring forces that are exerted on the floating structure are used as input for the solution of the equation of motion in both the frequency and time domains (Kim, 2010).

For the time-domain coupled floater–mooring line analysis, two distinct methods for obtaining the mooring line forces on the structure have been developed so far and are presented here: the first method is based on a static approach, and the second one on a dynamic approach taking into account the elasticity of the mooring line itself.

If we neglect the bending stiffness effects of the mooring line, a static method can be developed. Consider a mooring line (Figure 8.6a) that station keeps a floating structure if we examine a small element of this mooring line (Figure 8.6b), where w is the constant submerged line weight per unit length, T the tension of the mooring line, A the cross-sectional area, E the modulus of elasticity, and D and F are the mean hydrodynamic loads on the small element per unit length. Using the free body diagram of the segment in two horizontal directions, Equations 8.17 and 8.18 hold true:

$$dT - \rho g A dz = \left[w\sin\varphi - F\left(\frac{T}{EA}\right) \right] ds \qquad\qquad \text{Equation 8.17}$$

$$T d\varphi - \rho g A z d\varphi = \left[w\cos\varphi + D\left(1 + \frac{T}{EA}\right) \right] ds \qquad\qquad \text{Equation 8.18}$$

If we neglect the hydrodynamic forces D and F and the elasticity of the element of the mooring line, we can estimate the response quantities of the mooring lines as presented in detail in Faltinsen (1990). As a result, we can estimate the horizontal T_H and the vertical T_Z component of the tension:

$$T_H = T\cos\varphi_w \qquad\qquad \text{Equation 8.19}$$

$$T_Z = ws \qquad\qquad \text{Equation 8.20}$$

Moreover, the mooring line tension T can be estimated with the use of Equation 8.21:

$$T = T_H + wh + (w + \rho g A)z \qquad\qquad \text{Equation 8.21}$$

In the previous method, the elongation of the cable (e.g. elasticity of material) due to the tension force had not been taken into account. To solve the nonlinear equations for cases in which we do not neglect the elasticity of the mooring lines, advanced numerical techniques are required. With the use of the static configuration of the mooring lines as a starting point, the solution of the nonlinear equations can be solved easier (Triantafyllou, 1994) by a variety of numerical methods. Two main categories exist for the solution of these nonlinear equations: the finite difference approach and the finite element approach. The lumped mass approach that is based on finite differences is widely used for the numerical analysis of mooring lines. In this method, the mooring line is discretized into point masses that are connected by weightless inextensible elements; this leads to a set of ordinary differential equations for the dynamics of each lumped mass. The finite element approach uses a discrete number of finite elements to approximate a continuum (Kim, 2013). The finite elements retain the material properties of the continuum, and finite element models for cable dynamics are established.

8.1.2 Taut Moorings

TLPs are generally used for deep-water offshore installations. TLPs consist of a platform that is kept in place with the use of a taut mooring system (Figure 8.7). The taut mooring system consists of a number (e.g. three, four or more) of tension legs, each of them comprising multiple tendon members. With the use of the taut system, the heave, roll and pitch motions of the offshore structure are well restrained, but the horizontal motions, surge and sway, and the yaw motion are permitted and may be excited by the wave excitation loads. The hull buoyancy is larger compared to the total TLP weight, and as a result tension in the tendons exists. In general, the tendons can be fixed to the outside of the hull, or attached in internal tendons or "tubes." Various multi- and single-tendon configurations have been proposed so far depending upon the use of the offshore structure. The basic TLP components are the pontoons, the stability columns, the deck and the taut legs. To date, the four-column configuration dominates the rest of the design choices.

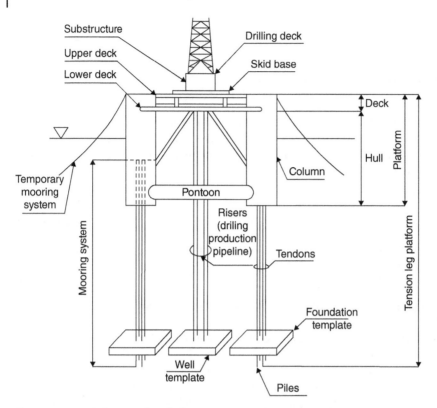

Figure 8.7 A typical tension leg platform (TLP) station kept with tendons.

For the numerical analysis of any type of TLP, the tendons are simulated mainly with the use of a restoring stiffness matrix that can be used from any possible solver. Two different types of analysis are performed to estimate the structural integrity of TLPs. The first one is based on a static consideration, and the second one on a dynamic consideration of the tendons. Based on static considerations, the pretension of the tendons T_i for the case that we ignore any possible lateral force (Figure 8.8) is given with Equation 8.22:

$$nT_i = \rho gV - W - T_{ris} \qquad \text{Equation 8.22}$$

where n is the total number of tendons, V is the displaced volume of the platform, W is the total weight of the platform and all the equipment and T_{ris} is the tension of the risers (for the case of a TLP in oil and gas industry). For the case that lateral forces exist (Figure 8.8) that result in a positive offset of the platform, x, the tendons that maintain their length as L_t cause the TLP to submerge by dz. As a result, a small increase at the tension exists dT_i:

$$\delta z = L_t \left[1 - \sqrt{1 - \left(x / L_t \right)^2} \right] \qquad \text{Equation 8.23}$$

$$\delta T_i = \rho gA_w \times \delta z / n_i \qquad \text{Equation 8.24}$$

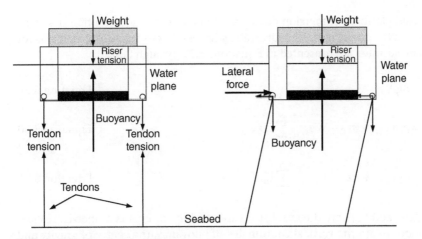

Figure 8.8 Free body diagram of a TLP.

TLPs operate in oceans, where intense dynamic environmental loadings exist. In general, the dynamic response of TLPs is nonlinear. Moreover, heave, roll and pitch motions of TLPs have high natural frequencies, while surge, sway and yaw motions have very low natural frequencies, as a result of the high axial stiffness and the low geometric stiffness of the tendons, respectively. So, three different frequency ranges exist that may lead the offshore structure to resonance: the low-frequency range, the wave excitation frequency range and the high-frequency range. A fully tendons-floater coupled analysis is required for the dynamic analysis of TLPs.

To model the tendons in the fully coupled numerical analysis, two different approaches exist (Senjanovic, 2013). The first one deals with the tendons that are behaving linearly. The total stiffness connected with the existence of the tendons, K_{tot} (which should be added in the numerical analysis), consists of three parts (Low, 2009): K_c, K_{hydr} and K_g:

$$K_{tot} = K_c + K_{hydr} + K_g \qquad \text{Equation 8.25}$$

where K_c is the stiffness provided by the axial stiffness of tendons, K_{hydr} is the hydro-static stiffness but with the influence of the tendons and K_g is the geometric stiffness of the tendons. For heave, roll and pitch motions, the axial stiffness of the tendons is estimated with Equations 8.26 through 8.28:

$$K_c^{33} = \frac{EA}{L} \qquad \text{Equation 8.26}$$

$$K_c^{44} = \frac{EI_x}{L} \qquad \text{Equation 8.27}$$

$$K_c^{55} = \frac{EI_y}{L} \qquad \text{Equation 8.28}$$

where A, I_x and I_y are, respectively, the cross-section area and the moments of inertia about the x- and y-axis of all tendons. With regard to the hydrostatic stiffness and due to the increased platform immersion due to the tendon forces, a correction has to be made in some of the coefficients of hydrostatic stiffness. For example, the formulas for the hydrostatic stiffness are presented in Equations 8.29 through 8.31:

$$K_{hydr}^{33} = \rho g A_{WL} \qquad\qquad \text{Equation 8.29}$$

$$K_{hydr}^{44} = \rho g I_{WLx} + B z_B - W z_G - \sum_{n=1}^{N} T_n z_n \qquad\qquad \text{Equation 8.30}$$

$$K_{hydr}^{55} = \rho g I_{WLy} + B z_B - W z_G - \sum_{n=1}^{N} T_n z_n \qquad\qquad \text{Equation 8.31}$$

where A_{WL}, I_{WLx} and I_{WLY} are the waterplane area and the moments of inertia, respectively; z_B, z_G and z_T are the vertical coordinates of buoyancy, the centre of gravity and the tendon top measured from the waterplane; B is the platform's total buoyancy; and W is the total weight. Regarding the geometric stiffness, this is attributed to the motions of the platform that are permitted. To numerically model this contribution, Equation 8.32 can be used for all the degrees of freedom:

$$K_g^{ij} = \iiint_V \sigma_{kl} h_{m,k}^i h_{m,l}^j dV \qquad\qquad \text{Equation 8.32}$$

where σ_{kl} is a stress tensor. For the case of surge, sway and yaw motions, the geometric stiffness is estimated with Equations 8.33 and 8.34:

$$K_g^{11} = \sum_{n=1}^{n} \frac{T_n}{L} = K_g^{22} \qquad\qquad \text{Equation 8.33}$$

$$K_g^{66} = \sum_{n=1}^{n} \frac{T_n}{L} \left(x_n^2 + y_n^2 \right) \qquad\qquad \text{Equation 8.34}$$

where x_n and y_n are the coordinates of the point where the tendons are placed at the TLP.

Due to all the possible motions of the platform of the TLP (e.g. setdown), nonlinear terms associated with the tendons should be added in the restoring stiffness that simulates the tendons. To estimate these nonlinear coefficients, an analysis is required. Based on this analysis, the following coefficients K_{nl} should be taken into account:

$$K_{nl}^{11} = \sum_{n=1}^{n} \frac{T_n}{L_z} + \rho g A_{WL} \frac{\delta^x}{L_z} \qquad\qquad \text{Equation 8.35}$$

$$K_{nl}^{22} = \sum_{n=1}^{n} \frac{T_n}{L_z} + \rho g A_{WL} \frac{\delta^y}{L_z} \qquad\qquad \text{Equation 8.36}$$

$$K_{nl}^{44} = \frac{\rho g I_{WLx}}{\cos^3 \varphi_x} + B \left(z_B - z_T \right) - W \left(z_G - z_T \right) + \rho g A_{WL} \left(z_B - z_T \right) + \frac{EI_x}{L} \cos^2 \varphi_x$$

$$\text{Equation 8.37}$$

$$K_{nl}^{55} = \frac{\rho g I_{WLy}}{\cos^3 \varphi_y} + B(z_B - z_T) - W(z_G - z_T) + \rho g A_{WL}(z_B - z_T) + \frac{EI_y}{L} \cos^2 \varphi_y$$

Equation 8.38

$$K_{nl}^{66} = r^2 \left(\sum_{n=1}^{n} \frac{T_n}{L_z} + \rho g A_{WL} \frac{\varphi^z}{L_z} \right)$$

Equation 8.39

8.2 Soil Mechanics

Offshore soil mechanics and marine geotechnical engineering study the equilibrium and movement of soil particles. Soil is the upper layers of the earth's crust. The nature of soil is different from that of artificial materials such as steel and concrete. Manmade materials are more uniform and present more linear mechanical behaviour (if their deformations and strains are not too large). However, offshore soil behaviour is highly nonlinear, and soil characteristics are quite dissimilar in different locations.

Soil represents irreversible plastic deformations under loading and unloading, even at low stresses. Soil often shows anisotropic behaviour, creep and dilatancy[1] (changing volume under shear). The soil structure is inhomogeneous due to geological history, and it is difficult to determine the soil's detailed behaviour even by tests. The presence of water makes the soil behaviour much more complicated. Due to these reasons, soil modelling is site dependent, and the soil properties should be set for a specific problem.

Depending on the problem under consideration, for example if it is the long-term stability of the structure (i.e. considering the soil–pile–structure interactions) or installation of the structure (i.e. hammering piles), the relevant specific data should be applied. Also, proper analyses implementing proper parameters should be used. Offshore geotechnics and soil mechanics help predict such soil behaviour. However, it is good to remember that the applicability of a specific parameter in soil mechanics is limited to a range of problems and several properties are not valid outside that range.

Site investigation and data collection from soil are needed to obtain proper soil information for planning, engineering, construction and installation of offshore structures. Geotechnical site investigations are normally performed, including: (a) surface investigations (topographic surveys), and (b) subsurface investigations by seismic surveys, cone penetration testing, vibracores and boreholes (Bakmar, 2009). The collected seabed material is tested in the laboratory to investigate the soil properties and load carrying capacity.

Offshore soil is made of solid particles ranging in size from clay to huge stones and water. The saltwater fills in the pore spaces, and soil sediments are in a saturated state. The size of soil particles is dependent on the earth's geological history. The soil material may be gravel, sand, clay, peat or loose granular (Verruijt, 2006). The soil sediment is mainly cohesionless (sandy) or cohesive (mud or clay).

The types and size of these particles influence the soil properties and soil capacity to carry loads. The sediments are detrital material and marine organisms' leftovers

---

1 Dilatancy is a non-Newtonian behaviour when fluid shear viscosity increases with shear stress, which is influenced by the size, shape and distribution of material particles.

(calcareous soils). In general, sediment thickness is higher close to shore; and in some sites, due to strong bottom current, the sediments are totally removed. The structure of soil particles is complex and affects the water flow and erosion. Soil conditions can be quite different, from soft clay to sand, rock and hard rock; hence, soil conditions can significantly affect the foundation and anchoring design.

There are different methods and devices to measure soil properties. Among the devices used for offshore geotechnical site investigations is piezocone. The piezocone device is pushed into the seabed, and by measuring the cone resistance, sleeve friction and pore pressures, the soil properties are derived.

There are various parameters defining the strength-deformation properties of soil, including the soil-bearing capacity, soil compressibility and soil modulus of subgrade reaction. Often, offshore structures are set on seabed, for example gravity-based structures or jack-ups. Also, jackets should sit temporarily on seabed before installation of piles is finished. Soil is a compressible material; hence, these structures experience settlement (downward movement). The settlement varies from small amounts to several metres (in extreme cases), depending on the soil characteristics and the structure/ foundation. Proper investigation of structure settlement is essential to ensure the structure stability and structural integrity.

In many problems, the settlement of the structure is the interest value. The elasticity of the soil, the boundary conditions, compressibility and applied stress (loads) affect the settlement. The theory of elasticity and continuum mechanics can be used to examine the associated settlements. In foundation engineering, if the applied stresses are much less than the failure values, soil may be considered as an elastic solid. However, this assumption is not always proper, in particular for soft clays and loose sandy soils.

A linear relation can be established to calculate the settlement based on applied load and subgrade modulus. In such case, the soil is modelled as a spring. This is a too-simplistic way of modelling soil behaviour; as was mentioned, soil exhibits nonlinear and irreversible characteristics in a real offshore world. But this can be the first step to understand the physics and phenomena.

Example 8.1   Derive the subgrade modulus for a soil under uniform structure/ foundation loads shown in Figure 8.9.

We assume that stresses in the soil are distributed homogeneously over an area that increases gradually with an angle (i.e. 45 degrees). At a depth Z below the seabed, the area carrying the load is $\pi/4(D+2Z)^2$ and the total force is $\pi/4D^2q$. q is the pressure upon soil. Hence, the stress is obtained by the following expression:

$$\sigma_Z = q/(1+2Z/D)^2 \qquad\qquad \text{Equation 8.40}$$

For linear elastic material with modulus of elasticity (E), the strain is $\varepsilon_Z = \sigma_Z/E$. We also know that $\varepsilon_Z = dW/dZ$, in which W is settlement.

$$dW = \frac{qdZ}{E(1+2Z/D)^2}$$

$$W = \int_0^\infty \frac{qdZ}{E(1+2Z/D)^2} = \frac{qD}{2E} \qquad\qquad \text{Equation 8.41}$$

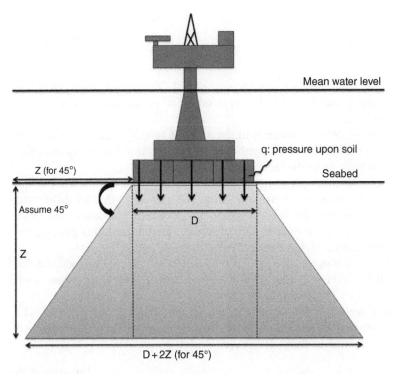

Figure 8.9 Schematic layout of a gravity-based structure on elastic soil.

If we write the settlement as a function of subgrade modulus ($c_{subgrade}$) and pressure upon the soil (q):

$$q = c_{subgrade}W \rightarrow c_{subgrade} = \frac{2E}{D}$$

Equation 8.42

In general, one may write $W = \dfrac{qD}{\beta E}$, in which β is replaced by 2 in Equation 8.41 to make a more general expression. Hence, the subgrade modulus may be written as:

$$c_{subgrade} = \frac{\beta E}{D}$$

Equation 8.43

The subgrade modulus of the soil is derived in Equation 8.43 in which the stiffness of the soil is presented as a function of the diameter of the foundation and soil modulus of elasticity. Increasing the diameter decreases the stiffness, and modulus of elasticity increases the stiffness.

If soil behaved like an elastic solid, most soil mechanic problems would be easier to solve. However, the compressibility of a soil deposit is generally not constant, and it decreases with increasing depth below the seabed. The modulus of elasticity (E) of soil can be related to compressibility constant (C_{comp}) in the Terzaghi formula (Kellett, 1974). Verruijt has shown that the modulus of elasticity is related to initial stress level (σ_0); refer to Verruijt (2006) for more information.

$$E = C_{comp}\sigma_0$$

$$c_{subgrade} = \frac{\beta C_{comp}\sigma_0}{D}$$

Equation 8.44

So, the stiffness of the soil increases with the stress level (increase of the depth below seabed). In engineering practices, the increase is not often as strong as the linear relation in Equation 8.44.

For sand, the compressibility constant is in the order of 100–500; for clay, it is in the order of 20–100. An oedometer test[2] can be used to determine the compressibility values of specific soil samples in the laboratory.

The values of compressibility constants given are assumed for virgin loading of soil for relatively large deformations. Under unloading and reloading the soil below the maximum stress ever applied, the soil is stiffer (e.g. 10 times). The soil under small deformations is much stiffer compared to that with large deformations passing the pre-consolidation stress level (Schanz *et al.*, 1999).

In Table 8.3, some typical values of modulus of elasticity are listed. The soil modulus of elasticity should be determined by laboratory or offshore field tests. When such test data are not available, Table 8.3 may be used.

Soil strength is a limit beyond which the soil cannot transfer the stresses. The Coulomb relation between the maximum shear stress (τ_{max}) and effective normal stress (σ_{en}) expresses the soil shear strength.

$$\tau_{max} = c_{coh} + \sigma_{en}\tan\phi$$

Equation 8.45

where c_{coh} is the cohesion and ϕ is the friction angle. For sand, cohesion is negligible and the friction angle is the only strength parameter. For clay, it is reasonable to consider the strength in undrained conditions, during which the effective stress (σ_{en}) remains constant (Verruijt, 2006). From Equation 8.45, it is clear that the soil strength is proportional to the stress level. Also, we have shown in Equation 8.45 that the soil stiffness is proportional to stress level. Hence, soil strength and soil stiffness are correlated.

Table 8.3 Typical values of soil modulus of elasticity.

Type	E(MPa)
Rock	2000–20,000
Weathered rock	200–5000
Dense sand and gravel	50–1000
Firm clay	5–50
Soft clay	0.5–5

Source: Yuen (2015).

---

2 An *oedometer test* is a geotechnical test to measure consolidation properties of the soil sample in which, by applying different loads, the deformation responses are recorded and analysed. *Consolidation* is a process in which soil decreases in volume. More specifically, consolidation occurs when a saturated soil loses water without replacement of water by air.

Among the different methods used in the laboratory, triaxial testing may be used to determine the soil's shear strength and stiffness characteristics. It is possible to control soil drainage and measure the pore water pressures in triaxial testing (Rees, 2013). The angle of shearing resistance, cohesion, undrained shear strength, shear stiffness, compression index and permeability are usually determined during the test.

Among different field-testing methods, the cone penetration test (CPT) is very practical and simple. A cone is pressed to the soil using hydraulic pressure devices, and the stress at the cone tip as well as the friction along the shaft are recorded. Also, the cone penetration test can be used for estimating the soil strength considering certain correlations.

The spaces between particles of offshore soil (such as sand and clay) are filled with water. So, offshore soil is saturated, or partially saturated. The stiffness of the porous material and the characteristics of the pore fluid influence the saturation. For saturated and partially saturated soils, deformation of the soil can be significantly affected by pore fluid, in particular if the permeability of the porous material is small. The flow of pore fluid and deformation of porous material are studied by consolidation theory. In this section, the theory of consolidation is briefly discussed (Terzaghi *et al.*, 1996).

As was mentioned in Chapter 5, equations of deformations involve equilibrium, compatibility and a stress–strain relation. As an example, a one-dimensional consolidation case is studied herein. Assuming that there are no deformations in horizontal direction, then:

$$\varepsilon_x = 0, \varepsilon_y = 0 \qquad\qquad \text{Equation 8.46}$$

So, any change in the volume (ε_{volume}) can happen just in the vertical direction:

$$\varepsilon_{volume} = \varepsilon_z \qquad\qquad \text{Equation 8.47}$$

Now, we need to propose some assumptions to relate stress and strain in the vertical direction by a simple relationship. It is practical to assume that the vertical strain is dominated by vertical effective stress. Karl von Terzaghi proposed the term *effective stress* in 1925. Effective stress acting on soil is related to total stress and pore-water pressure; that is, it is presented as follows:

$$\sigma'_z = \sigma_z - P \qquad\qquad \text{Equation 8.48}$$

$$\varepsilon_z = -C_{comp}\sigma'_z = -C_{comp}(\sigma_z - P) \qquad\qquad \text{Equation 8.49}$$

where C_{comp} is the compressibility of the soil.

Special sign conventions, strains positive for extension and stresses positive for compression are used in literature. Hence, the minus sign is introduced to account for this sign convention.

$$\frac{\partial \varepsilon_{volume}}{\partial t} = -C_{comp}\left(\frac{\partial \sigma_z}{\partial t} - \frac{\partial P}{\partial t}\right) \qquad\qquad \text{Equation 8.50}$$

Equation 8.50 provides a relation between pressure and volume. Another relation that exists between pressure and volume can be found based on the conservation of mass, which is called a *storage equation*, one of the most important equations from the theory of consolidation.

$$-\frac{\partial \varepsilon_{volume}}{\partial t} = nC'_{comp}\frac{\partial P}{\partial t} + \nabla.q \qquad \text{Equation 8.51}$$

where n is porosity and C'_{comp} is compressibility of pore water (fluid).

$$\nabla.q = \frac{\partial q_x}{\partial x} + \frac{\partial q_y}{\partial y} + \frac{\partial q_z}{\partial z} \qquad \text{Equation 8.52}$$

$$q = n(V_F - V_S) \qquad \text{Equation 8.53}$$

in which V_F is the velocity of the fluid and V_S is the velocity of the solid.

Based on experiments, Darcy (1857) found that the specific discharge of a fluid in a porous material is proportional to the head loss (Verruijt, 2006):

$$q = -\frac{k}{\mu}(\nabla P - \rho_F g) \qquad \text{Equation 8.54}$$

where k is the intrinsic permeability of the porous material (k depends upon the size of the pores), μ is the viscosity of the fluid, ρ_F is the density of the fluid and g is the gravity vector. It is possible to show that: $\nabla.q = -\nabla.\left(\frac{K}{\gamma_w}\nabla P\right)$, in which $K = \frac{k\rho_F g}{\mu}$ is hydraulic conductivity and $\gamma_w = \rho_F g$ is the volumetric weight of the fluid. Using Equation 8.51, we can write:

$$-\frac{\partial \varepsilon_{volume}}{\partial t} = nC'_{comp}\frac{\partial P}{\partial t} - \nabla.\left(\frac{K}{\gamma_w}\nabla P\right) \qquad \text{Equation 8.55}$$

Using Equations 8.50 and 8.55:

$$\left(nC'_{comp} + C_{comp}\right)\frac{\partial P}{\partial t} = C_{comp}\frac{\partial \sigma_z}{\partial t} + \nabla.\left(\frac{K}{\gamma_w}\nabla P\right) \qquad \text{Equation 8.56}$$

Equation 8.56 is a differential equation, and with initial conditions and boundary conditions it can be solved for a specific one-dimensional problem.

Example 8.2 Set up the one-dimensional differential equation for a layer of soil that is subjected to constant load (see Figure 8.10).

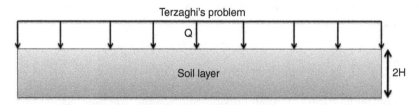

Figure 8.10 A layer of soil subjected to constant load.

If we assume that $\dfrac{K}{\gamma_w}$ is constant, then Equation 8.56 is written as:

$$\left(nC'_{comp}+C_{comp}\right)\frac{\partial P}{\partial t} = C_{comp}\frac{\partial \sigma_z}{\partial t} + \frac{K}{\gamma_w}\frac{\partial^2 P}{\partial z^2}$$

Equation 8.57

In Equation 8.57, $(nC'_{comp}+C_{comp})\partial P/\partial t$ is the loading rate that is very large at first but approaches zero afterward. To study the problem at loading time, if we integrate Equation 8.57 over a time interval and then let the time interval approaches to zero, then:

$$\frac{\Delta P}{\Delta \sigma_z} = \frac{C_{comp}}{\left(nC'_{comp}+C_{comp}\right)}$$

Equation 8.58

Based on Equation 8.58, if the stress quickly increases to the desired loading (Q), the pore-water pressure (P_0) is:

$$P_0 = \frac{C_{comp}}{\left(nC'_{comp}+C_{comp}\right)}Q$$

Equation 8.59

If $nC'_{comp} \ll C_{comp}$ or when the fluid is incompressible $nC'_{comp} \approx 0$, the initial pore pressure equals the external loading $P_0 = Q$.

As water is incompressible and no water has been drained out at the initial time, all the loads are carried out by the pore water, and no deformation occurs in the soil.

For constant loading, $\dfrac{\partial \sigma_z}{\partial t}$ is zero after the load has been applied, hence Equation 8.57 becomes:

$$\frac{\partial P}{\partial t} = \frac{K}{\left(nC'_{comp}+C_{comp}\right)\gamma_w}\frac{\partial^2 P}{\partial z^2} = C_{cons}\frac{\partial^2 P}{\partial z^2}$$

$$\frac{\partial P}{\partial t} = C_{cons}\frac{\partial^2 P}{\partial z^2}$$

Equation 8.60

$$C_{cons} = \frac{K}{\left(nC'_{comp}+C_{comp}\right)\gamma_w}$$

Equation 8.61

C_{cons} is the so-called consolidation coefficient.

The initial condition and the boundary conditions are:

$$P_{t=0} \equiv P_0 = \frac{C_{comp}}{\left(nC'_{comp}+C_{comp}\right)}Q$$

$$z=0 \;\rightarrow P=0$$

$$z=2H \rightarrow P=0$$

Equation 8.62

As the upper boundary ($z=0$) and lower boundary ($z=2H$) are drained, the pore pressure at boundaries will not change and remains constant (i.e. equal to atmosphere pressure). The above equations define the one-dimensional differential equation for a layer of soil that is subjected to constant load.

Equation 8.60 together with Equation 8.62 can be solved by the theory of partial differential equations (i.e. using the separation of variables; refer to Chapter 5 for more information regarding the separation of variables). The analytical solution of the problem can be found in many soil mechanics textbooks; for example, refer to Lambe *et al.* (1969).

$$\frac{P}{P_0} = \frac{4}{\pi}\sum_{j=1}^{\infty}\left\langle\frac{(-1)^{j-1}}{2j-1}\cos\left((2j-1)\frac{\pi}{2}\left(\frac{H-z}{H}\right)\right)\exp\left(-(2j-1)^2\frac{\pi^2}{4}\frac{C_{cons}t}{H^2}\right)\right\rangle$$

Equation 8.63

For practically large values of time, the series solution converges very quickly; and the solution can be approximated by the first term:

$$\frac{P}{P_0} \approx \frac{4}{\pi}\cos\left(\frac{\pi}{2}\left(\frac{H-z}{H}\right)\right)\exp\left(-\frac{\pi^2}{4}\frac{C_{cons}t}{H^2}\right)$$

Equation 8.64

As mentioned in this chapter, settlement of soil due to applied load should be assessed. Increase of settlement in time can be obtained using Equation 8.63, considering that the strain is given by Equation 8.49. By integrating the strain over the sample height, the settlement can be expressed as follows:

$$W = -\int_0^{2H}\varepsilon_z dz = \int_0^{2H}C_{comp}\left(\sigma_z - P\right)dz$$

$$= 2HQC_{comp} - \int_0^{2H}C_{comp}Pdz$$

Equation 8.65

where $W_\infty = 2HQC_{comp}$ is the final settlement. Right after the application of the load, the pore-water pressure is P_0. Hence, the immediate settlement (using Equation 8.62) can be found by:

$$W_0 = \frac{2HQnC'_{comp}C_{comp}}{nC'_{comp}+C_{comp}}$$

Equation 8.66

Assuming that the pore water is incompressible ($C'_{comp} = 0$), the initial settlement is zero. It is possible to relate the settlement to ratio of pore pressure (P/P_0) if we define a parameter called *degree of consolidation* (U_{cons}) as follows:

$$U_{cons} = \frac{W - W_0}{W_\infty - W_0} = \frac{1}{2H}\int_0^{2H}\left[1-\frac{P}{P_0}\right]dz$$

Equation 8.67

$$U_{cons} = 1 - \frac{8}{\pi^2}\sum_{j=1}^{\infty}\frac{1}{(2j-1)^2}\exp\left(-(2j-1)^2\frac{\pi^2}{4}\frac{C_{cons}t}{H^2}\right)$$

Equation 8.68

For $t \to \infty$, $U_{cons} = 0$; and for $t = 0$, $U_{cons} = 1$. Theoretically, the consolidation is finished when $t \to \infty$. Practically, when $C_{cons}t/H^2 \approx 2$, the consolidation is finished.

$$t \approx 2H^2/C_{cons}$$

Equation 8.69

Equation 8.69 enables us to estimate the duration of the consolidation process. It also allows us to study the influence of the various parameters on the consolidation process. If the permeability is increased, consolidation will be faster. If the drainage length is reduced, the duration of the consolidation process is reduced. This explains accelerating the consolidation by improving the drainage. Practically, consolidation can be accelerated by installing vertical drains. In a thick clay deposit, this may be very effective as it reduces the drainage length (Verruijt, 2010).

In this section, a brief introduction to soil mechanics has been provided to show how the basic mechanical theories are applied to geotechnical problems. The scope of the book does not allow us to go further into the development of equations and related mathematical approaches presenting the physics.

8.3 Foundation Design

Different structures like piles, caissons, direct foundation and anchors are used as interfaces between the support structure and the soil. There are similarities between design and function of foundation and anchors. As an example, for both shallow foundations and deadweight anchors, the main element is a footing that interacts with the soil. Both resist the sideward forces; however, shallow foundations mainly resist downward-bearing forces, while deadweight anchors resist upward forces.

Foundations are subjected to combinations of loads, including structure weight; environmental loading due to wave, current, wind and earthquake; as well as other external loads, such as drilling. The loads consist of overturning moments and lateral loads (the force components parallel to sea floor). Overturning moments due to environmental and external loadings result in tensile and compressive pressure (uplift and downward forces, respectively). Different foundations, depending on the type and magnitude of loads, sea floor conditions and soil type, can be considered.

If the resultant loading is downward, the design is mainly based on soil bearing capacity and soil friction on the embedded surfaces; while if the foundation is loaded upward, the design is mainly dependent on the submerged weight of the foundation, the soil friction on the embedded surfaces and suction beneath the foundation. In this section, some of the main types of foundations and anchors are discussed.

Figure 8.11 shows a schematic layout of a shallow foundation. The horizontal dimensions of a shallow foundation are normally much larger compared to the foundation thickness. Also, for shallow foundations, the depth of embedment (D_f) is less than the minimum horizontal dimension (B). In Figure 8.11, the base is below the grade $D_f > Z_s$. It is possible to design a shallow foundation with base at the grade $D_f \approx Z_s$ (which is more common). Z_s is the shear key height, and H is the foundation base height. The shear keys improve the lateral load resistance. In soft soils, shallow foundations are normally designed with skirts or shear keys.

The behaviour of a deadweight anchor is similar to the behaviour of a shallow foundation subjected to uplift forces (except for special deadweight anchors designed to dig into the soil when the anchor is dragged). In practice, a deadweight anchor is a heavy object placed on the seabed. Deadweight anchors are either set on the seabed or buried (partially or even fully). A deadweight anchor should resist uplift and lateral loading from a mooring line connected to a floating body (e.g. a buoy or a floating platform).

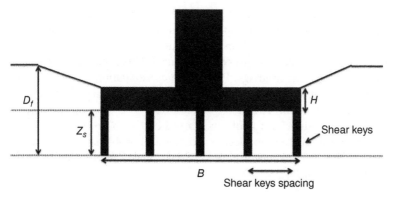

Figure 8.11 A layout of a shallow foundation with its main design parameters.

The uplift resistance is mainly from the net submerged weight of the anchor. However, some additional uplift resistance can be provided by special design. Deadweight anchors can be simple concrete clumps, or they can be more expensive, sophisticated anchors with shear keys (shear keys provide more lateral load resistance). The main parameters influencing the design of anchors and foundations are (Thompson *et al.*, 2012):

- Site specifications, including geotechnical information, water depth, bathymetry and slope, stratigraphy, potentially hazardous features and so on
- Structure or moored platform characteristics; responses and relative importance of the structure or moored platform
- Soil data, including the vertical and lateral extent of the soil data investigation
- Structural and environmental loading conditions
- Knowledge and experience based on similar foundations or anchors
- Cost and failure risk.

The design process of a foundation and anchoring system is interactive and involves the above considerations in varying degrees. Normally, safety factors accounting for the uncertainties in loading conditions, soils data and analytical procedure are applied. The safety factors are preliminary functions of loading conditions and soil properties. For long-term static loading and slow cyclic loading, a safety factor of 2.0 is recommended. The corresponding value for short-term static loading and rapid cyclic loading is 1.5.

The seafloor soil bearing capacity is a function of soil properties, foundation/anchor type and size, embedment depth, loading direction and seabed profile. The maximum bearing capacity of soil (Q_u) considering the side traction, represented as an equivalent base stress due to side adhesion and friction, is based on Thompson *et al.* (2012). The maximum bearing capacity should be greater than the sum of all normal forces to the seabed, considering proper safety factors.

$$Q_u = A'\left(q_c + q_q + q_\gamma\right) + P_{base}\, H_{ss}\left(\frac{S_{ua}}{S_t} + \rho_b\, z_{avg}\, \tan\delta\right) \qquad \text{Equation 8.70}$$

where A' (reduced foundation base-to-soil contact area) is the effective base area of foundation considering the eccentricity of load; when eccentricity is zero,

$$A' = A = LB(m^2) \qquad \text{Equation 8.71}$$

$q_c = S_{uz} N_c K_c$ is the bearing capacity stress for cohesion(N/m^2) Equation 8.72

$q_q = \rho_b D_f \left(1 + \left(N_q K_q - 1\right) f_z\right)$ bearing capacity stress for overburden$\left(N/m^2\right)$

Equation 8.73

$q_\gamma = \rho_b \left(B'/2\right) N_\gamma K_\gamma f_z$ is the bearing capacity for friction$\left(N/m^2\right)$ Equation 8.74

Due to intergranular friction, the portion of bearing capacity becomes attenuated at depth. The reason is that the high intergranular stresses cause particle crushing during shear failure. This is presented as a depth attenuation factor (f_z) for the frictional portion of bearing capacity stress that extends the soil bearing formulation (Equation 8.70) to any footing depth.

$P_{base} = 2B + 2L$ base perimeter of the foundation/anchor(m) Equation 8.75

$H_{ss} = \min\left[D_f, H + z_s\right]$ side – soil contact height(m) Equation 8.76

S_{ua} is undrained shear strength averaged over the side-soil contact zone (N/m^2).
S_t is the soil sensitivity, or the ratio of undisturbed to remoulded strength.
ρ_b is buoyant unit weight of soil above the foundation/anchor base (N/m^3).

$$z_{avg} = \frac{1}{2}\left[D_f + \max\left(0, D_f - H - z_s\right)\right]$$ Equation 8.77

average depth over side-soil contact zone (m).
δ is the effective angle between soil and foundation side (deg).

For rough-sided footings, $\delta = \varphi - 5$, and for smooth-sided footings (or when the soil is largely disturbed), $\delta = 0$. In these, φ is the soil friction angle.

S_{uz} is undrained shear strength effective for base area projected to shear key tip depth (N/m^2).
D_f is embedment depth of foundation/anchor (m).
z_s is depth of shear key tip below foundation base (m).
B' is effective base width depending on eccentricity (m).
B is base width (m), L is base length (m) and H is base block height.
N_c, N_q, N_γ are bearing capacity factors, and K_c, K_q, K_γ are bearing capacity correction factors; for more information, refer to Thompson *et al.* (2012).

Some external loads result in lateral forces on the foundation or deadweight anchor. These include wind loads, gravity force components due to non-horizontal seabed, current loads on foundation/structure, non-vertical mooring system loading, and wave and earthquake loadings. To increase lateral load capacity, shear keys may be used in the foundation base design. Hence, the surface on which the foundation will slide (failure surface) is forced into the seabed, where stronger soils can resist greater lateral loads.

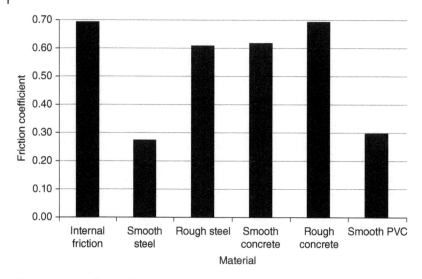

Figure 8.12 Coefficient of friction for sand (cohesionless soils), which depends on soil type, material type and roughness of foundation.

The maximum lateral load capacity parallel to the seabed (Q_{ul}) for foundations on cohesive soil (under undrained conditions) is given by the following expression:

$$Q_{ul} = s_{uz}A + 2s_{ua}H_{ss}L \qquad \text{Equation 8.78}$$

The parameters in Equation 8.78 are defined above in connection to Equation 8.70. If there are no shear keys, the short-term resistance on the foundation base is limited to a minimum of $s_{uz}A$ or $F_n\mu$.

μ is the coefficient of friction, which is roughly 0.2 for cohesive soil. F_n is total normal force. Coefficient of friction is between foundation and soil or between soil and soil. Figure 8.12 shows the coefficient of friction between cohesionless soils (sand).

The lateral load failure is a drained soil failure for cohesionless soils, and the maximum lateral load capacity in sliding is obtained by the following expression (Thompson *et al.*, 2012):

$$Q_{ul} = \mu\left[\left(W_{bf} + W_{bst} + W_b - F_{ve}\right)\cos\beta - F_h\sin\beta\right] + R_P \qquad \text{Equation 8.79}$$

where W_{bf} is the buoyant weight of foundation (N); W_{bst} is the buoyant weight of bottom-supported structure (N); $W_b = \rho_b A z_s$ is the buoyant weight of soil contained within the footing skirt (N); F_{ve} and F_h are the design environmental and mooring system loading in the vertical and horizontal directions, respectively (N); β is the seabed angle; and R_P is the passive soil resistance on the leading edge of base and footing shear key skirt.

Example 8.3 By using Equation 8.79, obtain a relation between the maximum lateral load, safety factor and loads cause sliding.

The maximum lateral load capacity (Q_{ul}) should be greater than the total forces cause sliding by a safety factor. The safety factor is recommended to be in the order of

1.5–2.0, depending on loading conditions and soil properties. The total forces cause sliding is found by:

$$F_{sliding} = \left(W_{bf} + W_{bst} + W_b - F_{ve} \right) sin\beta + F_h \cos\beta$$

Equation 8.80

Hence, the safety factor is defined by:

$$F_s = \frac{Q_{ul}}{F_{sliding}} = \frac{\mu \left[\left(W_{bf} + W_{bst} + W_b - F_{ve} \right) cos\beta - F_h \sin\beta \right] + R_P}{\left(W_{bf} + W_{bst} + W_b - F_{ve} \right) sin\beta + F_h \cos\beta}$$

Equation 8.81

Piles are relatively deep foundation structures that are widely used in offshore technology. Among different installation methods, hammering (driving) and drilling are the most common approaches. Normally, the cost of installation of piles limits their application when a surface foundation (or anchor) can be used. However, shallow foundations and deadweight anchors may not supply the support needed, and hence piles are used in such cases. The most common pile material for offshore applications is steel, while wood and pre-stressed concrete are also used in nearshore and harbour constructions. Circular sections and H-profiles are the most common types for use as foundations/anchors in deep offshore. Piles have been used for anchoring as well, in some cases with a special design to improve lateral and uplift capacity. The lateral load capacity of a pile-anchor may be increased by (Thompson *et al.*, 2012):

1) Lowering the attachment point along the pile length
2) Lowering the pile head beneath the soil surface into stronger soils
3) Attaching fins or shear collars near the pile head to increase the lateral bearing area.

Piles are subjected to different loads, such as (a) compression, (b) uplift force, (c) lateral and (d) bending loads. For normal piles, the axial forces are taken by soil friction developed along the pile as well as by the bearing capacity at the pile tip for compression loading. The lateral and bending loads are taken by the pile shaft bearing. Design of a foundation-pile is usually governed by compression and lateral loads, while moment and uplift govern the anchor-pile design. However, significant moments may appear depending on the action point of the lateral load.

In practice, design can be performed by a trial-and-error procedure, in which the pile will be checked against design criteria. The main point is determining the pile length, width and stiffness to be capable of resisting applied loads without excessive movement of the pile and without exceeding the allowable stresses for the pile material. In a simplified manner, the pile may be considered as a beam surrounded by elastic material (with an elastic modulus that increases linearly with depth).

In this section, the mechanics of pile-foundation and its response to axial and lateral loads are described. Generally, the response of pile-foundation is nonlinear and usually involves large deformations. Some examples are presented for problems involving nonlinear material behaviour. The soil surrounding the pile is represented by nonlinear spring, and the local displacement of the pile determines the spring (soil) response.

First, an axially loaded pile with a constant cross-section (A) in linear elastic material with modulus of elasticity (E) is discussed. The pile circumference is denoted by $O = 2\pi rl$, in which r is the radius and l is the length of the pile element. The friction along the circumference is related to normal axial force (N_{axial}). Figure 8.13 shows an

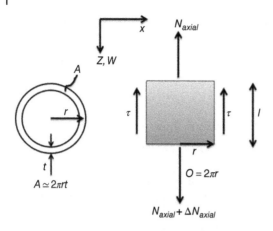

Figure 8.13 Element of a pile subjected to axial load.

element of the pile subjected to axial load; the equation of equilibrium for this element is written in Figure 8.13.

From equilibrium of forces, we can write:

$$\frac{dN_{axial}}{dz} = \tau O$$
Equation 8.82

Also, the stress is related to normal force by:

$$N_{axial} = \sigma_z A$$
Equation 8.83

Using the relation of stress and strain, the normal force is related to vertical displacement by:

$$N_{axial} = EA\frac{dw}{dz}$$
Equation 8.84

$$EA\frac{d^2w}{dz^2} = \tau O$$
Equation 8.85

In Equation 8.85, E is the modulus of elasticity of the pile material. If the shear stress and vertical displacement are linearly related, $\tau = cw$, in which c has the character of a subgrade modulus (Verruijt, 2010), the basic differential equation for an axially loaded pile supported by continuous linear springs is written as:

$$EA\frac{d^2w}{dz^2} - cOw = 0$$
Equation 8.86

If the parameters are constant, the solution of Equation 8.86 can be written as:

$$w(z) = a\exp(z/l_{ch}) + b\exp(-z/l_{ch})$$
$$l_{ch} = \sqrt{\frac{EA}{cO}}$$
Equation 8.87

where l_{ch} is the characteristic length. By applying the boundary conditions, the constants a and b are found. For an axially loaded pile, the boundary conditions are:

$$N_{axial} = -F \quad at \quad z = 0$$
$$N_{axial} - 0 \quad at \quad z - L$$

Equation 8.88

Hence, the vertical displacement (w) and corresponding normal force (N) in the pile are found as follows:

$$w = \frac{Fl_{ch}}{EA} \frac{\cosh\left[(L-z)/l_{ch}\right]}{\sinh(L/l_{ch})}$$

$$N = -F \frac{\sinh\left[(L-z)/l_{ch}\right]}{\sinh(L/l_{ch})}$$

Equation 8.89

If $L \rightarrow \infty$, then the displacement and normal force formulation are reduced to:

$$w = \frac{Fl_{ch}}{EA} \exp(-z/l_{ch})$$
$$N = -F \exp(-z/l_{ch})$$

Equation 8.90

Example 8.4 By using Equation 8.87, calculate the characteristic length of a steel pile with a length of 10 m, diameter of 1 m and thickness of 5 cm (see Figure 8.13).

Hint: Assume that the subgrade constant is related to the elasticity of the soil (E_{soil}), and it can be written as $c = E_{soil}/D$, in which D is the diameter of the pile.

Considering the pile circumference (O) and area of the pile cross-section (A), Equation 8.87 is rewritten as:

$$l_{ch} = \sqrt{\frac{EA}{cO}} = \sqrt{\frac{E(\pi Dt)}{(E_{soil}/D)\pi D}} = \sqrt{\frac{EtD}{E_{soil}}}$$

Equation 8.91

Considering Table 8.3, it is clear that for most soil types, the ratio of E/E_{soil} is more than 1000. For the given pile (with a diameter of 1 m and thickness of 5 cm), the characteristic length (l_{ch}) is in the order of 7 m.

In Example 8.4, it is assumed that the shear stress is linearly related to the displacement, $\tau = cw$. In Equation 8.89, the normal force and displacement are decreasing exponentially by depth. Hence, the maximum shear stress occurs at the top of the pile, and it decreases with depth as well. This assumption is reasonable for the current example; however, it conflicts with the soil shear strength characteristics. For example, considering a cohesive material (e.g. clay), the shear stress along the pile is limited by a value (τ_p); see Figure 8.14. w_p is the displacement necessary to generate the maximum shear stress (τ_p). For small displacements, the relation

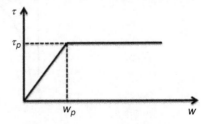

Figure 8.14 Shear stress and displacement relationship for cohesive (clay) material.

is linear in the elastic region; and, after passing such a limit, even if the displacement increases, the shear stress is constant in the plastic region.

As was mentioned, it is reasonable to consider the clay shear strength in undrained conditions. Although the maximum shear stress is influenced by the roughness of the pile and clay behaviour, it is practical to assume that the maximum shear stress is close to the undrained shear strength (s_u). Also, the maximum shear stress depends on the direction of the normal force (compressive or tensile).

For engineering purposes, it is practical to assume that the $w_p \approx 0.01D$, in which D is the diameter of the pile. Hence, the subgrade modulus is $C = 100s_u/D$, which means that for clay, the modulus of elasticity is about 100 times the undrained shear strength ($E_{soil} = 100s_u$).

Considering an axially loaded, infinitely long pile in a homogeneous cohesive material; there is a depth (d_p) over which the plastic deformation occurs. At the top of the pile where the maximum displacement occurs, the shear stress is constant. Below d_p, the shear stress has a linear relation with displacement; see Figure 8.15. In the elastic region, the solution is similar to Equation 8.90, and can be written as (Verruijt, 2006):

$$w = w_p \exp\left(-\left(z - d_p\right)/l_{ch}\right)$$
$$N = \left(-EAw_p/l_{ch}\right)\exp\left(-z/l_{ch}\right)$$

Equation 8.92

In the plastic region, the differential equation governing the displacement is:

$$EA\frac{d^2w}{dz^2} = \tau_p O$$

Equation 8.93

The solution using the boundary conditions, given load at the top, and the known force and displacement at the bottom of the plastic region, is given in the soil mechanics literature:

$$w = w_p\left(1 + \frac{d_p - z}{l_{ch}}\right) + \frac{\tau_p O}{2EA}\left(d_p - z\right)^2$$

Equation 8.94

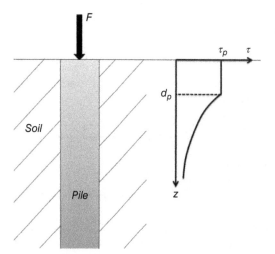

Figure 8.15 Shear stress for a pile in cohesive soil.

The maximum displacement is at the top of the pile:

$$w_{max} = w_p\left(1+\frac{d_p}{l_{ch}}\right)+\frac{\tau_p O}{2EA}d_p^{2}$$

Equation 8.95

The normal force in the plastic region (N) and the force at top of the pile (F) are as follows (Verruijt, 2006):

$$N = -EAw_p \,/\, l_{ch} - \tau_p O(d_p - z)$$
$$F \equiv -N\big|_{z=0} = EAw_p \,/\, l_{ch} + \tau_p O d_p$$

Equation 8.96

The force–displacement relation of the pile when $d_p > 0$ is nonlinear, which is shown by:

$$\frac{w_{max}}{w_p} = 0.5\left\langle 1+\left[\frac{Fl_{ch}}{EAw_p}\right]^{2}\right\rangle$$

Equation 8.97

If $d_p = 0$, then, based on Equation 8.95, $w_{max} = w_p$. So, Equation 8.97 is simplified to:

$$F = EAw_p \,/\, l_{ch}$$

Equation 8.98

For an axially loaded pile with finite length in clay (cohesive material), it may be assumed that the deformation in the top region, $0 < z < d_p$, is plastic, while the elastic deformations appear in the lower region, $0 < z < L$.

For a pile loaded and unloaded, permanent displacements appear and stresses remain in the pile. So, the stresses and deformations of such a previously loaded-unloaded pile are dependent on load-displacement history. Herein, the analysis is limited to an initially unstressed pile.

In the elastic zone, Equation 8.84 is the governing differential equation, and Equation 8.86 shows the related solution. Knowing that the normal force at the bottom of the pile is zero, the solution is written as:

$$w(z) = 2a\exp(z\,/\,l_{ch})\cosh\big((L-z)\,/\,l_{ch}\big)$$

Equation 8.99

Also, applying the displacement (w_p) at $z = d_p$, Equation 8.99 is further modified to:

$$w(z) = w_p\frac{\cosh\big((L-z)\,/\,l_{ch}\big)}{\cosh\big((L-d_p)\,/\,l_{ch}\big)}$$

Equation 8.100

Based on Equation 8.83, the normal force is found to be:

$$N_{axial} = EA\frac{dw}{dz} = -\frac{EAw_p}{l_{ch}}\frac{\sinh\big((L-z)\,/\,l_{ch}\big)}{\cosh\big((L-d_p)\,/\,l_{ch}\big)}$$

Equation 8.101

In the plastic region, the differential equation governing the displacement is based on Equation 8.93, which can be written as $d^2 w \,/\, dz^2 = w_p \,/\, l_{ch}^2$, with the solution given by:

$$w = 0.5w_p z^2 \,/\, l_{ch}^2 + a_1 z + a_2$$

Equation 8.102

where a_1 and a_2 are found using the boundary conditions (i.e. the normal force) and displacement at $z = d_p$ is given from Equation 8.100 and Equation 8.101. This results in:

$$a_1 = -w_p d_p / l_{ch}^2 - \left(w_p / l_{ch}\right)\tanh\left((L-d_p)/l_{ch}\right)$$
$$a_2 = w_p + 0.5 w_p d_p^2 / l_{ch}^2 + \left(w_p d_p / l_{ch}\right)\tanh\left((L-d_p)/l_{ch}\right)$$

Equation 8.103

If the pile is loaded by a dynamic action such as wave loads, the basic differential equation will be modified and written as:

$$EA\frac{\partial^2 w}{\partial z^2} - cOw = \rho A\frac{\partial^2 w}{\partial t^2}$$

Equation 8.104

where ρ is the pile material density. Normally, inertia effects during pile driving are important and should be carefully investigated.

Offshore structures are subjected to extensive lateral loads due to wave, wind and current loads. Hence, pile foundations' capacity to carry lateral loads should be checked as well; although piles have to carry axial loads, still they are subjected to lateral loads. This subject is briefly discussed herein; for more information, readers may refer to Poulos (1988), Kooijman (1989), and Verruijt et al. (1989).

Let us consider an element of a pile subjected to lateral loads, see Figure 8.16. From Chapter 5, we know that the equilibrium equations for a Bernoulli beam result in:

$$\left.\begin{array}{c} \dfrac{dQ}{dz}=-f \\[2mm] \dfrac{dM}{dz}=-Q \\[2mm] EI\dfrac{d^2 u}{dz^2}=-M \end{array}\right\} \rightarrow \left.\begin{array}{c} \dfrac{d^2 M}{dz^2}=-f \end{array}\right\} \rightarrow EI\dfrac{d^4 u}{dz^4}=f$$

Equation 8.105

where Q is the shear force, M is the bending moment, f is the transversal loads (distributed), E is the modulus of elasticity of the pile material, I is the second moment of the pile cross section and u is the transversal (lateral) displacement. For more information concerning the theory of bending of beams, refer to Chapter 5.

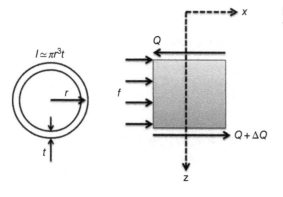

$I \approx \pi r^3 t$

Figure 8.16 Element of a pile subjected to lateral load.

If we assume that the load is proportional to the displacement ($f = -ku$), Equation 8.105 is written as follows:

$$EI\frac{d^4u}{dz^4} + ku = 0 \qquad \text{Equation 8.106}$$

where k is subgrade modulus; when the displacement is positive, the soil reaction is negative, resisting against the external applied load on the pile (Figure 8.16). The general solution of Equation 8.106 is in the form of:

$$u = a_1 \exp(z/\beta)\cos(z/\beta) + a_2 \exp(z/\beta)\sin(z/\beta)$$
$$+ a_3 \exp(-z/\beta)\cos(z/\beta) + a_4 \exp(-z/\beta)\sin(z/\beta) \qquad \text{Equation 8.107}$$

where $\beta = \sqrt[4]{4EI/k}$ is dimension length; and a_1, a_2, a_3, a_4 are determined from the boundary conditions.

Example 8.5   Consider an infinitely long pile loaded with a constant force at the top in transverse direction, and find the maximum displacement.

For an infinitely long pile, the term $\exp(z/\beta)$ is approaching zero; hence, by using Equation 8.107:

$$u = a_3 \exp(-z/\beta)\cos(z/\beta) + a_4 \exp(-z/\beta)\sin(z/\beta) \qquad \text{Equation 8.108}$$

Using boundary conditions at top of the pile:

$$\left.\begin{array}{l} M\big|_{z=0} = -EI\dfrac{d^2u}{dz^2} = 0 \\[2mm] Q\big|_{z=0} = -EI\dfrac{d^3u}{dz^3} = -F \end{array}\right\} \Rightarrow u = \frac{F\beta^3}{2EI}\exp(-z/\beta)\cos(z/\beta) \qquad \text{Equation 8.109}$$

$$u_{max} = \frac{F\beta^3}{2EI}$$

Displacement is periodically zero along the pile at $z = 0.5\pi\beta$, $z = 1.5\pi\beta$,.... Practically, a pile with length larger than 5β can be considered infinitely long. Also, from Equation 8.109, we may say that the stiffness of the soil (subgrade modulus) usually has a larger influence on the displacement rather than the pile stiffness.

References

American Petroleum Institute (API). (2015). *Design and Analysis of Stationkeeping Systems for Floating Structures*. Washington, DC: API.

Bakmar, C.L. (2009). *Design of Offshore Wind Turbine Support Structures: Selected Topics in the Field of Geotechnical Engineering*. Aalborg, Denmark: Department of Civil Engineering, Aalborg University.

Barltrop, N. (1998). *Floating Structures: A Guide for Design and Analysis*. London: Oilfield Publications.

Det Norske Veritas. (2013a). *DNV-OS-E301: Position Mooring.* Oslo: Det Norske Veritas.

Det Norske Veritas. (2013b). *DNV-OS-E302: Offshore Mooring Chain.* Oslo: Det Norske Veritas.

Faltinsen, O. (1990). *Sea Loads on Ships and Offshore Structures.* Cambridge: Cambridge University Press.

International Organization for Standardization (ISO). (2013). *ISO 19901-7: Stationkeeping Systems for Floating Offshore Structures and Mobile Offshore Units.* Geneva: ISO.

Jeon, S.C. (2013). Dynamic response of floating substructure of spar-type offshore wind turbine with catenary mooring cables. *Ocean Engineering*, 72:356–364.

Kellett, J.R. (1974). *Terzaghi's Theory of One Dimensional Primary Consolidation of Soils and Its Application.* Canberra: Department of Minerals and Energy, Australian Government.

Kim, B. S. (2010). Finite element non linear analysis for catenary structure considering elastic deformation. *Computer Modeling in Engineering and Sciences*, 63:29–45.

Kim, B. S. (2013). Comparison of linear spring and nonlinear FEM methods in dynamic coupled analysis of floating structure and mooring system. *Journal of Fluids and Structures*, 42:205–227.

Kooijman, A.P. (1989). *A Numerical Model for Laterally Loaded Piles and Pile Groups.* Delft, the Netherlands: Delft University of Technology.

Lambe, T.W. and R.V. Whitman. (1969). *Soil Mechanics.* New York: John Wiley & Sons.

Larsen, K.S. (1990). Efficient methods for the calculation of dynamic mooring line tension. In *Proceedings of The First European Offshore Mechanics Symposium.*

Low, M. (2009). Frequency domain analysis of a tension leg platform with statistical linearization of the tendon restoring forces. *Marine Structures*, 22:480–503.

Matsumoto, K. (1991). The influence of mooring line damping on the prediction of flow frequency vessels at sea. In *Proceedings of the OTC*, Paper 6660.

Nakamura, M.K. (1991). Slow drift damping due to drag forces acting on mooring lines. *Ocean Engineering*, 283–296.

Poulos, H.G. (1988). *Marine Geotechnics.* London: Unwin Hyman.

Rees, S. (2013). *What Is Triaxial Testing?* Retrieved 12 September 2015 from http://www. gdsinstruments.com/__assets__/pagepdf/000037/Part%201%20Introduction%20to%20 triaxial%20testing.pdf

Senjanovic, I.T. (2013). Formulation of consistent nonlinear restoring stiffness for dynamic analysis of tension leg platform and its influence on response. *Marine Structures*, 30:1–32.

Schanz, T., P.A. Vermeer and P.G. Bonnier. (1999). The hardening soil model: Formulation and verification. *Beyond 2000 in Computational Geotechnics-10 Years of PLAXIS.* Rotterdam, Netherlands: Balkema.

Terzaghi, K., R.B. Peck and G. Mesri. (1996). *Soil Mechanics in Engineering Practice.* New York: John Wiley & Sons.

Thompson, D. and Beasley, D.J. (2012). *Handbook for Marine Geotechnical Engineering.* Washington, DC: NAVFAC.

Triantafyllou, M. (1994). Cable mechanics for moored floating systems. *BOSS*, 57–77.

Verruijt, A. (2006). *Offshore Soil Mechanics.* Delft, the Netherlands: Delft University of Technology.

Verruijt, A. (2010). *An Introduction to Soil Dynamics.* New York: Springer.

Verruijt, A. and A.P. Kooijman. (1989). Laterally loaded piles in a layered elastic medium. *Geotechnique*, 39:39–46.

Yuen, S.T.S. (2015). *Compressibility and Settlement.* Retrieved from http://www.unimelb. edu.au/: http://people.eng.unimelb.edu.au/stsy/geomechanics_text/Ch9_Settlement.pdf

Index

Offshore Mechanics: Structural and Fluid Dynamics for Recent Applications, First Edition.
Madjid Karimirad, Constantine Michailides and Ali Nematbakhsh.
© 2018 John Wiley & Sons Ltd. Published 2018 by John Wiley & Sons Ltd.